KW-191-611

DISPOSED OF
BY LIBRARY
HOUSE OF LORDS

ON THE NATURE OF WAR
Swedish Studies in International Relations 8

To Janka

SWEDISH STUDIES IN INTERNATIONAL RELATIONS is a series published by the Swedish Institute of International Affairs. It comprises reports on research carried out at the Institute or otherwise published under its auspices. The Institute, a non-profit organisation, is a centre for information and research on international relations. The views put forth in its publications do not necessarily represent those of the Institute itself.

1   Kjell Goldmann, *International Norms and War Between States: Three Studies in International Politics,* Stockholm: Läromedelsförlagen, 1971.

2   Katarina Brodin, *Finlands utrikespolitiska doktrin: En innehållsanalys av Paasikivis och Kekkonens uttalanden aren 1944–1968,* Stockholm: Läromedelsförlagen, 1971.

3   Daniel Tarschys, *Beyond the State: The Future Polity in Classical and Soviet Marxism,* Stockholm: Läromedelsförlagen, 1972.

4   Kjell Goldmann, *Tension and Detente in Bipolar Europe,* Stockholm: Esselte Studium, 1974.

5   Harold Hamrin, *Between Bolshevism and Revisionism: The Italian Communist Party 1944–1947,* Stockholm: Esselte Studium, 1975.

6   Thomas G. Hart, *The Cognitive World of Swedish Security Elites,* Stockholm: Esselte Studium, 1976.

7   Gunnar Sjöstedt, *The External Role of the European Community,* Farnborough, Hants: Saxon House, 1977.

8   Julian Lider, *On the nature of war,* Farnborough, Hants: Saxon House, 1977.

On the nature of war

JULIAN LIDER

SAXON HOUSE

© Swedish Institute of International Affairs 1977

All rights reserved. No part of this publication may be reproduced, stored in a retrieval system, or transmitted in any form or by any means, electronic, mechanical, photo-copying, recording, or otherwise without the prior permission of Teakfield Limited.

Published by
SAXON HOUSE, Teakfield Limited,
Westmead, Farnborough, Hants., England

 British Library Cataloguing in Publication Data

Lider, J
  On the nature of war. - (Swedish studies
  in international relations ; 8).
  1. War
  I. Title   II. Utrikespolitiska institutet
  III. Series
  355.02'01   U21.2

  ISBN 0-566-00178-0

ISBN 0 566 00178 0

Typeset in 10/12 Times by
Preface Ltd, Salisbury, Wilts
Printed in Great Britain by
Biddles Ltd, Guildford, Surrey

Contents

Acknowledgements

I have found the Swedish Institute of International Affairs to be an ideal place for research, reflection, and writing. I am particularly indebted to the Institute's Research Director, Dr Kjell Goldmann, for his continuous support and advice over the years.

It is almost impossible to express adequately my gratitude to Mr Donald Lavery whose work with my manuscript went far beyond the normal duties of an editor. Many ideas, which were unclear in the original draft, now appear in more comprehensible language due to his efforts; this book could hardly have been completed without his cooperation. Of all the experiences I have had in the course of writing, our friendship is the one I most cherish.

Thanks go also to Mr Manne Wängborg and to my colleagues at the Institute who sacrificed so much of their time to read the long manuscript and make valuable comments.

I should also like to thank Sonja Johnson, Elsa Lindberg and Ann-Marie Bratt who patiently provided material for my research.

1 Introductory remarks

My purpose in this study is to compare Soviet and Western views of the nature of war. In doing so, I shall examine, if but briefly, a great variety of approaches and schools of thought. In a sense such a general review constitutes a comparison, since differences and similarities are, at least by implication, pointed out. But I shall devote special attention to comparing analytically views representative of the two opposing ideological camps, the two antagonistic socio-political systems of today.

The reason for this emphasis seems obvious: the threat of war between these systems has until quite recently constituted the greatest threat to mankind; relations between them influence all conflicts in the world, even remote ones, and conversely, each local conflict affects world-wide competition and struggle and might escalate into a Third World War.

Although the climax of the Cold War is past, and the need to solve international problems through negotiation and cooperation is generally accepted, to make an inventory of the differences in views of war, and to point out in what direction they are changing, is perhaps worthwhile.

The ideas the opposing camps have about such problems as the nature of war, the relation between war and politics, and the aims of conceivable future wars, are of great importance for the occurrence of war itself. The debates between, for example, the behaviourists and Freudians, or between those who believe that war is a manifestation of human nature and those who maintain that war is a regulating mechanism of the international system of sovereign countries, that prevents anarchy and imbalance – all these competing claims have contributed to our understanding of mankind and its social activity through politics, international relations, and war. But it is the difference in outlook between Western and Marxist–Leninist science, between political and military leaders in the Soviet Union and the West, between their advisers and consultants, that is of crucial importance for the outbreak or prevention of war in our time. This is true because today, as often in history, the influence of ideas on current events owes less to their inherent validity than to the material strength of the socio-political forces that believe in them and, being prepared to fight for them, shape their foreign and military policies in accordance with them.

War is obviously too vast a subject to be adequately dealt with in one book. To limit the material requires a criterion by which relevant topics can be chosen. By declaring in the title that this is a study of the nature of war, I have tried to indicate that I am mainly concerned with very general ideas about war — those features which are most essential to war and distinguish it from other similar social phenomena. In principle my ambition is to deal with only those features which are common to all historical instances of war and to leave aside such questions as the description of the manifest characteristics of particular wars

1

and types of wars, their aims and immediate causes, and their purely military aspects. What in turn is meant by the most essential feature of war is of course a controversial matter reflecting border ideological issues; here the author must be allowed some discretion. I shall devote most of my attention to war as organised armed violence carried on by social groups with conflicting interests; as a form of struggle in the whole process of resolving social conflicts; and as a continuation and instrument of policy. This choice is not completely arbitrary, however. My reasons, which are related mainly to the comparative aim of this study, are given more fully in the course of the text.

I had originally intended this study to be a comparison of Marxist and non-Marxist views of the nature of war. In the process of researching and analysing the different theories, however, I have modified my initial aim. In some important ideological and political matters there are now different views in what was once called the Marxist camp, a fact which makes it necessary to choose between them. At the same time I wanted to present views representative of the Warsaw Pact states, as the main politico-military opponent of the Western camp. Both reasons have contributed to the choice of the Soviet variant of the contemporary Marxist point of view; this variant is called Marxism–Leninism in the Soviet Union.

To represent the governing non-Marxist views, the Western political approaches, and especially the variant called political realism, have been chosen. This choice is motivated by the following considerations. In the first place, the political and military doctrines of the leading Western countries, particularly those of the United States, seem to be influenced to a significant degree by ideas usually regarded as stemming from the tradition of political realism. Moreover, the intellectual impact of this tradition is very far-reaching. It is the basis of a mode of thinking about politics, in terms of the state, interests, and power and about war as a political tool, that is shared by a great number of Western scholars who study war, whether these be military advisers or peace researchers. It seems that in combining a system of abstract concepts with guides for practical action, political realism comes closer than any other Western intellectual tradition to what Marxists would call an ideology.

There is a difference, however, between the impact the two theories to be compared have upon policy-making. While Marxist–Leninist ideas on war serve directly to explain, justify, and support political decisions, and are often included as integral parts in political statements (resolutions of Party Congresses, pronouncements of political and military leaders, etc.), the way in which the ideas of political realism are reflected in the official Western policy statements (Presidential State of World Messages, Defense Secretary Posture Statements in the USA, British, French, and West German White Papers, NATO Council resolutions, etc.) can only be hypothesised, although with a reasonable degree of accuracy.

However, this asymmetry does not constitute an obstacle to fulfilling the principal aim of this study, which is to examine theories of war, and not political decisions or policy-making processes.

2

Part I

THE NON-MARXIST ANALYSIS OF WAR

2   Approaches

War, the most many-sided and complex of social phenomena, commanding tremendous material and moral resources from entire nations, has been studied from many points of view, and by representatives of many disciplines. An analysis of war may be only a part of a broader study of social evolution or international relations; the purpose of the analyst may be to describe fully the nature of war, but he may also examine only the causes or consequences of war, or confine himself to one particular dimension or one particular type of war.

The familiar approaches to the topic are the biological, psychological, anthropological, political, and sociological. A scholar who restricts himself to one of these approaches need not regard war as a one-dimensional phenomenon; he may have one of many reasons for concentrating on just one aspect of war; he may not feel competent to treat characteristics of war that fall outside his field of scientific training; he may wish to point out the explanatory power that the conceptual apparatus of his discipline has in disclosing part of the essence of war; and even when he believes that the direct causes of war lie outside his own field, he may show that important indirect causes lie within. [1] Students of war often point out that a full understanding of the nature of war requires that its many facets be examined, even if war is first of all a socio-political phenomenon. [2]

Studies of war may also differ with regard to the level of analysis chosen. [3] The choice often depends on the assumptions the analyst makes about the nature or roots of war. For example, if the scholar believes war to be an expression of man's inherent nature, or if he believes that wars have their origin in the ideas men hold, then he focuses his attention on the behaviour of individuals. If on the other hand he regards war as one kind of relation between groups of men, then his analysis is likely to be conducted on a more aggregated level. Moreover, the particular social group chosen will give the analysis a definite character; if the analyst defines war as class-war, or as inter-state war, or as a balancing mechanism of the international system, the questions he asks and the answers he provides will be quite different in each case.

In consequence of the variety of assumptions about the nature and roots of war, and of the range of possible dimensions and levels of analysis, a great array of definitions of war can be found. These can in general be divided into two groups: those which emphasise the primacy of the socio-political nature of war, and those which do not. Since the latter group falls outside the socio-political perspective, that is basic to both the Marxist–Leninist and non-Marxist traditions, I shall devote less space to them than to the others. They must nevertheless be included, for both the Marxist–Leninist and non-Marxist approaches have been developed through confrontation with the whole gamut of theories of the nature of war.

The biological approach

Underlying the biological approach can usually be found one or some combination of two basic assumptions.

The first is the alleged law of survival of life. The biologist usually assumes as a basic premise that all forms of life, from plants to man, are engaged in a constant struggle in which only the fittest survives. To him, war is simply one of the forms in which this struggle becomes manifest. The struggle itself is understood to have two interrelated functions. In the first place, it is the means by which the fittest survive and the species evolves to some higher form of life. In the second place, it is the way in which individuals compete for the best positions in a group, at the same time as different groups, such as tribes, nations, and classes, fight for the best living conditions or the highest status in the system of groups. In these theories, social groups are often treated as living organisms. [4]

Secondly, the biologists may assume that it is part of the nature of man that he is driven by instinct to struggle and to war. [5] Human behaviour is said to be determined by aggressive drives developed through the process of natural selection and inherited by all members of the species. War is regarded as one of the forms in which such behaviour is expressed.

One extreme variant of the biological approach is based on ethology, which is the science of the behaviour of animals in their natural environment. When studying man, the ethologists thus emphasise those aspects of human behaviour which are not learned. They argue that much social and political behaviour, including war, originates in, and can therefore be explained by, man's animal nature.

The 1960s marked a revival in the biological approach, including ethology. [6] A new version of these theories was presented in a number of publications, which received a great deal of attention. [7] Ethologists asserted that man inherited his aggressive instincts from the animal kingdom, where they served useful functions such as dividing the land available to a species to ensure an adequate supply of food. In primitive societies these instincts filled a similar role, but as man's social organisation became increasingly complex, these instincts were diverted from their original tasks.

Two ethological hypotheses especially relevant to the nature of war might be mentioned:
(a) Man has an innate tendency to kill members of his own species.
(b) Man has a 'sense of territory', which is manifested by a deep emotional attachment to a particular territory. Related to this feeling is man's need of a certain 'personal space', an area not occupied by any other person; without such a space man cannot function normally.

6

Due to these irrational and functional needs, man has developed the tendency to fight for his territory. Ardrey calls this the 'territorial imperative'; he contends that it is an instinctive trait that is genetic and ineradicable. [8]

The claims biologists make for the power of their approach in explaining war vary greatly, from the modest assertion that war cannot be explained without taking the biological factor into consideration, to the most extreme contention that war is both the form and the result of the biological struggle. [9]

The biological approach, and especially its ethological variant, is not without its critics. [10] Most social scientists deny the basic assumption of ethological theories, namely that the behaviour of man, like that of all animals, is governed by genetically programmed predispositions; they deny consequently that man is inherently aggressive. To support their criticism, they maintain that the evidence of an instinct of aggression in lower animals is not reliable, and even if it were, it would be naive to extrapolate from other animals to man. Such a method of reasoning would rather put the 'territorial imperative' in doubt, for man's nearest relatives amongst the primates do not exhibit 'territorial' behaviour. [11] The notable feature of human behaviour, these critics claim, is that it is learned. For the most part, men react aggressively when irritated by particular, not general, causes, most commonly when they are frustrated. What men experience as averse, they learn to experience as averse through socialisation. The critics therefore consider the ethological theories to be harmful; they provide relief from the heavy burden of guilt that many nations and individuals bear after two world wars, and they absolve men of the sin of preparing for new wars; for if man is born innately aggressive, then he cannot be held responsible for wars.

The psychological and socio-psychological approaches

The psychological approach to the nature of war generally emphasises the influence of social or group psyche on the outbreak and course of war. War is thus viewed as a manifestation of the aggressive tendencies of the masses; aggressiveness is a social expression of such human feelings as possessiveness, hatred, and frustration. In this approach, war is regarded primarily as a behaviour pattern of human groups. 'War represents', to paraphrase Clausewitz, 'the continuation of behaviour in another context. It is the translation of personal aggression to the cultural sphere.' [12]

This approach has four main variants, which differ in their explanation of the way the social psyche is formed.

Aggressive behaviour as an evolved trait

The first variant is essentially a modification of the biological approach. Aggression is seen as the product of natural selection, that is, as an evolved trait. [13] As the struggle between individuals in the form of bodily combat

7

became impossible, it was replaced by collective combat between communities and by more refined forms of combat within communities. [14] Since the conditions of life favoured those who could fight in groups, the pugnacious instinct gradually evolved into a collective trait.

Aggression through transformation

According to the psychoanalytical interpretation, aggression also has its roots in innate elements of the individual human psyche. However, the social context in which men live requires them to suppress their natural aggressive reactions. As a result, aggression becomes transformed into part of a social psyche, and thus causes or at least influences the outbreak of war. [15]

Aggression through frustration

The third explanation, which is connected to some degree with the previous one, is based on the idea that aggression is a product of frustration, an emotion men feel when a goal they desire cannot be reached. If men become very frustrated at the same time as they are aware of the fact, they may act aggressively. [16] They are more likely to do so when not inhibited by, for example, the threat of punishment. [17] War may become an institutionalised pattern of behaviour, resolving the personal tensions of the members of society and maintaining their psychic well-being while enabling them to avoid other undesirable consequences of frustration. [18]

The original thesis of this interpretation, that frustration is a necessary and sufficient condition for aggression, has been subsequently modified: frustration does not always lead to aggression, [19] nor is aggression always the product of frustration alone. [20]

Studies have also been made of the way in which collective frustrations that lead to aggression arise. Two main mechanisms have been indicated: they may constitute a social aggregation of frustrations held by members of the group, or they may derive from such structural (also called environmental) characteristics of collectivities as socio-economic inequality. [21]

Aggression as learned behaviour

The behaviouralists hold that group aggressiveness, if it occurs at all, is learned and not innate. They assume that man is by nature neither aggressive nor submissive, neither warlike nor peaceful; he is capable of developing in either direction depending on what he learns through interaction with his environment and culture. [22] In other words, people do not act and react in a simple, mechanistic way according to their instincts or frustrations, but deliberately, according to the views and emotions they have learned through past experience. [23]

All four variants of the psychological interpretation of the nature of war assume that man and groups of men become aggressive, but they differ greatly in the significance they attribute to aggressiveness. At one extreme is the view that war is fundamentally a manifestation of a social psychic condition; at the other is the opinion that such collective feelings may have a considerable impact on the outbreak of war, but they are not its main cause. Between these extremes comes the contention that the clash between the psyches of conflicting groups or nations is one of the main fronts of war. Moreover, the opinion has been expressed, that no one psychological theory has proved sufficient to explain the nature of aggression, for there are a number of psychologically and functionally different kinds of aggression [24] with disparate roots. [25] The ways in which the social psyche is formed, and especially the ways in which the hostile views, emotions, and behaviour of individuals are transformed into collective ones, are not directly relevant to this study however; the question belongs rather to a study of the causes of war.

The anthropological approach

In contrast to the biologists, who trace war to man's animal instincts, and the socio-psychologists, who stress the importance of the interplay between the psychic needs of individuals and the social forces binding them into groups, the anthropologists see war as a cultural phenomenon. The basic point of departure for this approach is thus that war is in some way a product of man's cultural experience. Several alternative or complementary hypotheses have been put forward to illuminate this connection; these will be reviewed briefly below.

The anthropological approach raises the important question whether war as a product of cultural evolution is also a means by which this evolution proceeds, on the assumption that nature and society generate phenomena, which further their evolution. This question will also be given a short examination.

War as a product of culture

All theories holding that war is not an eternal phenomenon, that it first appeared at some definite stage of man's development, imply thereby that war is the product of this cultural evolution, that earlier periods were both warless and less-developed. Sometimes this thesis is expressed directly: Margaret Mead's contention that 'war is an invention' is a well-known example of this type. [26] Often the thesis is an unstated corollary, as when it is suggested that the first wars occurred between tribes, the first definite forms of culture, or in another version between nation-states.

A second group of hypotheses, which are usually derived from studies of more recent historical periods, relates the occurrence of war to rapid cultural transformations. For example, Quincy Wright underlines (at many points in his collection of theories and data) the correlation between increased cultural sophistication and the greater frequency of war, and especially between the growth of political organisations in a cultural entity and its bellicosity. Another scholar has contended that in times of political, social, cultural, and territorial growth and expansion, war occurs at least as often as, if not more often than, in periods of decline. [27] Others have posited that a greater command of technological resources has always influenced the occurrence of war, [28] and that the political structure, especially of the capitalist and imperialist systems, has also had such an impact. [29]

Hypotheses belonging to a third category try to explain war in terms of special cultural characteristics. One of the most simple of these is the theory that certain cultures tend to generate more aggressiveness than others. More complicated is the hypothesis that cultural differences between socio-political units may hinder communication, give rise to suspicions and misinterpretations of each other's intentions, and ultimately lead to wars. [30] Some scholars put forward the opposite thesis: the more similar two units are culturally, the more similar are also their needs and interests, which thereby conflict; the danger of violent conflict increases in consequence.

Finally, in a category by itself comes Malinowski's view that the essence of war is a clash between cultures in the form of independent tribes or nations. [31] He defines war as an 'armed contest between independent political units, by means of organized military force, in the pursuit of a tribal or national policy'. The cultural entities that wage wars are thus, in his view, first the tribe, and later, the nation; war is for them a manifestation of their cultural independence. Malinowski's thesis is based on two assumptions: first, that wars are culturally and not biologically determined, that they constitute complex cultural responses formed by collective sentiments and values; and second, that wars are always fought by cultural units to achieve their political goals. Wars between or within tribes were thus waged to create primitive tribe-states, whereas the wars fought later between culturally differentiated groups were an instrument of national policy.

There are also many 'mixed' typologies which connect the occurrence of war in primitive society with several variables – cultural, political, psychological, etc. [32]

Does war promote the evolution of man?

To establish a correlation between the frequency of war and cultural development does not answer the question of whether war promotes such development. Nor does a correlation between the incidence of war and a period of rapid change prove that war is beneficial. It could be argued that the death and devastation that has always followed in the wake of war more than offset any conceivable gains. Many would agree with the conclusion Toynbee has come to in his analysis of the relation between war and culture: that war has always been the direct cause of the decline of each culture, even if it has been an institutionalised phenomenon in all cultures. [33]

Finally, those who subscribe to the 'destructive hypothesis' state categorically that war adversely affects civilisation by imposing identifiable costs and hindering development.

But opposite views are also often expressed. [34] On the basis of their study of the early stages of man's social development, many anthropologists conclude that war has indeed made a contribution. [35] It has always destroyed what was effete and made way for what was viable. [36] The role war has played in the creation of large social units (the state in particular), [37] their political, social, and economic institutions, and their culture, has been given special emphasis.

Such conclusions might be illustrated by the extreme view taken by Andreski. [38] He contends that war was a necessary condition for both the emergence of civilisation and the biological ascent of man; the two processes being closely interrelated. By the process of natural selection, of which war was an important part, man appeared as a species different from the apes, with a much improved brain. What this implied was that the evolutionary process accelerated remarkably since brain power now became the decisive quality for selection. War ensured that those with better brains and better weapons could improve their conditions of life, by expanding their territory, for example. They could thus multiply more quickly and win in the struggle for existence. The higher forms of social organisation were also dependent on war, for it was only by conquest that small tribes could be welded into states, and small states into larger ones. War was thus ultimately a necessary condition for science, technology, art, and all other aspects of culture that require comparatively large and organised societies.

A second anthropological interpretation considers the evolution of man to be inseparably connected with the development of tools of warfare.

This theory should be seen as an alternative to the perhaps more familiar view that man's evolution has been indistinguishable from his progress in developing tools of production. [39] In one variant, war and military needs are said to stimulate the productive capabilities of society, particularly industrial organisation, commerce, and finance, which are the bases of economic growth. [40]

Finally, in a more general form, war is regarded as a coping adaptive mechanism, which can be used in response to many problems of survival and thereby can ultimately make some contribution to cultural evolution. [41]

Some scholars take no position in the reviewed question, either stating that since war sometimes helps and sometimes hinders development and many even do both simultaneously, no general assessment can be posited, or contending that nations improve their material and cultural conditions regardless of war. [42]

Adherents of the anthropological interpretation of war have been criticised for ignoring value judgements, and for attempting to justify war. [43]

The ecological approach

The ecological approach to the study of war regards war as a manifestation of the struggle for better environmental conditions, and/or as an instrument in this struggle, [44] or, in a more moderate form, as something that is greatly influenced by the environment.

The environment (or milieu) can be defined as the 'whole spectrum of environing factors, human as well as non-human, intangible as well as tangible'. [45] The prevailing view combines the products of human culture and the physical features of the earth in the definition of the environment. [46] In the history of man, the combination of these two elements has gradually changed; in the earlier stages, the environment comprised mainly natural forces that exerted a determining influence on the social activity, but as human life has developed, social laws have come to take a dominant position. Nevertheless, nature does still limit to some degree man's *potentialities*. [47]

The ecological approach to the study of war has a long history. The main idea of many earlier works was that the geographical environment directly determines the policy of nations, and consequently conflicts and wars. Men were thus compelled to strive for the best environment possible, but in the process, the milieu indirectly moulded their character and affected their will to act. Some of the most modern ecological theories claim that the environment has at least some effect on policy: it promotes certain policies, it provides opportunities and sets limits for action, and it influences the way a policy is conducted. As man attempts to exploit the environment and improve it, he is driven into conflicts with his fellow man, especially in periods of a rapid increase of population density. [48] The environment may therefore be said to affect both the cause of, and the reason for war.

Since there are many aspects to the environment, [49] a political unit that seeks to improve its milieu may choose one of several means of doing so. If, for example, it suffers from a shortage of land, it may resort to war to acquire new territory or to open new trade routes. [50] If, in a period of rapid technological growth, supplies of energy run short, war may be one way of obtaining new sources. [51] It has also been suggested that, at least in primitive societies, war is one way of regulating the size of population in relation to the available land and other resources. [52] Some scholars contend that a war-generating ecologic crisis can soon come about; it would be caused by contamination or waste of the earth's surface and the depletion of the earth's raw materials. [53]

Nowadays the geopolitical approach to the study of war is sometimes regarded as a variant of the ecological approach. The struggle for more space, for more secure boundaries, or for an improved geographical situation is considered to be a variant of the struggle for a better environment in general. Since geopolitics has had a long history of its own, however, it will be treated separately in this review.

Ecological ideas have sometimes been borrowed by representatives of other disciplines and approaches as arguments in their theories of social evolution. Toynbee's view of history as a cyclic pattern of challenge and response is one example: civilisations are born and grow in response to challenges posed by the environment. [54] Another example is Andreski's contention that war, which is generated by the tendency for a population to outgrow its resources, leads to the evolution of society as a whole. [55] The ecological approach is also closely connected with the new version of Malthusian theories. There has been a modern revival of theories of overpopulation, largely due to the need for tools to analyse the population explosion in the Third World. One of the main themes in this literature is the danger that overpopulation will in the future be a fundamental cause of violent conflicts.

According to some advocates of neo-Malthusianism, the struggle for a material base sufficient for satisfying human needs may take the form of war, or may be carried out by means of war. [56] Some contend that war may thus fulfil two functions: it may provide the victor with needed space and resources and at the same time reduce the number of people the given resources must support. Some even assert that however cruel or inhumane war may be, it is nevertheless a necessary solution to the problem of overpopulation. [57]

The main objection raised against the ecological approach is that rapid evolutionary changes have occurred in the absence of war-like activity; on the other hand, it has not been proved that war always contributed beneficially to human evolution or to human adaptation to the environment.

The geopolitical approach

Of all the ways of studying the nature of war, the geopolitical is perhaps the one with the longest tradition, and the one that has had greatest impact on the way statesmen themselves have regarded war. It can no longer be seen as a single unified approach, however. [58] In its classical form, the geopolitical situation of a state is defined as the space required for a nation to live prosperously and securely, as well as its boundaries and geographical location, including its relation to the crucial geographical areas of the world. All these conditions are therefore considered to be of crucial importance for the state's whole history and destiny, enabling it to expand and achieve a position of power or even world dominance. War in this scheme is a manifestation or a form of the struggle for better geopolitical conditions. [59] In the more moderate form of the approach, the geopolitical situation of a state is regarded as a very important but not the only item in the set of factors that influence the origins, the outbreak, and the course of war. [60]

Classical geopolitics

One of the most frequent themes of the older geopolitical analyses was the struggle for living space. Influenced by Darwinism, the classical geopoliticians

13

regarded the state as an organism that occupied space, grew, contracted, and eventually died. [61] The struggle for space was thus equivalent to the struggle for life, and a nation's land area was an indication of its power and vitality. [62] The life-cycle of states was expressed in various sets of laws in which expansion, [63] including the conquest of strategically important territories were important concepts. [64]

A second group of geopolitical theories laid less stress on the geographical size of a country *per se*; the key organising concept was the need for secure boundaries which each nation had in order to ensure its economic, political, and cultural development. War was seen in this perspective as a struggle for such boundaries. The aim of geopolitics was therefore to indicate which sort of boundaries were the easiest to defend. [65] In this regard a sea coast, that marked the limits of a country's interests and strength at the same time, was considered to be the ideal boundary. [66] In one concept it was held that control of the seas had been a decisive factor in all wars since the seventeenth century and that control of the oceans was crucial for the power-status of a state. The theory that world domination could be achieved by capturing the strategic points that provide control of the Pacific Ocean was a variant of this basic idea. [67]

In contrast, a common theme in the writings of many American and British geopoliticians was that land powers, especially those located in the 'heartland' of the world, as central Eurasia was called, had the best geographical situation. [68] Such land powers as Russia and Germany tended to exploit their geographical position not only to rule the heartland, but also to conquer the countries on its periphery and eventually the whole world.

Parallel to the heartland theories, the theory of *Lebensraum* was developed by German geopoliticians. They maintained that great countries experienced the need to push back their frontiers, to obtain *Lebensraum*, in order to ensure economic self-sufficiency and provide for a large growth in population. [69] Here too geographic size was equated with strength and greatness, and much attention was devoted to the problem of securing boundaries by creating a sparsely populated zone at the frontiers. These ideas, together with such theses as the need for natural frontiers and control over the central Eurasian–African landmass, were exploited by Nazism to justify its policy of expansion through war.

Modern geopolitical theories

After World War II, a need was evidently felt to reinterpret, or supplement, the classical geopolitical positions. Mackinder, who had been the father of the heartland theory, now suggested that the command the Atlantic Community had gained over the seas, which was of decisive importance in the postwar world, gave it a position of strength. [70]

More important for American geopolitics was the appearance of Spykman's

14

'rimland' theory, however. [71] If one places the geopolitical position of a country in a global perspective, [72] he maintained, then the crucial areas of the world were the 'rimland' of the Eurasian land mass, the region located between the heartland and the sea coast. In Spykman's view, the great wars of history, like World War II, were struggles for control of this 'rimland', [73] and future wars would also be fought with this aim; if it were conquered by Communism, the United States would be encircled. [74] The 'rimland' theory has influenced the thinking of American political and military scholars, and the formation of American foreign policy. Traces of it can be found in the policy of containment, [75] including the formation of NATO and the network of global anti-Communist alliances, in Nixon's 'Asian Doctrine', and in NATO's 'forward strategy'. In one variant, emphasising air power, a British military writer contends that the control of crucial territories from which the enemy can be destroyed is decisive for victory in war, and consequently for strategy and defence policy. [76] In another variation Rear Admiral Lepotier of the French Navy suggests that between the continental Eurasian bloc and the complex maritime bloc there exist 'zones of friction'. [77] These zones are of great geopolitical and geostrategic importance, since control over them is decisive for the relative strength of the main geopolitical blocs. The zones he mentions are spread throughout the world, chiefly along the periphery of Eurasia. What is unique in the modern geopolitical situation is the great importance of the polar regions: the Arctic, Lepotier contends, is becoming a front of air contact between two leading powers, and the Antarctic the decisive rear platform for sea and air communication of the Western bloc.

A second French analyst, General F. Gambiez, who also views war as a struggle for positions of geostrategic importance, puts more emphasis of effect the use of outer space will have on the age-old rivalry between sea powers and land powers. Since the conquest of outer space has become the decisive element in the world-wide geopolitical struggle, the object of geostrategy is to gain control of those spots in the world from which the conquest of space is most easily accomplished. [78]

The revolution in strategy caused by the advent of nuclear-armed missiles has also led to a reappraisal of some of the basic tenets of geopolitics. It has been suggested for example, that a very large territory with a dispersed population and industry is even more of an asset to a country because it is less vulnerable to nuclear attack. [79] Theories focusing on the crucial importance of seas and seapower have also been reinterpreted and repeated.

A revival of geopolitical ideas could be noted in West Germany in the 1960s and 1970s. [81] Some scholars have advanced the argument that because of the specific location of West Germany, both geographically (in the centre of Europe) and politically (between the two antagonistic blocs), the geopolitical (or as it is often called 'geostrategic') factor has become an extremely important link between foreign policy and military strategy. They posit that geographic realities greatly influence the security policy of all countries involved in the

15

East–West struggle and that they may even bring about situations that invite states to go to war. In their view one of the main reasons why West Germany and the other Western powers decided to include West Germany in NATO was the country's geographical location. They contend that the foreign policy of the United States is based on a desire to gain control over the opposite shores (*Gegenküste*) – a theory that can be seen as a variant of the 'rimland' theory. The United States therefore regards West Germany as an important ally because of its location. [82] Geopolitical concepts are also applied in historical studies of German policy, in which it is maintained that German defeats have been rooted in the policy, mistaken for a central power, of expanding both to East and West at the same time. [83]

Critique of the geopolitical approach

The underlying assumptions of the pure geopolitical theories – that war is in some sense the effect of geopolitical causes – has been attacked from many angles. [84] There seem to be four main objections:

  1 The geographic conditions of a country cannot be a stable factor determining its policy because many of them can be changed by the man. [85]
  2 The influence that geographic conditions may have on the policy of a country can only be exerted through the complex mix of historical, political, spiritual, and cultural elements that also constitute the environment.
  3 The determining influence of boundaries is questioned on two grounds: in the first place, the importance of frontiers has historically diminished owing to technological development including the development of armaments; [86] in the second place, the boundaries of a particular society are never natural in some sense independent of the values or desires of that society, and they therefore change as the goals of the society change. [87]
  4 Geopolitics makes the error of eliminating the ethical aspect of human behaviour. If war is necessarily tied to the geopolitical situation of the state, the question of its desirability becomes irrelevant. Geopolitics provides those who advocate war with an apology for shirking responsibility for the consequences of their policies. [88] It can be used by advocates of any ideology to give their desired goals an air of scientific authority that obviates criticism. [89]

Use of the geopolitical terminology

The geopolitical perspective has been adapted by some scholars as a tool of analysis in a somewhat looser form. One way was to use it for the investigation of various problems of political geography, in particular for the analysis of the peculiarities of the political attitudes and the behaviour of different regions. [90]

Hosmer has suggested, for example, that the American military posture has to be based on such geopolitical postulates as continental defence, protection of lines of communication, maintenance of alliances, protection of American interests overseas, and a capability to participate in local conflicts outside the USA. Changes occurring in the 'geopolitical climate' should be analysed and answered with corresponding changes in the military establishment. [91]

Some scholars who are not themselves geopoliticians sometimes find it useful to apply geopolitical terminology to their studies of war. To cite one example, Duchènes does this to denote the two main regions where he considers the outbreak of war to be most likely: the 'geopolitical periphery', comprising the states on the border of systems of developed countries, where armed conflicts are an expression of a large number of interacting and struggling forces in a period of rapid change; and the 'geopolitical backyards' of the superpowers, such as Eastern Europe and the Caribbean, where armed intervention is a constant possibility. [92] Moreover, the term geopolitical is often used together with the term situation in studies of international relations. [93] For journalists also the term seems to have some appeal, as when they write about the 'oil age in geopolitics'. [94] Such adaptations do not seem to have much in common with the geopolitical approach, and are intended simply to indicate some connection between the policy of a country and its geographic location. They may neverthe-less indicate some sort of projection of geopolitical ideas.

The legal approach

Despite appearances, the legal approach does not treat war as a purely legal phenomenon, [95] but rather emphasises the legal aspects of war. Two variants will be discussed below, one in which legal terms are important in defining the nature of war, and one in which the legal term 'sovereignty' is used to describe the main goal of war.

A definition of war in legal terms – Quincy Wright

Those who stress the legal aspects of war maintain that a belligerent status implies sovereignty. [96] A struggle can be considered a war only if the contenders are sovereign political units (tribes, fiefs, empires, nation-states, etc.). A rebellion against a sovereign authority may assume the character of a war, an internal one, only if the rebellious party succeeds in establishing a structure for asserting the sovereign power it claims. A certain amount of ambiguity is involved however in defining the warring parties as both sovereign and political. The point can be illustrated in the work of Quincy Wright.

In *A Study of War,* Wright tries to combine the legal, sociological, military, and psychological views of war and offer a synthesis. The resulting definition holds that war is a state of law and a form of conflict involving a high degree of

17

legal equality, of hostility, and of violence in the relations of organised human groups; in a simpler wording, war is the legal condition which equally permits two or more hostile groups to carry on a conflict by armed force. [97]

Similar definitions are presented at other points in his study. In one, he asserts that war may be regarded 'from the standpoint of each belligerent' as an extreme intensification of military activity, psychological tension, legal power, and social integration, and 'from the standpoint of all belligerents' as an extreme intensification of simultaneous conflicts of armed forces, popular feelings, jural dogmas, and national cultures; he also repeats here the definition of war as a legal condition. [98] In another place he writes that war is at the same time an exceptional legal condition, a phenomenon of intergroup social psychology, a species of conflict and a species of violence. [99]

In all three definitions, each of the four viewpoints is represented, but prominence is given to the legal aspect. What is more surprising is that no connection between war and policy is mentioned.

In another place, Wright suggests that wars occur as the result of rapid change in the social, technological, psychic, or intellectual life of a society. Again, political change is only implied, unless the term social is taken to mean socio-political. [100]

Of course, in such an encyclopaedic work, the role of politics could not be neglected. Wright does briefly describe war as a method for achieving major political changes, and the goals for which he suggests wars are fought are essentially political: building nation-states, speeding up or slowing down the development of civilisation, preserving or destroying peace and stability, maintaining or destroying the international system of states. [101] In his commentary on war since 1942 (and up to 1965), Wright repeatedly mentions the political causes of war, and devotes a special passage to 'war as an instrument of policy in the atomic age'. [102]

In a later study Wright distinguishes between 'war in the legal sense' and 'war in the material sense'. The legal conception, which he calls the narrower of the two, describes situations in which the participating groups are equally permitted to combat each other with the use of armed force. If war is seen in a material sense as armed struggle of considerable magnitude without regard to the legal status of the contenders, then, Wright concludes, 'war has usually been employed by states as a method of international politics; it has also been employed by insurgents and rebels to gain independence, by governments to suppress domestic and colonial revolts, and by international organizations to suppress aggression'. [103] This is a description of war as a political instrument. In an encyclopaedic article, however, he once more repeats the assumption of the primacy of the legal concept of war. [104]

It can be suggested that the ambiguity of Wright's position derives from his combining the constituent elements of war as a social phenomenon (war as a phenomenon of intergroup psychology, as a species of conflict, or a species of violence) with certain legal regulations to which war was submitted as it became institutionalised. [105] The general impression that remains from his very valuable contribution to the study of war is nevertheless that he was most concerned with what legal conditions had to be met before an armed conflict could be called war. Wright himself asserts that every scientist must make a choice between many points of view; it seems that he chose the legal viewpoint. [106]

War as a transfer of sovereignty

In a more obviously legal approach, Reves proposes the 'social law' that war, which he defines as fighting between units of equal sovereignty, takes place whenever and wherever such units come into contact; the corollary to this law is that war ceases the moment sovereign power is transferred from the fighting equal units to a larger or higher unit. His conclusion is that the problem of the transfer of sovereignty is the sole cause and object of war. [107]

This is surely an excellent example of the dangers of legalism. Reves's definition of war is so narrow that it excludes the great majority of, what are commonly called, wars. Moreover, his contention that the transfer of sovereignty to a higher unit is the only result of each war is clearly invalidated by any fair test. His thesis may imply that the only way war can be prevented is by merging all nations into a world-state, the single and highest sovereign power. [108]

The moral approach

There are two moral problems related to the phenomenon of war. One of these concerns the moral character of particular wars, how to decide whether they are just or unjust. This aspect will be looked at in a later chapter. The second treats war as an instrument in some moral system, as a means of perfecting man or society. This might be called a moral interpretation of the nature of war.

By this approach, the explanation of war lies in the goals of the moral system: these aims determine the characteristics of war, and war serves these aims. Quite often the approach is part of a teleological outlook on nature, man, and society, in which all motion is considered to be a striving towards some final end. Those who adopt the moral approach to the study of war must therefore define the ends to which wars lead (what is perfected through war?) and the motive force that brings about war (who are the agents of goodness?).

The perfection of man

First there are those theories which hold that war has a perfecting influence on the nature of individual men. In some theories this is an effect independent of the origins of war; [109] in others the effect is the cause. An example of the latter type are the religious theories that have appeared in many cultures at many times, in which warriors become virtuous by fighting for the just cause. In the ancient Greek and Roman religions, the road to becoming a god-like hero passed through the battlefield, and according to Mohammedanism, participation in wars against 'unbelievers' was the shortest way to salvation. Similar ideas can be found in Japanese religious beliefs and in Christianity. Such religious ideas are often intertwined with other social values: [110] the age of chivalry is one example from Europe, [111] but participation in war as a means of proving manliness was also common amongst American Indian tribes. [112] Some recent anthropological research has also suggested that in some cultures war is fought as an end in itself, as the highest form of living. [113]

In the less teleological form of the positions, war is regarded as a phenomenon that gives men (or both men and the state) an opportunity to demonstrate their high moral qualities and even to heighten them. [114] This theme recurs in German philosophical thought between the 17th and 19th centuries. Hegel's analogy, in which the influence of war upon men and nations is compared to the motion of the winds on the seas, both preventing a foulness which a constant calm would otherwise produce, was well known and often quoted. Nietzsche was another who praised the benefits of war as a means for moral perfection, without which humanity would lose its culture and its very existence. [115] Similar views were expressed by, amongst others, Max Scheler, [116] von Moltke, [117] and Ludendorff. [118]

Although the intellectual tradition to which these thinkers belong is no longer fashionable, the germ of thought they express has been developed in more modern versions. Sorel has described internal violence, of which civil war is the highest form, as a means of creating new moral values, as an expression of a sense of honour and an indifference to material advantage. [119] And Coser, a prominent representative of modern social functionalism, writes that participation in revolutionary violence is an act by which an individual affirms his identity, manifests his maturity, and participates in active political life. Man is 'reborn' through the act of violence. [120] Such ideas are not the property of any one political ideology but, contrary to the traditional variants, they are often expressed by representatives of radical groups (Franz Fanon, for example), [121] for whom civil violence is a means of social revolution, and only rarely by exponents of the established order. [122]

The modern apologists of violence have been given a penetrating criticism by many scholars, who have emphasised the danger of finding a justification for violence in the realm of nature and then transferring the justification to the social realm. Hannah Arendt asserts that whereas violence in nature destroys the old and thus helps the new to flourish, the collective violence made possible by man's social life is not a creative but a destructive force. [123]

The perfection of the human race

The second variant of the moral approach is the one regarding war as a means by which the human race may be perfected. Some of those connected with this view include Houston Stewart Chamberlain, Arthur de Gobineau, Daniel G. Brinton, [124] Giovanni Gentile, Alfredo Rocco, Benito Mussolini, Alfred Rosenberg, [125] and Adolf Hitler. [126] Racism can be considered to have been discredited, but a reminder in the form of a few voices from pre-Nazi Germany might not be out of place: 'Only war can demonstrate which nation can win in the competition of the Aryan peoples'. [127] 'War is the only just law court. It appears as a means of natural selection in which the perfect German peoples triumph over the less valuable peoples, imperfect and weak'. [128] And from the first years of the Nazi's rule: 'War has been and will always be indispensable for the Nordic race as a means of solving the problem of having enough room to live in.' [129]

The perfection of human culture

The third end to which it is sometimes indicated that war contributes is the cultural development of man. The term culture can be used to mean different things. Ruskin sees war as a stimulus to culture in the narrower sense when he says that 'great art is only possible in those countries which often wage war'. [130] Using the word in a broader sense, sociologists and anthropologists, often influenced by Social Darwinism, have interpreted war as a means by which man has been able to progress to higher forms of social life. [131] These views have already been mentioned above; it is sufficient to note here their moral implications. [132]

The perfection of the state (nation)

Finally, perhaps the most influential of these moral theories contended that war was a manifestation of the moral strength of the state (or the nation or nation-state) and at the same time a means by which the state improved its character. The roots of this idea go back to ancient times; in Sparta, for example, the over-riding ideal to which all citizens sacrificed their lives was the strong state (*polis*), and war was a means by which this goal was achieved and the virtues of the state demonstrated. This tradition had influence on a great number of later scholars, but the more immediate source of modern variants is Hegel, in whose philosophy the state is again given a position of unquestionably supreme moral importance. By strengthening the state, war is thus an instrument of moral progress. Among other writers of the German national-Romantic epoch were Spengler, [133] who did not advocate war but recognised the role it played in enabling states to strengthen themselves and conquer weaker peoples, Ratzenhofer, who developed the idea that war strengthened the structure of the state, [134] and von Treitschke, [135] who regarded war as majestic and sublime because it allowed states to maintain and strengthen their independence.

Out of this philosophical tradition developed the war ideas of German fascism, which are the most extreme of the theories reviewed in this section. The theory of total war formulated by General Erich Ludendorff had a special place as a forerunner of Nazism. In his view, war is a permanent force in the life of a nation; it helps the nation survive, awakens its soul, and stimulates its will to live. War demands of a nation the most difficult yet the most ennobling of tasks. War is total; it requires the mobilisation of all physical and spiritual resources, which is precisely why it serves to perfect the nation. War is not an instrument of policy: it fulfils its creative functions independently of the aims of policy, although it does so better when aggressive than when defensive. What is more, policy is an instrument of war: it is the role of policy to prepare the nation for total war through a mobilisation of all resources. It must indicate the enemy, or invent him if he does not exist. Policy does in time of peace what war does in times of fighting: perfecting the nation. [136]

The motive force of war

Thus far the discussion has dealt with the ends to which different moral approaches have seen war as a means. Since in teleological theories, events are interpreted as steps towards the fulfilment of a goal, the defined goal becomes in some sense a cause of the event. The motive force identified in these various theories as the mainspring of war is therefore closely related to the ends they claim war serves. In religious theories, war is often explained with reference to the will of God, who turns his wrath on sinners or, alternatively, leads the faithful to victory. Theories that view the dynamics of the cosmos as movement toward the perfection of nature usually claim that nature creates the means for its own perfection; in this perspective men may be said to go to war because of their natural instinct to do so. On the other hand, theories that, like Ludendorff's, are more prescriptive than explanatory, presume that the moral worth of the ends to which they believe men attain is sufficient to account for the fact that men use war as a means to achieve them. [137]

The military-technical approach

In the military-technical approach, emphasis is placed on war as a violent clash between any two hostile social entities. War is in other words seen, or analysed, as a phenomenon whose main features and laws of conduct are independent of the socio-political conditions that have led the conflicting parties to be hostile to each other. This is the approach that professional military men have tradition-ally adopted, but it also appears in scholarly analysis, [138] two examples of which will be given below.

The outlook of the professional

In many countries emphasis on the technical aspects of war has had a great influence on the development of military theory and practice. From this perspective, policy may be granted a role in the process leading up to the outbreak of war, but it is considered to have very little influence on its conduct. In this sense, war is a continuation of policy, but not a controlled instrument of policy.

The attitude of the German General Staff is often cited as an example of how the professional soldier tries to free himself from the shackles of his political masters. The liberal-democratic ideal of a professional civil servant who can serve equally well whichever government the people elect also fosters a politically neutral view of war amongst the military. Up until World War I the military leaders in the United States considered war to be a technical matter in both theory and practice, [139] and some critics have maintained that the tradition of neglecting the primacy of policy influenced the way American troops were used in both World Wars. [140] According to these analysts, the American military did not recognise the political bounds within which these wars were fought, but considered it obvious that war was waged for total victory and the unconditional surrender of the enemy. [141] Evidence that the theory of 'pure war' had a hold on the thinking of American military leaders even later can also be found. [142]

The double approach

In most modern military science, the view that the nature of war is essentially socio-political seems to be the prevailing one. Nevertheless, the technical approach is still used in many strategic studies, especially concerning the role of nuclear weapons.

The ambivalence of this position is reflected in a number of studies in which two separate analyses are conducted side by side, one treating war as a military-technical phenomenon, the other examining the political side of war. A well-known exponent of this double approach is the English military scholar Air Vice-Marshal E. J. Kingston-McCloughry. [143] He defined war as an 'armed conflict between nations concerned to impose their will on each other', [144] and his discussion of the evolution of war deals with changes in its technical nature; the milestones are the invention of new armaments. [145] He reminds the reader however that war 'was once a national instrument of policy used to achieve national purposes or more power'; [146] this is a political definition of war. [147] The analysis of the political side of war is kept parallel (in some sense secondary) to the military-technical analysis, and even then deals mainly with the influence of the political situation in the world (the formation of alliances, blocs, and politico-strategic zones) on the conduct of war and the management of military forces.

Military-technical factors as the cause of war

The military-technical approach is also basic to some theories in which the state of military technology is an independent factor influencing state behaviour. In some models, the growth and improvement in a nation's arms is regarded as something which follows its own laws; in this view, wars occur semi-automatically or else they are to some extent self-generating. Theories in which the balance of power or the arms race are factors explaining the outbreak of war imply to some extent the assumptions of this approach. [148] In modern literature these ideas have gained renewed status due to the realisation that competition in weapons-technology involves risks of war of inconceivable devastation. All theories of accidental war, catalytic war, and other sorts of unwanted wars are partial reflections of this tendency.

Typologies of war

As a footnote it could be pointed out that the military-technical aspect of war has had a greater influence on the terms in which war is discussed, and thus the idea of war, than any other. In almost any dictionary definition, more space is devoted to war as armed conflict than war as an instrument of policy. [149] More importantly, it is the technical aspect that defines the tools with which war is classified and analysed in many non-Marxist studies: war is limited or total, nuclear or conventional, etc. Only recently have the socio-political characteristics of war been admitted as grounds for differentiation in non-Marxist classifications.

The sociological approach

While all of the foregoing approaches have contributed to the study of war as a multidimensional phenomenon, the sociological, political (or socio-political), and economic dimensions of war seem to predominate in contemporary war theory. A brief preliminary description of these three approaches follows; a broader analysis will be undertaken later.

The schools of thought that could be grouped together under the heading of sociological approach to the study of war are many and varied. In general, what links them together is their treatment of war as a kind of socially recognised conflict in which the combatants are groups who resort to armed violence. War is thus assumed to fall under those laws which govern all social conflicts. These sociologists derive their theories from an analysis of the structure of society; conflicts are in this perspective rooted in the formation of groups, which pursue their own interests. The problems investigated with this approach include the conditions for the occurence of social conflicts, the way they occur, and the nature and dimensions of different kinds of conflicts. The chapter on 'War as a kind of conflict' provides some comments on these theories. [150]

The political approach

In what might be called the political approach to the study of war, emphasis is placed on the political (or socio-political) nature of war, its essence and role as an instrument of policy. By this definition the approach includes a great many theories, which may differ sharply in their interpretation of the role war plays in international relations and in the domestic life of societies, in their evaluation of its moral worth, and in their prescriptive conclusions.

One source of variety lies in the level of analysis chosen. [151] Most often the nation-state is taken as the unit whose action is to be explained. At the subnational level, the interest of the analyst is focused on social groups subordinate to the state, as classes, political parties, religious sects, and labour groups. At the supranational level, including even the whole international system groups of states figure as the acting units. Finally, at the transnational level, the actors are associations such as the Roman Catholic Church. War within or among the political units at each level may be one of the topics of such analyses. Until now, however, the focus has been on the nation-state as the main actor in war, and war has been regarded predominantly as an interstate phenomenon. [152] Wars between coalitions of states are also studied from this point of view.

Of the many variants and combinations of the sociological and political approaches, the socio-political one, in which the role of a conscious political decision to resort to war and the political aims of war are emphasised, is the most important. It is examined in more detail in the chapter entitled 'War as an instrument of policy'.

The politico-economic approach

Studies belonging to the politico-economic approach also underline the political motives for war, but they see a struggle between political interests as a reflection of an underlying conflict between economic interests. The Soviet treatment of war might be considered as an example of this approach (although it could also be regarded as a sociological or socio-political theory) since it emphasises that war has its roots in the division of society into economic classes, while at the same time it asserts that war is one of the means by which classes implement their policy. This will be discussed more thoroughly below.

A somewhat different emphasis is given in studies where the analytical focus is more purely economic, that is, where war is regarded as the direct manifestation of a struggle between economic interests. It seems justifiable to include the economic approach as a variant of the politico-economic approach, however, for the scholars who adopt it realise that economic forces can only act through policy; the different terminology merely indicates a different emphasis.

One of the pioneers of this variant was Norman Angell, who contended that wars were fought for economic reasons. He concluded from his study that war in the industrial age could not be profitable, and that it should therefore be banned. [153]

Since the beginning of the century a rich literature devoted to the economic factors influencing the outbreak of wars has been produced. Four main theses have been put forward:

1 Wars occur between states because the societies have conflicting economic interests, which they try to advance or protect. [154] To digress, the critics of this hypothesis object that it is difficult to speak of the interests of the whole society, for various groups within the society have differing interests and what they gain from war consequently varies. Moreover, in modern times, war between equal opponents is profitable for neither owing to the destructive nature of modern weapons. [155]

2 Wars between states are a means by which domestic economic conflicts, which may derive from systemic inconsistencies, are redirected and given an outlet less harmful to the society.

3 Wars between states arise because of the influence powerful economic groups, who profit by war, have on governments. All non-Marxist theories of imperialism as a source of war include this contention. [156]

4 Internal wars are forms of the struggle between groups with conflicting economic interests.

A general criticism of this approach is that it is too one-sided; war, it is argued, is a very complex phenomenon with many dimensions and many causes. It is also pointed out that a conflict of economic interests is not a sufficient condition for war since such conflicts can also be resolved in other ways, for example through economic development.

The importance of economic motives as a cause of war is recognised by many representatives of various other approaches. Anthropologists have suggested, for example, that in the early stages of social history, men took to armed fighting mainly for economic reasons. [157] Often they prefer not to call such fighting war, however, but reserve that term to describe fighting that takes place mainly for political reasons or that has become institutionalised. [158]

Multi-dimensional approaches

Whatever the strengths and weaknesses of all these various approaches to the study of war, there can be no doubt that the nature of war has changed profoundly within the last two or three generations. The experience of the two world wars, the prospect of a nuclear war, and the course of the local wars fought in the past decades stand as evidence that modern wars are carried out with a complex of military, political, economic, psychological and other means simultaneously, and that the relative importance of each dimension can vary from war to war and in the various stages of any one war. War has spread into all areas of human activity. Political leaders treat preparation for war, whether defensive or offensive, as an all-encompassing enterprise in the firm conviction that future wars will be fought on many fronts: military, economic, scientific, and ideological. Moreover, and perhaps more importantly, there is an awareness shown by the public to the fact that war in modern times can reach cataclysmic proportions and have consequences of the utmost significance in all fields of human endeavour. In short, war has become a much more many-sided social phenomenon than ever before; this is a fact which cannot fail to result in changes in the approaches employed in the study of war.

In recent years a number of scholars have tried to develop ways of incorporating as many aspects as possible into their analysis. Often their ambition is to create a general theory of war, [159] but the common trait is a deliberate adoption of a multi-dimensional approach. There is, however, a notable lack of conceptual clarity in the use of the term dimension. Properly speaking, a dimension of war is one of its characteristic qualities as distinct from one of its causes, whether basic or proximate. Since there is a conceptual connection between dimensions and causes or factors, however, the multi-dimensional approach is usually considered to include studies in which war is treated as a social phenomenon with many attributes as well as studies in which war is seen to have many causes. [160] The relation between the roots of war and the dimensions of war is a problem that goes far beyond the theme of this study and forms a large field of separate research. For the present it may suffice to indicate four different directions multi-dimensional approaches have taken with respect to the nature of war.

War has many dimensions, but only one is studied

Most scholars recognise today that war is a many-sided phenomenon, even if they choose to concentrate on one or two particular aspects of it. Many of those who confine themselves to one of the approaches described above do so for practical reasons – their academic background, the need to keep their study within manageable bounds, and so on. Others (the political realists, for example), justify their use of one basic approach on the grounds that the dimension studied (power) is dominant over the rest and determines the main course of war. [161] Nevertheless, accompanying these decisions to study only one dimension is a basic awareness, often unexpressed, that the total complexity of war is not captured by their analysis.

Multi-dimensional analysis

In the second line of development, the multi-dimensional character of war is the point of departure, [162] and an effort is made to encompass the totality of war within the analytical framework. The French school of polemology serves as an example. [163]

In one variant, three levels at which all conflicts can be studied are discerned. On the surface are the apparent causes of war: the shot at Sarajevo, for example. At the third level, the underground, are the basic factors (demographic, economic, psychological, political, and sociological) that make up the structure of society and lead to possible aggressiveness. Between these comes the middle level, at which the competition and contradictions between different societies are worked out in political relations and the balance of power.

Multi-causal analysis

A third example of an attempt to develop a synthetic view of war is the theoretical analysis made by Kenneth Waltz entitled *Man, the State, and War*. [164] Here the task which Waltz sets himself is to come to a better understanding of the causes of war. This he does by analysing several key theories in the history of political thought, some of which trace war to man's nature, others to the internal structure of states, and a third group to the nature of relations between states. Waltz does not delve directly into the essence of war, but since each of the three images of war implies a different perspective on the nature of war, his attempt to integrate these images leads to a multi-dimensional view of war itself.

Waltz's study is a conceptual analysis, but examples of empirical research in which a multi-causal analytical approach implies a multi-dimensional conception of war can also be cited. The most ambitious of these is unquestionably the 'Correlates of War' project conducted by Singer and Small. [165] Their long-range aim is to contribute to a theory of modern international war. In the meantime, they are busily engaged in testing the correlation between war as a dependent variable and as many independent variables as possible. These comprise two main categories, ecological and behavioural variables. The ecological variables are in turn subdivided into structural attributes (alliances, trade patterns, the socio-economic structure of states in the system, etc.), physical attributes (geographical features, technological development, etc.), and cultural attributes (popular beliefs, élite values, etc.). Singer and Small believe the ecological variables to have the most explanatory power, for the patterns in conflict behaviour seem also to be outcomes of the ecological configurations. Regardless of the results of this project, however, the general approach seems to imply that war is seen as a multi-dimensional phenomenon.

Agnosticism

Doubts about the possibility of ever revealing the true nature of war or of discovering the common causes of all wars in history have also been expressed by prominent scientists. Anatol Rapoport even goes so far as to question whether there is a class of events involving human behaviour that can be legitimately subsumed under a single term war, i.e. whether war is a discretely identifiable event. War is dependent upon what we think about it; it can be perceived as a recurring disaster akin to disease, but it can also be regarded as a normal state of affairs, an institutionalised recurring phase of international relations; tribal wars, wars of political unification, colonial conquests, religious wars, or national wars of the sort described by Clausewitz, require separate explanation.

Rapoport states that an argument about the true nature of war would lead nowhere, as do many other similar philosophical disputes'. [166]

Such a position can perhaps be called agnostic, and it is in a sense a multi-dimensional approach. Those who might be called agnostics do not deny that the causes (or some of them) of particular wars can be known, but they do question whether there are causes common to all wars. [167]

The position of Raymond Aron might also best be described as agnostic. Critical of abstract sociology, he argues that wars must be analysed 'in the forms in which history presents them', a method he calls 'historical sociology'. [168] Only when this has first been accomplished can one 'compare . . . and possibly in the end, trace out common features and underlying causes'. Thus, he does not preclude the possibility of ultimately finding common causes, but his doubts could hardly be more strongly expressed. [169]

Notes

[1] Therefore some scholars apply terms like 'biopolitical', 'sociopolitical', etc. For example, the term biopolitical is considered to be useful, because biological concepts may help to understand political phenomena, and biological factors play a significant role in political behaviour (Somit, 1972, p.209).

[2] Bramson, Goethals, 1964.

[3] For a review of different levels of analysis, see Waltz, 1959, Midlarsky, 1975 (pp.2–8).

[4] Donald G. MacRae, 'Darwinism and the Social Sciences', in: Barnett, 1958; Banton, 1961; Ali A. Mazrui, 'From Social Darwinism to Current Theories of Modernization', *World Politics* XXI, 1968; James Davies, 'Biology, Darwinism, and Political Science: Some New and Old Frontiers', American Political Science Association, Annual Meeting, Chicago 1971. According to a biopolitical theory, relationships between states 'are treated as an inevitable outcome of the "struggle for survival" to which all living organisms are presumably condemned' (Somit, 1972, p.209).

[5] Ronald Fletcher, *Instinct in Man,* International Universities Press, New York 1957; McDougall, 1960; N. Tinbergen, *The Study of Instinct,* Clarendon Press, Oxford 1951; Storr, 1968; Thorson, 1970. The term often used is 'instinct theories'.

[6] Some scholars observed that such a revival was expected, because war was connected with irrational elements in human nature. 'Current thermo-nuclear technology and delivery systems make irrationality a key element in the possibilities associated with the causes of war' (Bramson, Goethals, 1964, p.7). Cf. Röling, 1970, about the growing impact of the 'blind' forces on the occurrence of wars (ch.19, pp.248 ff.).

[7] Lorenz, 1966; Ardrey, 1961, 1966, 1970; Morris, 1969; Tiger, 1969; Davies, 1963, 1970; Arthur Koestler, *The Ghost in the Machine,* Macmillan,

New York, 1968; R. F. Ewer, *Ethology of Mammals,* Plenum Press, New York 1968.

'. . . man is a predator whose natural instinct is to kill with a weapon' (Ardrey, 1966, p.316).

[8] In a variant of this theory, N. Tinberger contends that the so-called group-territorialism (the common fighting of members of a group to acquire or defend its territory), which is one of the behavioural characteristics of man's animal ancestors, and which is still innate in him, is one of the main causes of war. Man's limited behavioural adjustability has been outpaced by the culturally determined changes in his social environment. This has increased the probability of wars, because recruitment to the human population surpasses losses through mortality. We now live at a far higher density than that in which genetic evolution has moulded our species ('On War and Peace in Animals and Man, An ethologist approach to the biology of aggression', *Science,* vol. 160, 28 June 1968, pp.1411–18). The scholar concludes that one has to do the utmost to return to a reasonable population density as well as find other ways and means of keeping our intergroup aggression in check.

[9] 'War is an overt action resulting from man's innate aggressiveness. Like other social animals, groups of men defend their home territories at definite boundaries and aggressively seek to own or control larger territories in keeping with the aggressiveness and fear of their leaders and power of their weapons and those of their friends compared to those of their foes. Friends and foes are identified by the leaders by means of similarities or differences of national cultures, religious, class, or political ideology. Aggression is increased by social stress and by authoritarianism, and decreased by distance of a disputed territory from home' (Alcock, 1972, p.199).

For a detailed argument in support of the biological approach, see Norman Macdonald, 'The Biological Factor in the Etiology of War: A Medical View', in Nettleship, et al., eds, 1975, pp.209–31. He contends that of the five factors in the etiology of war – biological psychological, social, economic, and political – the biological is the basic one and its influence has been greatly underestimated. Considering the human affinity with the weapons (which he regards as one of the most powerful features of man's existence), as basically biological he explains this term as follows: 'the term "biological" implies that our use of weapons is not purely cultural in origin but that our genetic code conveys a message which renders resort to weapons a likely response to stimuli which threaten, or appear to threaten, the individual or group' (pp.209 ff.).

[10] Bernard, L. L., 1924; Montagu, 1952, 1968; Hofstadter, 1961; Mackenzie, 1967.

[11] John Hurrell Crook, 'The Nature and Function of Territorial Aggression', in Montagu, 1968, pp.141–78.

[12] Bramson, Goethals, 1964, Introduction, p.14.

[13] Corning and Corning, 1971; Corning, 1971.

[14] McDougall, 1960, ch.11.

[15] 'The authors hold that war – an organized fighting between large groups of adult human beings – must be regarded as one species of a larger genus, the genus of fighting. Fighting is plainly a common, indeed a universal, form of human behaviour ... Wars between groups within the nation and between nations are obvious and important examples of this type of behaviour ... The causes of simple aggression – possessiveness, strangeness, frustration – are common to adults and simpler creatures. But a repressive discipline drives the simple aggression underground – to speak in metaphors – and it appears in disguised forms. These transformations are chiefly those of displacement and projection. These mechanisms have as their immediate motive the reduction of anxiety and the resolution of the conflicts and ambivalence and guilt. They result in the typical form of adult aggressiveness – aggressive personal relations of all kinds – but above all in group aggression: party conflict, civil war, wars of religion, and international war. The group life gives sanction to personal aggressiveness. The mobilization of transformed aggression gives destructive power to groups. Aggression takes on its social form' (E. F. M. Durbin, J. Bowlby, 'Personal Aggressiveness and War', in: Durbin et al., after Bramson, Goethals, 1964, pp.81, 98).

[16] 'The proposition is that the occurrence of aggressive behaviour always presupposes the existence of frustration, and contrariwise, that the existence of frustration always leads to some form of aggression'. Frustration occurs, 'when a goal response suffers interference' (Dollard et al., 1939.). Cf. Corning, Corning, 1971. Frank, 1967, contends that the demands of war have always stimulated a feeling of social solidarity; war has satisfied important psychological needs of man, e.g. a feeling of manhood, and has been an outlet for aggressiveness (pp.35–6).

[17] Ross Tagner 'The Psychology of Human Conflict' in McNeil, 1965; cf. Sheriff, Sheriff, 1953.

[18] War is one of the most effective devices ever invented for releasing hatreds, frustrations, thwarted ambitions, unfulfilled wishes. 'War is the great trigger-release of pent-up emotions' (Turney-High, 1971, pp. 141 ff.).

[19] N. E. Miller, 'The Frustration-Aggression Hypothesis', *Psychological Review,* July 1941; McNeill, 1959.

[20] Berkowitz, 1962; 'Basically I believe a frustrating event increases the probability that the thwarted organism will act aggressively soon afterwards' but 'the existence of frustration does not always lead to some form of aggression, and the occurrence of aggressive behaviour does not necessarily presuppose the existence of frustration' (Berkowitz, 1969).

[21] Gurr analysed the correlation between the outbreak of internal group violence and socio-economic deprivations (1968).

[22] 'if ... man's environment and his culture are organized so that fighting is more rewarding than peaceful pursuits, he will acquire habits of aggression and attitudes of hostility' (Mark A. May, 'War, peace and social learning', in: Bramson, Goethals, 1964.). Cf. his: *A Social Psychology of War and Peace*, Yale

University Press, New Haven 1943, p.200. He contends that four types of social learning are 'mobilized' in war-time: learning to hate and to fight, to fear and to escape, to love and to defend, to follow the leaders.

[23] Hadley Cantril, *The Human Dimension: Experiences in Policy Research,* Rutgers University Press, New Brunswick, New Jersey 1967, p.16.

[24] For example predatory, intermale, fear-induced, irritable, territorial, maternal, instrumental, sex-related.

[25] Review of the psychological theories, which state or imply the above conclusion: Peter A. Corning, 1973 Human Violence: Some Causes and Implications, in Beitz, Herman, 1973.

[26] 'Warfare by which I mean recognized conflict between two groups as groups in which each group puts an army (even if the army is only fifteen pygmies) into the field to fight and kill . . . is an invention like any other of the inventions in terms of which we order our lives, such as writing, marriage, cooking our food. . . .' (Margaret Mead, 'Warfare is Only an Invention – Not a Biological Neccessity' in Bramson, Goethals, 1964).

[27] Sorokin, Band III, also Band II.

[28] Gjessing, 1967.

[29] See: Hobson, 1902; Sombart, 1913; Weber, 1921/22; Schumpeter, 1968.

[30] Wright, 1965, pp.829, 1231; Gantzel appreciates this view as one of the main hypotheses in the study of war (1972, p.197).

[31] Malinowski, 1936: also his: 'War, past, present, and future', in Clarkson, Cochran, 1941.

[32] For a review of studies on the dependance of the primitive militarism and occurrence of wars on culture and personality, see William Eckhardt, 'Primitive militarism', *Journal of Peace Research*, 1975. The scholars concluded that the anthropological findings suggested three basic variables contributing to war in primitive cultures as well as modern societies: private property, frustrated personality, and egoistic morality (p.62).

[33] Toynbee, 1950, p.V ff.

[34] Barbera, 1973, quotes many views of this kind.

[35] For a brief review of the anthropological literature, see Schneider, 'Primitive Warfare: A Methodological Note', in: Bramson, Goethals, pp.275–83. Clarence W. Young demonstrates that war is a product and a tool of cultural evolution, a highly evolved aspect of human political organisation which has proved its viability. The mechanism of evolution is in fact biological, consisting in a combination of intentional mutation with preferential and natural selection; the trait to fight evolved because it was advantageous for human and social development ('An Evolutionary Theory of the Causes of War', in Nettleship, et al., eds, 1975, pp.199–207).

[36] William Graham Summer: 'War', after: Bramson, Goethals, 1964.

[37] '. . . the great watershed of human history, the organization of the state and subsequent archaic empires, was an adaptive process that closely involved predation, conquest, assimilation, and so on – in short, new and intensified

forms of warfare' (Elman R. Service, in Fried, Harris, and Murphy, p.169). 'War lies at the root of the state' (Robert L. Carneiro, 'A theory of the origin of the state', *Science*, no. 169, 1970, pp.733–8); War is the mechanism of state function and it is the necessary although not sufficient cause of the rise of the state (Park, 1964, p.242); 'War has been the medium of the birth and growth of modern nations. All major nations have periodically engaged in war, and the fact that they exist today means that they have won many of them; resort to war has been reinforced by success' (Frank, 1967, pp.35–6).

[38] Andreski, 1965, 1968, 1971. 'War has been throughout history the most effective stimulant of cohesion of states and efficiency of government ... without war neither technology nor the art of organizing could have progressed very far' (1971, p.92).

[39] 'The history of human beings is a history of adaptation to warfare. Man's first tools were weapons – the jawbone of an animal that could be used as a means of defense or offense' (Anthony Harrigan, 'Men Over Weapons. The Chinese Concept', *Military Review*, 1965, vol.1, p.13).

In his review of some findings of psychologists, sociologists, and anthropologists (Ardrey and others), Clarke, 1971, writes that their ideas lead to a reassessment of man the tool-maker: 'What man really made was not a tool but a weapon' and 'It was the weapon that fathered man' (pp.200–2). Ardey is quoted: 'Our history reveals the development and contest of superior weapons as *Homo sapiens* single, universal cultural preoccupation'. Norman Macdonald combines both ideas: 'weapons were in fact tools of acquisition of means of existence' (loc.cit. – cf. note 9, pp.211–3).

[40] Werner Sombart, *Krieg und Kapitalismus,* Duncker and Humblot, Leipzig 1913, after Barbera, 1973, who terms such concept of the role of war 'the productive hypothesis' (pp.4–7).

[41] Corning, 1973, p.128. He quotes Wallace's expression 'revitalization movement' which is defined as a 'deliberate, organized, conscious effort by members of a society to construct a more satisfying culture' ('Revitalization movement', *American Anthropologist,* April 1956).

[42] Barbera, 1973, terms these views 'the unpredictable hypothesis' and 'the irrelevant hypothesis' respectively (pp.10–14).

[43] In a review of anthropological ideas: 'they entertain the view that war is an inevitable adjunct, and in many ways the architect of the civilization that man has built' ('The Case for War', *Time*, 9 March 1970).

[44] For a detailed exposition of the ecological approach, see Peter A. Corning, 'An Evolutionary Paradigm for the Study of Human aggression' in Nettleship, et al., eds, 1975, pp.377 ff. In his description, the main assertion of the ecological approach is that warfare represents a form of collecting coping behaviour clearly associated with basic, survival-related pressures. Anthony F. C. Wallace describes revolutions and other social movements for change as coping or adaptive mechanisms ('Revitalization movements', *American Anthropologist*, vol.58, 1956, p.2).

[45] Harold and Margaret Sprout, *The Ecological Perspective on Human Affairs with Special Reference to International Politics,* Princeton University Press, Princeton 1965, p.27.

[46] Dougherty, Pfaltzgraff, 1971, p.47. Environment is 'the aggregate of surrounding things, conditions, or influences' (Lynton K. Caldwell, 'Environment: A New Focus for Public Policy', in D. L. Thompson, 1972, p.19).

[47] Man has become 'nature limited' instead of 'nature directed' (Kristof, 1960, p.16).

[48] Anderson Nettleship, 'A Description of Certain Features of Prodromal War', in Nettleship, et al., eds, 1975, pp.401–15. The main theme is that sudden local population growth produced war, or, in other words, the need for securing natural resources for the growing population was the main source of war. Cf. Andrew P. Vayda, 'Hypotheses about functions of war', in Fried, et al., eds, 1968.

[49] The term environment has become a vogue word of such attraction that writers uncertain of the intrinsic merits of their ideas are greatly tempted to use it to give their work an air of novelty and freshness. Moreover, since the term is now used in a very general sense to include human as well as physical factors, its misuse is often not immediately apparent. For example, in a recent description of the nature of modern war, an American military monthly considers the following to be environmental factors of crucial importance: (a) The presence of militant communism, which makes war unavoidable; (b) the differing stages of development, reached by the states of the world, which creates a need to form larger and more efficient political and economic units; (c) the revolutionary explosiveness of the aspirations for a better life held by those who live in the backward areas of the world, which must be exploited by the West and not by Communism; and (d) the scientific revolution, which endangers the survival of man at the same time as it offers the prospect of a better life. Of these, only the second and fourth bear any resemblance to actual ecological factors (Robert K. Cunningham, 'The Nature of War', *Military Review*, 1959, vol. 1).

[50] '. . . wherever it exists, warfare serves a cultural adaptive purpose in that it results in a more advantageous relationship between people and their cultural ecology . . . cultural systems define and regulate the circumstances under which expressions of aggression are permitted, what form they take, against what or whom they are directed and the legitimate means of such expressions. . . . The stimuli evoking these responses are culturally determined and extremely varied, defense of the territory being only one such stimulus and not a universal one at that' (Napoleon A. Chagnon, 'Yanamamö Social Organization and Warfare', in Fried, Harris, Murphy, 1968, pp.112, 113).

[51] Gjessing, 1970, also his *Kriehn og kulturene*, Oslo 1950. For a review of the school of American cultural anthropology which sees in scarce resources (including energy) one cause of war (Leslie A. White, Robert L. Carneiro, Marvin Harris, Andrew P. Vayda), see Introduction to Part Four: 'The Ecology and Etiology of War', in Nettleship, ct al., 1975, pp.395–9.

A variant of such a theory holds that one can trace the origin of the war to the consequences of different paces of national growth – economic, demographic, and technological (Choucri, North, 1975).

[52] A. P. Vayda, 'Hypotheses about functions of war', in Fried, Harris, Murphy, 1968. In a more general form, G. Bouthoul contends, that because each war involves some killing of men, wars activated by population pressure in principle are mechanisms for demographic regulation (1951, 1961, 1968).

[53] Erwin R. Brigham, 'Ecostrategy: The Ecological Crisis and National Security', in *New Dynamics in National Strategy,* 1975, pp.37–8.

[54] Toynbee, 1950.

[55] Andreski, 1971. '. . . wars might cease to be a permanent factor of social life only after the restoration of the demographic balance whose disappearance made them inevitable' (p.91).

[56] In a recent version, lack of food and space calls forth aggression in order to obtain a balance to the resources. The contention is illustrated by the behaviour of monkeys and related also to men. Internal and international wars result from lack of food and extensive crowding (Claire and W. M. S. Russell, *Violence, Monkeys, and Man,* Macmillan, London 1968).

[57] For a detailed analysis of the function of war consisting in 'perpetuation of the species through the death of a number of individuals', see Bouthoul, 1951; cf. Fornari, 1974. The 'explosive structure' of society produced by a 'surplus of young men' is treated as reinforcing other causes of war. Against such ideas many objections have been raised, e.g. that the unavoidable costs of war are mistaken for its essential function (Lang, 1972, p.136).

[58] Bernard, 1946; Strausz-Hupé, 1942; Earle, 1943; Dougherty, Pfaltzgraff, 1971, ch.2: J. P. Scott, 'Comment' to P. A. Corning, in Nettleship et al., 1975, pp.385–7.

[59] 'Geopoliticians hold that international relations and, therefore, internal conditions . . . are governed by geographical factors. Political ideologies . . . have little influence on international policy' (T. Hammer, 'The Geopolitical Basis of Modern War', *Norsk Liftmilitaert Tidskrift*, Norway, April 1955); 'Geopolitics is the study of political phenomena in their spatial relationship with, dependence upon, and influence on, earth as well as on all those cultural factors which constitute the subject matter of human geography (anthropogeography) broadly defined' (Kristof, 1960, p.34).
It should be pointed out, however, that some scholars present the geopolitical or geographic factor as a very broad and complex one; it includes many factors from space relationships and climate to industries, population size, communication and transportation networks, etc. (Collins, 1973, pp.167 ff.; Walters, 1974, pp.26–7).

[60] Some scholars consider the term geopolitics as outmoded and contend that it has been replaced by the term political geography defined as the study of political phenomena in interconnection with goegraphical aspects (Lewis Alexander, 'The new geopolitics, a critique', *Journal of Conflict Resolution,*

Dec. 1961); 'Political geography . . . may be described as the study of the variation of political phenomena from place to place in interconnection with variations in other features of the earth as the home of the man' (Richard Hartshorne, Political geography in the modern world, *JCR,* March 1960, p.52).

[61] Kjellen, 1917, held that geography is 'a political categorical imperative'. He defines geopolitics as the study of a country as a geographical organism, from the point of view of its location, form, and area.

[62] Ratzel, 1899; Bernard attributes to the geopoliticians the view that if culture, race, nation, and state don't expand, they die (1946, p.418).
Strausz-Hupé describes the main idea of the theory as follows: since nations grow continuously, territories and frontiers of states are the expression of a transitory power situation (1942).

[63] Ratzel formulated seven laws of expansion: (a) Space increases with the development of culture. (b) The growth of a nation is a result of many tendencies. (c) The growth of a state consists of adding smaller entities. (d) It can result from moving boundaries or from conquering distant territories. (e) State aims at acquiring valuable territories. (f) The incentive for expansion of primitive societies comes from outside. (g) The incentive for expansion passes from country to country.

[64] Some of the ten laws of the influence of boundaries on policy: states tend to acquire access to sea; they tend to conquer territories on the opposite counter-side of the sea; the stronger countries absorb the weaker ones; the number of the independent strong countries decreases (Stephen S. Visher, 'Territorial Expansion', *Scientific Monthly,* May 1935).

[65] Huntington, 1940, 1945.

[66] Mahan, 1897.

[67] Lea, 1909, 1912, Mitchell, 1925.

[68] Mackinder, 1904. His well-known dictum: 'Who rules East Europe commands the Heartland. Who rules the Heartland commands the World Island (Eurasia). Who rules the World Island commands the world' (p.150). He held that also the evolution of the 'coastal peoples' was affected by their geopolitical situation. The need to resist the permanent pressure of the 'heartland powers' increased the internal cohesion of the 'coastal peoples', stimulated the development of the industry and of the marine transportation. The history of both land powers and coastal powers was called by him 'geography in motion'. Crowes' memorandum on British foreign policy is quoted by Carr (1964, p.66): '. . . The general character of England's foreign policy is determined by the immutable conditions of her geographical situation' ('British Documents on the Origin of War', Goods and Temperley, III, p.397).

[69] Writings of Karl Haushoffer, the founder of the German Academy and the chief-editor of *Zeitschrift für Geopolitik,* active supporter of Hitler and one of his advisers.

[70] Mackinder, 1962.

[71] Spykman, 1942, 1944.

[72] 'Modern states can preserve their power position only if they do their strategic and political thinking in war and peace on a global scale. The basis of all sound geopolitical analysis is, therefore, a world map expressive of the location on the earth of the state or states concerned' (Spykman, 1944).

[73] A new dictum has been presented: 'Who controls the rimlands rules Eurasia; who rules Eurasia controls the destinies of the world'; this is because 'It is clear that the oceans play a most significant role in the economic, cultural, and political relations of the states of the old and new worlds' (Spykman, 1944). Rimland is an intermediate region, located between the heartland and sea-countries; if it is conquered by Communism, the USA will be encircled.

[74] 'Since World War II we have had to formulate our foreign policy approximately in Spykman's terms, i.e. to prevent the communist states from expanding into the rimland nations' (William D. Franklin 'U.S. Strategic Policy and the Eurasian Rimland', *Army*, March 1968). 'Spykman inverted Mackinder's thesis, placed the United States of America at the centre of the map, and urged that this country should seek to avoid the unification of the opposing shores of the Atlantic and Pacific Oceans under one or two hostile powers' (Prescott, 1975, p.16). In a recent variant, Cohen, who divided the world into geo-strategic and geopolitical regions, boundaries between them extending across oceans, and the prime division being that between the maritime world and the Eurasian continental world, stressed the importance of the rimland countries and urged a selective involvement in their affairs with the aim of securing dominance over the sea-contacts between the regions (after Prescott, 1975, pp.16–17). Another scholar traces contemporary nuclear deterrence theories to an acceptance of the geopolitical 'heartland' theories of Mackinder. The fear of the Soviet attack is however based on wrong assumptions. Since the West is now geographically and militarily in a position of advantage over the USSR, it has no need to rely so much on nuclear deterrence. Nor did it have to do so in the past after 1945, it exaggerated the possibility of a Soviet massive lightning land attack. Exploiting the terrain, Europe could be successfully defended. The author concludes his geopolitical analysis with a contention that USSR is now vulnerable to an attack on the northern flank, from the Arctic Ocean, and in the south, China; these are the geopolitical causes of its weakness (Walters, 1974).

[75] George E. Lowe analyses attempts of the 'oceanic powers' (Britain and Japan) 'continental powers' (France and Germany) and 'ambivalent powers' (the USA) to conquer the whole 'heartland' or to penetrate it deeply, and concludes that each such attempt always led to abatement of the given country. These 'great mistakes' are the indirect evidence of the value of rimland strategy ('Rimlands Strategy Seen As Basis of New Nixon Asian Doctrine', *Navy Magazine* Nov. 1969, p.35): 'A Rimland Policy is a strategic compromise between two strategies: Fortress America and Pax Americana'. A re-orientation of the American military forces towards a total Grand ocean strategy is therefore required.

[76] Writings of John Slessor, A. P. de Seversky and others. Seversky holds

that America, Africa, part of Europe, and Asia are within the circle of the USA air dominance, the remaining part of Europa, North Africa and almost the whole Asia are within the ellipse of Soviet air dominance. The overlap of the American circle and of the Soviet ellipse is the 'area of decision', the control of which is crucial for the mastery of the air and the dominance over the world (1950).

[77] Rear Admiral Lepotier, 'Geopolitique et géostrategie', *Revue de Défense Nationale,* 1958:2.

[78] Gen. F. Gambiez, 'Geostrategie et géopolitique mondiales', *Revue Militaire Generale*, January 1973, pp.13–18. He contends that the geopolitical importance of the countries in what he calls the seabord world, which like Spykman's rimland includes the countries lying between continental Eurasian bloc and the maritime bloc centred in the America and Oceania, and thus the countries of Western Europe, has therefore declined; Europe would, however, regain its importance in the geopolitical struggle if it were to become a link between the two main forces in the world.

[79] 'Therefore geographic conditions and territorial expanse have acquired new significance' (Hammer, 1965, p.82).

[80] The main theme of Walters, 1974. He criticises the 'heartland theory' and the strategy of nuclear deterrence based on it ('the nuclear fallacy').

[81] Adolf Grabowsky, Raum, Stadt und Geschichte, Köln 1960: Peter Schöller, art. 'Geopolitik', in: Stadslexikon vol.3, Freiburg 1959, pp.776 ff.; Kurt Kipp, 'Die politische Bedeutung der Gegenküste', *Wehrkunde* 1967, pp.397–409; Friedrich Ruge, Seemacht und Sicherheit, 3.erweiterte Ausgabe, Frankfurt am M., 1968; his 'Politik und Strategie', 1967; Boris Meissner, Gotthold Rhode eds, *Grundfragen sowjetischer Aussenpolitik,* Stuttgart 1970; Heinz Brill, 'Zentraleuropa: Glacis oder Cordon Sanitaire? Eine geostrategische Betrachtung', I–IV, *Wehrkunde* 1974:4, 5, 7, 8; geography plays the same role in history as a hereditary feature in the development of an individual: it constitutes the disposition to act in a determined direction (Möschel, 1975, p.390).

[82] Brill, p.190. Following geostrategic aims are pointed out as the most frequent: the access to high seas (for the land-locked countries), the domination over the opposing shores (for the sea powers), the possession of foregrounds etc.

[83] And a 'pendular policy' (Schaukelpolitik) was interrelated, which was sharply criticised by Konrad Adenauer, who waged the policy of a stable alliance with the West.

[84] Bernard, 1946; Dougherty, Pfaltzgraff, 1971; Strausz-Hupé, 1942; Kristof, 1960; Jean Gottman, 'Geography and International Relations', *World Politics,* 1950; Harold and Margaret Sprout, *Man-Milieu Relationship Hypotheses in the Context of International Relations,* Center of International Studies, Princeton 1960.

[85] One of the themes of Prescott, 1975.

[86] Harold and Margaret Sprout write that ballistic missiles, thermonuclear warheads, nuclear submarines, and other changes in weapons and communi-

cations 'have profoundly altered the relative military value of heartlands, marginal lands, rimlands, and islands' (*Foundations of International Politics,* Van Nostrand, New York 1966, p.338).

[87] Kristof points out, that the Pacific Coast was once proclaimed America's 'natural boundary', but afterwards the latter was extended over the Pacific to other countries (1960, p.30).

[88] 'Maps drawn in the service of a given ideology can always tell certain things and conceal many others' (Kristof, p.45). Geopolitics was 'the pseudo-science of the dictatorships, such as was practiced under Hitler, Mussolini, and the men in the Kremlin' (C. London White, 'Vigilance–Yes: Fear–No!' *Military Review,* 3/1957, p.4). Bernard contended, that geopolitical theories were used by reactionary, bellicose ideologies, and combined with the German Fascist theory, thereby they accelerated the outbreak of World War II (1946, p.415).

[89] Critique of Spykman's views has been reviewed in Kristof, ibid. But Spykman also has admirers. Editors of the translation of *Voennaya Strategiya,* H. S. Dinnerstein, L. Goure and T. W. Wolfe maintain that Spykman doesn't praise power, but only regards it as the main means of foreign policy. They quote Spykman's statement: 'Power means survival, the ability to impose one's will on others, the capacity to dictate to those who are without power, and the possibility of forcing concessions from those with less power', such a statement is not to be identified as a direct call for the use of power (p.133).

[90] Muir, 1975, Part 7, *Political geography and the international system.* Cf. Russett, 1967; Cohen, 1964; N. J. G. Pounds, *Political Geography,* New York 1964; Kristof, 1960.

[91] Craig Hosmer, 'The New Geopolitics', *U.S. Naval Institute Proceedings,* Aug. 1973, pp.19–23; Prize Essay 1973.

[92] Francois Duchenes, Introduction, in: 'Civil Violence and International System', Part I, *Adelphi Papers* no. 82, London, December 1971, pp.4–6.

[93] For example the strategic choice between the triangles: Washington–Moscow–Peking, and Washington–Tokyo–a capital representing Western Europe, is a choice made 'in geopolitical terms' (Stanley Hoffman, 'Choices', *Foreign Policy,* Autumn 1973, p.12).

[94] Peter Grose, 'A New Geopolitics', *International Herald Tribune,* 13 October 1974.

[95] Some Marxist–Leninists contend, however, that war is seen by the representatives of the legal approach as a legal condition justifying the resolution of conflicts by use of armed violence.

[96] Paul Kecskemeti, 'Political Rationality in Ending War', in *The Annals of the American Academy of Political and Social Science,* 'How Wars End', November 1970, p.106.

[97] Wright, 1965, p.13.

[98] Ibid., p.698.

[99] Ibid., p.700.

[100] The author of the preface to the second edition of Wright's study

contends, that the four listed factors may be interpreted as technology, law, social organisation and distributions of opinions and attitudes concerning basic values; they correspond to the technological, legal, socio-political and bio-logical–psychological–cultural levels of human life (Karl W. Deutsch, Quincy Wright's contribution to the study of war in Wright, 1965, p.xiii).

[101] Wright, 1965, ch.X.

[102] Ibid., pp.1512 ff., 1521 ff.

[103] Wright, 1955. A general definition of war has been added: 'War is the art of organizing and employing armed force to accomplish the purpose of a group' (p.148).

[104] 'War', in *International Encyclopaedia of the Social Sciences,* Macmillan Free Press, New York 1968, vol.16, p.453.

[105] See Kotzsch, 1956, pp.22–4.

[106] In his encyclopaedic article he states, however, that the legal concept of war was chosen because sociologists and lawyers sought criteria sharply separating war from peace; war 'in the ordinary sense' i.e. as armed conflict among political groups, is not sharply distinguished from peace (1968, p.453).

[107] Reves, 1945, 1946.

[108] Wright presents a different impact of war on processes of integration. In the earlier stages of human history war favoured political disintegration, in the first period of the modern civilisation it acted on behalf of integration (states and civilisations have been built up by wars). Afterwards the modern empires disintegrated through wars, and wars of our times may result in the disintegration of our civilisation, although the end of one civilisation may be the beginning of the next one (1965, pp.257 ff.).

[109] For an analysis of this problem, see Deutsch, 1967, pp.91–4.

[110] War was said to summon forth human efforts of heroic proportions (William James, *A Moral Equivalent for War,* Carnegie Endowment for International Peace, New York 1926).

[111] R. Benedict, *Wzory kultury,* Warsaw 1966; H. I. Marrou, *Historia wychowania w starozytnosci,* Warsaw 1969; Paul Lafargue, *Pisna Wybrane,* Warsaw 1961; A. Ossowska, 'Ethos rycerski w legendach średniowiecza', *Studia Socjologiczne* 1968:2; A. C. Knudson, *The Philosophy of War and Peace,* New York 1947.

[112] Some anthropologists indicate that in the primitive tribes fighting was often a means of proving one's virility and manliness.

[113] An expedition, led by Roberg G. Gardner, examined the Willigiman–Wallalua people in New Guinea. The people under examination lived, warred, and died as Stone Age men may have. They lived in a state of a permanent warfare with their neighbours, although none of the usual reasons for waging wars were detected, neither winning a territory, nor seizing goods or prisoners. The weapons were never improved, no bit of strategy was devised. Obviously, the people were not interested in a victory in a common sense, it treated fighting as a way of life. 'They fight, because it is to them the vital

function of the complete man, because they feel they must satisfy the ghosts of the slain companions' (A review article, 'The Ancient World of a War-torn Tribe', *Life*, 28 September 1962, p.78).

[114] J. Huizinga writes that war was often motivated by strife for glory. 'The contemporary outbursts of worship of war ... return to the Babylonian–Assyrian conception of war as a dictate of God for the saint glory) (*Homo Ludens: A Study of the Play Element in Culture,* Beacon Press, Boston 1955).

[115] 'Human, All Too Human', vol. 1, 1878, in Geoffrey Clive, ed, *The Philosophy of Nietzsche,* New American Library, New York 1965, pp.372-3.

[116] He called war the way to truth, a medicine given by God, and badly needed by man (*Der Genius des Krieges und der deutsche Krieg*, Verlag der weissen Bucher, Leipzig 1915: *Die Idee des Friedens und der Pazifismus,* Der neue Geist Verlag, Berlin 1931: *Ursachen des Deutschenhasses,* Wolff, Leipzig 1917). See: Ossowska, *Socjologia moralnosci,* Warsaw 1963.

[117] For an analysis of Moltke's and Bernhardi's views, see Andrzej Jozef Kaminski, *Militaryzm niemiecki,* Wyd MON, Warsaw 1962, ch.III Czciciele wojny.

[118] Ludendorff's theory of total war, see section (d) and note 130. Moltke held that such virtues as courage, unselfishness, fidelity and readiness to self-sacrifice manifest themselves in war. His dictum has often been quoted, 'Eternal peace is a dream, and even not a nice one, war is a link of God's order in the world' (In a letter to Professor Bluntschli, 11 December 1880, in *Moltke in seinen Briefen,* Mittler und Sohn, Berlin 1902, v.ii, p.252-3). Another well-known opinion connected with the biological approach to war: 'War is a biological necessity ... it is as necessary as the struggle of the elements in the Nature. . . It gives biologically just decisions, since they rest on the very nature of things' (Friedrich von Bernhardi, *Germany and the Next War,* Longmans, New York).

[119] 'Social war, by making an appeal to the honour which develops so naturally in all organized armies, can eliminate those evil feelings against which morality would remain powerless. If this were the only reason ... this reason alone would, it seems to me, be decisive in favor of the apologists for violence' ('Apology for violence' 1906. Appendix 2). See also Vilfredo Pareto, *The Mind and Society*, transl. by A. Bongiorno and A. Livingston, Harcourt, Brace, Co., New York 1935.

[120] 'Some Social Functions of Violence', in 1967, pp. 78–81. This function of violent internal conflict is qualified as 'achievement'. Hedley Bull observes, that some scholars, expressing such opinions, consider violence as a 'way of life', they enjoy violence 'as a culture or life-style' (1971, p. 29).

[121] *The Wretched of the Earth,* 1961. (Grove Press 1968). Sartre in the preface writes that irrepressible violence is 'man-recreating himself'.

[122] Bellicose military writers expressed such views in extreme wordings: 'War is the supreme test of man, in which he rises to heights never approached in any other activity' (Gen. George S. Patton, in *Time,* 9 March 1970).

[123] Arendt, 1969, p.30. She states that such a view, expressed by the newest apologists of violence as a life-promoting and morally perfecting force means a revival of Bergson's and Nietzsche's philosophy of violence, life, and creativity.

[124] 'In spite of the countless miseries which follow in its train, war has probably been the highest stimulus to racial progress' (D. McKay, *Races and Peoples*, Philadelphia 1901, p.76).

[125] *Blut und Ehre, Reden und Aufsätze, I–III*, Thilo v. Trothe, München 1937–1940, IV, Karlheinz Rüdiger, 1943; *Mythos des XX Jahrhunderts*, 1930.

[126] Philip Bouhler, *Der grossdeutsche Freiheitskampf Reden*, Franz Eher Nachf., München 3 vols. 1940–43; *Mein Kampf*, 464–8 Auflage, Zentralverlag der NSDAP, München 1939.

[127] Friedrich Lange, *Reines Deutschtum*, IV ed, 1914, p.237.

[128] Klaus Wagner, *Krieg*, 1906, p.129.

[129] Karl Zimmermann, *Deutsche Geschichte als Rassenschicksal*, 6 ed, Quelle u. Mayer, Leipzig 1933.

[130] J. Ruskin, Galazka dzikiej oliwy, Cztery odczyty: O pracy, handlu, wojnie i przyszłości Anglii, Warsaw 1900.

[131] For example Ottmar, Spann, *Zur Soziologie und Philosophie des Krieges*, Berlin 1913; Heinrich Gomperz, *Philosophie des Krieges in Umrissen*, Gotha 1915; '. . . those intellectual and moral factors which ensure superiority in war are also those which render possible a general progressive development' (Friedrich von Bernhardi, quoted in Dougherty, Pfaltzgraff, 1971, p.163); war is 'the great force of culture', a servant to mankind (Steinmetz 1929); 'Die Bedeutung des Krieges bei den Kulturvölkern' (in *Zeitschrift für sozialwissenschaft*, 1914, p.396); 'all higher cultures developed from war' (Picht, 1952, pp.32–3, 97, 115).

[132] 'When men were fighting for glory and greed, for revenge and superstition, they were building human society, discipline, cohesion, cooperation, perseverance, fortitude and patience were products of "war-education"' (Summer, in Bramson, Goethals, 1964, p.212). And: 'war operated as selection, destroying social prejudices, privileges, orthodoxy, tradition, popular delusion, false ideas and doctrines, furnishing a stimulus to thought and knowledge, to higher moral and progressive principles' (pp.222–23).

[133] Spengler, 1926–28, 1934.

[134] Ratzenhofer, 1881.

[135] 'We have learned to perceive the moral majesty of war through the very processes which to the superficial observer seem brutal and inhuman. The greatness of war is just what at first sight seems to be its horror – that for the sake of their country men will overcome the natural feelings of humanity, that they will slaughter their fellow men who have done them no injury. . .' (*Politik*, Vorlesungen gehalten an der Universität zu Berlin, Hirzel, Leipzig 1897–8, 2 vols.; 5th ed. 1922, transl. Politics, Macmillan, New York 1916, p.395–6). See also: *Aufsätze, Reden und Briefe*, Karl Martin Schiller ed, Hendel, Meersburg, 1959, 5 vols. and *Deutsche Geschichte im neunzehnten Jahrhundert*, Hirzel,

Leipzig 1927, 5 vols. Treitschke's writings and teachings were well known, learned and quoted in Germany. He was for more than twenty years the chief-editor of the *Preussische Jahrbücher* and of the weekly *Grenzboten*. Alfred Rosenberg in an introduction to the one-volume edition of *Geschichte* stated, that Treitschke was one of the 'fathers' of Nazism.

[136] *Meine Kriegserinnerungen 1914–18,* E. S. Mittler und Sohn, Berlin 1919; *Der totale Krieg,* Ludendorffs Verlag, München 1937; Ludendorff wrote that by war the nation could become conscious of its aims and destiny (Der Krieg bringt das Volk sich selbst).

[137] 'Some scholars contend, that there is a small chance to abolish wars, unless some substitute could be found to improve human nature and to liquidate the undesirable social, political, and economic conditions. A premature step might only lead to unbearable stagnation and preservation of the untenable *status quo*' (Anthony E. Sokol, 'Disarmament – Is It Possible?', *U.S. Naval Institute Proceedings,* May 1962, p.62. 'Mankind can find no better way to overcome poverty than war' (p.63).

[138] Wright, 1965, pp.108–9; Nieburg 1963; 'The Threat and Use of Force', in *The Theory and Practice of International Relations,* p.363.

[139] 'American military leaders did not become acquainted with ... the relationship between national policy and military strategy until World War I' (Ginsburgh, 1965). 'The horizon of war was strictly limited to the theatre of combat, and battle was the end–all, the final objective. Almost no connection was made between war and a nation's political objectives. War was a separate and distinct phenomenon from statecraft and politics' (Justin H. Smith, *The War with Mexico,* New York 1919, after Ginsburgh, op.cit.). Some other opinions cf. ch.8.

[140] Reviewed in J. Lider, 1971, pp.54–62.

[141] Cf. ch.3.

[142] Smith, 1955; Gavin, 1958; Ginsburgh, 1965. Gen. Mark Clark stated before a Senate subcommittee: 'We should fight to win, and we should not go in for a limited war where we put our limited manpower against the unlimited hordes of Communist manpower. If fight we must let's go in there and shoot the works for victory with everything at our disposal' ('The Korean War and Related Matters, Report of the Subcommittee to Investigate the Administration of the Internal Security Act and Other Internal Security Laws to the Committee on the Judiciary US Senate', 84 Congress, I Session, 21 January 1955, Government Printing Office, Washington, 1955, p.7). Dean Acheson quotes this opinion and writes, that the American military were 'entranced by Clausewitz's theory of pure war' and certain generals 'became the poor man's Clausewitz' (1958, p.35).

[143] *War in three dimensions,* London 1949; *The Direction of War,* London 1955; also 1960, 1964. Cf. Wylie, 1967; he sees war as a phenomenon independent of policy, governed by its own laws. War is regarded as a 'radically different world' from the peace world the latter governed by policy.

[144] 1957, p.17. Therefore he divides wars according to the amount of violence applied to 'impose the will' which leads to classification based on military-technical indicators: total, limited, local and cold wars (1957, pp.17–18; 1960, pp.14–18; 1964, Ch.1). Beaufre defines war as a competition of power (Machtprobe) in which one side attempts to impose its will upon the other side. Strategy is a dialectical, violent conflict of wills. The aim of war is the capitulation of one side to the will of the opponent. This is a common feature of all wars, which differ in the way, how they are conducted (1975, pp.13–14).

[145] Some other scholars relate all evolution of the nature of war to changes in the technology. Beaufre writes that the beginning of the classical wars and the emergence of armies were greatly influenced by the development of fortifications (1975, p.17); Coats's analysis of the development of war from primitive warfare to modern fighting is based on the development of military technology. Modern warfare developed concurrently with the rise of industrialisation and technology (1966, pp.36–7) and is distinguished from its earlier forms by expansion of technology (p.80).

[146] 1957, pp.19, 26.

[147] However, although war begins as a fight for definite political aims, after some time the influence of policy greatly decreases: '. . . the aims which maintain hostilities, once declared, are apt on both sides to become nebulous . . . the war may continue long after the original cause has been forgotten, by both sides being anxious even to fanaticism, to avoid failure or defeat. Nations are even capable of redoubling their efforts when they have forgotten their original purpose' (1957, p.22).

[148] '. . . the most serious wars are fought in order to make one's own country militarily stronger, or more often, to prevent another country from becoming militarily stronger, so that there is much justification for the epigram that "the principal cause of war is war" (Carr, 1951, p.111). The question whether or not arms race is a cause of war has been discussed in many books and articles, e.g. Huntington, 1958, Clarke, 1971 (ch.VII, section: 'The Arms Race'), Blainey, 1973 (pp.134–150); John C. Lambelet, 'Do Arms Races Lead to War?' *Journal of Peace Research,* 1975. In a collective study, in which twenty historical times and places from 225 BC to 1776 AD have been investigated, the theory that an arms race increases the probability of war has been disproved (as well as the theory that deterrence diminishes the war frequency) (Narroll, Bullough, Narroll, 1974).

[149] For example war has been defined as: 'A conflict among political groups involving hostilities of considerable duration and magnitude' (*Encyclopaedia Britannica,* Enc.Brit.Inc., 1964, vol.23, p.321); 'Hostilities between states or within a state or territory undertaken by means of armed force. . . De facto war exists . . . whenever one organized group undertakes the use of force against another group' (Jack C. Plano, Roy Olton, *The International Relations Dictionary,* Western Michigan University, Holt, Rinehart, and Winston, Inc., New York 1969, p.77); 'A conflict between two states carried on in general under

the recognized rules for such conflicts, and including both armed hostilities and the intention to carry on such a conflict' (White, 1948, p.309); '1. Conflict carried on by force or arms, as between nations or states (international war or public war), or between parties within a state-civil war)' (*The American College Dictionary*, Claurence L. Barnhart, ed, Random House, Inc., New York 1947). 'War means hostilities carried on by armed forces between nations, states, governments or groups within a nation or state' (*The Shorter Oxford English Dictionary on Historical Principles,* 3rd rev. ed, vol.2, Oxford 1965, p.2383); this definition is assessed as generally accepted and quoted in: Sovjetsystem und demokratische Gesellschaft, Eine vergleichende Enzyklopädie, Herder, Freiburg-Basel-Wein, Band III, p.1027.

[150] Robert C. Angell, 'The Sociology of Human Conflict', in: McNeil, 1965; cf. Dahrendorf, 1962.

[151] J. David Singer considers the nation-state as the main actor of each international conflict ('The Political Science of Human Conflict', in: McNeil, 1965, pp.140–2); in a later study, together with Melvin Small (1972) he presents five levels of analysis: (a) the global system; (b) the international system; (c) the interstate system; (d) the central system; (e) the major power system. The state remains the basis of all these systems (pp.16 ff.). K. J. Holsti sees three levels of political analysis: the individual (actions and attitudes of individual policy makers), state activity, and systemic (international) politics viewed from the perspective of entire systems of states (1972, p.17). Cf. note 3 in this chapter.

[152] Midlarsky applies a combined approach; he uses three levels of analysis: the international system, the boundary between the system and the nation-state, and the state. On the last level, the possible effects on war by domestic institutions are considered. He excludes investigation of the 'individual as actor' and deals with aggregate behaviours leading to war over a relatively long period of time and a larger number of cases (1975, Ch.I, section 'Levels and theories', pp.2 ff.).

[153] Angell, 1910.

[154] '... the most constant cause of war is the determination of a people to raise national living standards or to protect existing standards' (Wallace, 1957, p.278). War is 'a non-economic activity of social institution purposefully directed to the maintenance of the conditions of a society's economic life' (Frederick E. Emery, in Wallace, 1957, p.41).

[155] '... the elementary population pressures and food shortages are not potentially within the control of human technology and need no long drive to war. Economic forces in general are less direct causes of war than used to be thought' (Geoffrey Sawer, in Wallace, 1957, p.383). Quincy Wright reviewing the collective work edited by Wallace (1957), states: 'With this the reviewer agrees and is happy that the somewhat categorical assertion by some of the authors of the priority of economic conditions and motives in causing war is thus played down' (A review of Victor H. Wallace, Path to Peace, *JCR,* December 1958, p.350).

[156] Hobson, 1902, Schumpeter, 1968. Hobson dispraised the 'financial capitalism' for organising wars for economic reasons. Imperialism was 'the use of the machinery of government by private interests, mainly capitalist, to secure for them economic gains outside their country'. Wars were of an economic nature because all behaviour of capitalists that waged them was primarily motivated by the desire to gain economic (financial) profits. Schumpeter stated that economic reasons played a great part in wars waged by the contemporary developed countries. Both contended that war could be eliminated, or at least its likelihood greatly diminished, by improvement of the capitalist system. Of those scholars who recently have described socioeconomic roots of wars Richard J. Barnett may be taken as an example (1972); he recognises three roots of wars waged by the USA in the twentieth century: the pressure of the capitalist economy toward economic expansion and military presence abroad – the main root, the concentration of power in a national security bureaucracy, and the vulnerability of the public to manipulation on national security issues. Interests of the corporations seeking economic opportunities abroad are presented by them as identical, or at least not conflicting, with the national interest.

[157] Cf. the typology presented by Prof. Malinowski in the 'anthropological approach'.

[158] Kotzsch observes that war was at first 'armed robbery' and only afterwards, when it was converted into institution and when the original socio-logical conception of war became mixed with theological moral and quasi-legal elements, it was called war (1956 p.25).

[159] One of the main conclusions in Dougherty, Pfaltzgraff, 1971, pp. 138 ff. Also the general point, on which converge the contributors to Wallace, 1967.

[160] The theory which states that the outbreak of war is affected by a combination of societal, attitudinal, and structural factors may serve as an example; societal factors refer to variables characterising the society as a whole (the geographic position, the economic, technological, political conditions, the size of population, etc.); attitudinal refer to variables characterising individual attitudes, values and motivations toward other nations, of which the attitudes of the decision-makers are of special importance; and structural factors relate to the power structure (what groups have control over foreign policy decisions); for the exposition of this theory see H. C. Kelman, 'Societal, attitudinal, and structural factors in international relations', *Journal of Social Issues,* 1955:1.

[161] The literature is obviously too vast to be presented here. To exemplify, Blainey in his analysis of the causes of wars, 1973, presents war as a test of power. 'War is a dispute about the measurement of power', or in other words, 'wars usually begin when two nations disagree on their relative strength'.

[162] For a review of the multi-dimensional theories see Bernard, 1944.

[163] Gaston Bouthoul, 'Les baromètres polémologiques', *Etudes Polémologiques,* 1972:4; Rene Carrere, 'Die Zukunft der Gefahr und die Polemologie', *Beiträge zur Konfliktforschung* 1971:1/2.

[164] Waltz, 1959; some typical opinions expressed by other scholars: 'No

major conflict has ever had one simple or single cause that has been adduced' (R. M. Crawford, in Wallace, 1957, p.16); war as a complex multi-dimensional social phenomenon has so many sources and causes that no theory of a single cause can explain its nature. One cannot find a single necessary condition and a single sufficient condition, one can only try to find sources, factors, conditions important for the occurrence of wars (Robin M. Williams, Jr, 'Conflict and Social Order: A Research Strategy for Complex Propositions', in Collective Violence and Civil Conflict, *The Journal of Social Issues,* 1972:1).

[165] Singer, 1972.

[166] 'Approaches to Peace Research' in Nettleship et al., 1975, pp.44 ff.; and 1974, p.154; cf. 'Systemic and Strategic Conflict', in Falk, Mendlowitz, 1966.

[167] 'No two wars will manifest precisely the same configuration of causal variables' (Corning, 1973, p.129). And: 'For instance, an incompetent leader may stumble into war that could have been avoided; a skilful leader, seeking desperately to avoid war, may be overwhelmed by events; or an ambitious and highly aggressive leader may precipitate war that is unjustifiable on any ecological ground'. Many diversified causes of war were analysed by Turney–High in his excellent study on primitive war (1971). He stated: 'Every war is fought for more than one motive, spurious or real, appreciated or unrealised' (p.141), and pointed out that the relative importance of particular causes changed in the course of history.

[168] 'Conflict and War from the Viewpoint of Historical Sociology', in McNeil, 1965, p.201. Kingston–McCloughry contended that 'No two wars take the same form', because of variety of their military–technical correlates. As to the future war, it 'may be something quite different from any which mankind has so far experienced' (1957, p.17).

[169] A more or less agnostic approach can be recognised in many contemporary studies on war, 'Why wars take place is not at all clear . . . even more problematic is why a given war occurs between the given states at a given time' (Levi, 1974, p.200). After examination of various theories of the causes of war, all of them considered by him as unsatisfactory, Bernard Brodie concludes: '. . . any theory of the causes of war in general or of any war in particular that is not inherently eclectic and comprehensive, that is which does not take into account at the outset the relevance of all sorts of diverse factors, is bound for that very reason to be wrong' (1974, p.339).

3 War as an instrument of policy: the non-Marxist theories

In the political approach, which was only sketched briefly in the foregoing chapter, war is regarded as a way in which political competition may occur, distinct from other forms of political struggle in that military means are used. Moreover, war is the ultimate arbiter of political conflicts and is applied only when peaceful means have not proved effective.

An assumption common to the various political views of war is that it is the continuation of and an instrument of policy. The similarity is, however, deceptive for no two theories mean exactly the same thing by the principal terms, policy, state, and war. These are given different meanings in accordance with the different ideological and political values that underly the theory of war. Since the aim of this book is to confront the Soviet and non-Marxist theories of war, it is therefore necessary first to identify the non-Marxist interpretations of such terms as policy, state, and war.

The terms policy, state, and war

Policy

Policy is most often interpreted as activity taken by one group in relation to others, in order to realise its interests and exert influence on the other groups. The activity includes both the elaboration of guidelines and plans and their realisation. [1] In non-Marxist studies of war, the political unit chosen by the overwhelming majority of scholars is the state. Thus war is considered to be an instrument of that policy which expresses relations between states, and policy is defined either as the activity by which a state realises its interests, or more simply as the activity of the state. It has consequently been used most often in the sense of foreign policy. Domestic policy has been treated as a factor influencing foreign policy, but its connection with war is usually regarded as indirect and indecisive. The school of political realism, which will be described below, shares this traditional view of policy. Policy in this school is closely connected with power; the conduct of policy is the acquisition, defence, and application of power. [2] Thus search for power distinguishes policy from other kinds of human activity.

State and its functions

In the non-Marxist interpretation, the state is a form of social organisation, acting on behalf of a nation. Its existence and its activity is based on its recognition by the nation as the legitimate form of its organisation and on voluntary submission to its authority. Within the normal state, people live in relative harmony with each other, they respect the law, largely out of voluntary habit, but partly out of fear of punishment, for the state has the right to punish those who break the law. [3]

The state acts as the nation's agent, both inside and outside the country. Amongst the internal functions the state is considered to have are the maintenance of the unity of the state, the regulation of economic activity, the maintenance of law and order, reconciliation of classes, and the regulation of conflicts by non-violent means. The external functions include the protection of the population and the defence of national interests, and even, it is sometimes suggested, the conquest and exploitation of other states. [4]

The convention of treating the state as the agent of the whole nation, as well as using the word nation to denote the people who make up the state, has made such an impact on the terminology that in the study of state politics, the terms state and nation are now used almost interchangeably. [5]

War

According to international law, war, in principle, can only take place between sovereign political entities, that is, states. War is thus a means for resolving differences between units of the highest order of political organisation. The majority of those who have been concerned with war as a socio-political phenomenon have also adopted as their basic premise that there is a fundamental difference between domestic conflicts, for which there are normally mechanisms for peaceful resolution, and international conflicts, which occur in a state of anarchy. Wars have been seen to involve directly state institutions, such as the foreign office and the armed forces. Since war is put in an international context, the stakes of war may be the life and death of states. [6]

This general outlook on war as an international or inter-state phenomenon has been shared by many students, [7] regardless of their professional background as political scientists, [8] historians, sociologists, [9] psychologists, [10] or military analysts. [11] The school of political realism (see below) maintains [12] that the nation-state can only realise their national interests by demonstrating their willingness to fight and by making use of wars of various degrees of magnitude as an instrument of national policy to achieve legitimate ends. [13]

In recent years, many attempts have been made to place war in some other perspective in which the state as actor is not the only unit of analysis. One approach has been to choose another main actor, [14] for example, the system. Another has been to achieve a synthesis by combining kinds of actors – men, state, and the international system – or by including internal wars as part of a synthetic concept of war. [15]

One of the most important challenges to the concept of the state as an actor has come from the contention that the behaviour of states is really the behaviour of the men who act in the name of the state. Those who have power may not represent the national interests but rather the interests of those groups whose views the decision–makers consider relevant. The role that interest–groups have in the making of foreign policy of developed countries is a subject of recurrent interest. [16]

Three different groups have been indicated. The first group is big business, which it is maintained, gains by war through the close ties between business magnates and political leaders. Secondly, special interest has been devoted to the arms industry, which with its direct contact with the military establishment constitutes an important lobby. [17] The literature concerning the activity of the 'merchants of death' or the 'industrial–military complex as this combination is called, is quite large. [18] Finally, in some analyses those who carry out the foreign policy are mentioned as having many opportunities to influence it, and as liable to act in the interests of the ruling class. Moreover, it has been suggested that the rules by which international diplomacy is carried out give diplomats an interest in creating tensions, crises, and even war. [19]

On the other hand, some scholars deny that the influence of interest groups, although important, is decisive for the explanation of war. Karl Deutsch, for example, presents a classification of different interest groups: special interest groups, more general interest groups, and all-purpose interest groups. [20] He suggests that the decision-making process and the definition of the goals of foreign policy are determined by the interaction between many different interest groups, which often compete with each other. Moreover, these interest groups are in turn affected by elements of randomness, discontinuity, and unexpected change, and they are also to some extent controlled by opinions from the social and economic élite, the governmental élite, the mass media, and others. The result of these interactions and counter-influences is that the policy finally adopted reflects the interests of the nation. The influence of the large economic interest groups on decisions about war is as a result thinned out. Deutsch concludes therefore, that the most powerful and cohesive interest group is the nation itself, which he also defines as the people in control of the state. [21]

In spite of many differences in the way various theories and analyses define, classify, or evaluate the role of interest groups, one feature they share is that they do not relate the existence and activity of such groups to the socio-economic and political system as does the Marxist thesis.

Within the non-Marxist tradition and especially within the field of peace research, there are, however, many other theories in which the role played by interest groups towards preparations and instigation of war is emphasised and placed in a structural setting. The two classical scholars are John Hobson and Joseph Schumpeter, who indicated that to understand war it is necessary to see 'who stood to gain', and searched for the sources of wars in the imperialist system. More recent examples include studies of the armed interventions taken by France and Great Britain in their colonies after the Second World War, the influence of large corporations and military bureaucracies in the conduct of the war in Vietnam, [22] and the role of interest groups in the arms race and policy of deterrence, especially a study made by a group of West German researchers which included Dieter Senghaas and Fritz Vilmar. [23]

Studies dealing with the war in under-developed countries have also taken up the role of privileged powerful élites. It is suggested that these groups are responsible for internal wars that often lead to international ones, since they can take advantage of dominance imposed by force to increase their spoils. [24]

War as a means of regulating international relations

Schools and approaches

It was mentioned in the introduction that amongst the various non-Marxist political theories of war the school of political realism and in the first line its American variant will be chosen for comparison with Marxist theory. In order to sketch some general scheme of the political approaches, and to place the school of political realism in relation to other ones, some scheme of classification is necessary. In this section three different kinds of classification often used will therefore be presented.

Political orientation There are many ways of classifying theories of war on the basis of political orientation. Here the three most general forms will be reviewed. The first is to place the theories according to the common parliamentary configuration from left radical to conservative right. The difficulty with such a classification is however, that the terms 'left' and 'right', 'radical' and 'conservative', 'democratic' and 'anti-democratic', have been put to so many different uses that their meaning has become ambiguous. Moreover, views of the nature of war and the role it plays are often only indirectly connected with party politics.

A second classification according to political criteria is based on the general aims of military policy. Those who consider the primary goal of American policy to be to stem or reverse the advance of communism constitute one wing; those who regard the prevention of war, especially thermonuclear war, as the most important end are on the other wing; and those who assert that the real interests of the USA lie between these two make up a middle group. [25]

The third common classification of this type is a twofold division according to basic political values. On the one hand are those who believe that international relations are, can, and should be a field of cooperation because the interests of particular states coincide with the interests of the community of nations. They

51

believe consequently that war is a mistake and that it may be eliminated and replaced by cooperation through rational appreciation of common interests. On the other hand are those who assert that competition dominates international relations, that the aims towards which states strive conflict, and that states can only assert and defend their interests by using, or threatening to use, their power. In this view, war is a natural means of resolving conflicts. [26]

Fields of interests, methodological orientation, and association with universities One can find in the literature, examples of different classifications that group researchers use according to their fields of interests, and methodological orientation, [27] or, in some cases, according to the research institute to which the scholars have been attached. [28] However such classifications do not reflect the real differences in the views of the individual researchers. Each methodological or theoretical orientation can encompass a wide range of different views, and one research institute may support research representing various schools of thought. Besides, many researchers are connected with work in two or more places at the same time, or move from one to another.

Politico-military views A third common method of classification, one that is adopted by the present author, [29] is based on the politico-military views of the analysts. Others have made use of such single criteria as the predicted nature of a future war, the principle aims of a future war, and the postulated strategy of deterrence as a basis for foreign policy. It seems more purposeful, however, to arrange American politico-military thought according to the main assumptions of the politico-military doctrine taken as a whole, that is, to the views of the conceived or postulated nature of a future war, and of military policy in time of peace. At one end of the scale come what might be called the most bellicose (in Marxist-Leninist terminology the most right-wing) school, which holds that war with world communism is inevitable and that the possibility of a preventive war must be taken into account. [30] Next come those who for similar reasons are proponents of preemptive war. [31] Then follows the school of 'forward strategy', associated with the Foreign Policy Research Institute in University of Pennsylvania (Philadelphia), which advocates an active policy of 'rolling back' communism even at the risk of a nuclear war. [32] In the centre come three variants of the flexible response. In this group belong the most influential politicians, political scientists, sociologists, and military men, most of whom may be included in the school of political realism. [33] The common denominator of the views presented by this group seems to be the assumption that the position of the West in the world-wide struggle between the two opposing systems must be based on strength, it must be led and directed by USA, but it must be flexible. The policy of the West must be carried out with great care so that a global nuclear war does not break out, but limited wars and, when necessary, measures of arms control are considered to be possible means. In principle, the position is based on the strategy of deterrence, in which a complex of political, ideological, and military means are deployed. The next category is the school of minimum deterrence, which advocates diminishing the danger of world war by decreasing the nuclear arsenals to the minimum necessary to real deterrence. [34] Then come the various schools of thought that present different forms of disarmament proposals, from graduated reduction of international tension (GRIT), [35] to total nuclear disarmament (nuclear pacifism), and, at the exteme end of the scale, total disarmament (integral pacifism).

Political realism

Power The characteristic tenant of theories belonging to the school of political realism is that international politics is a struggle for power. [36] Power is said to be necessary to promote the interests of the state. The political realists thus admit that nations may pursue other goals with their foreign policy but, they contend, the attainment of all other goals is dependent on the amount of power at the nation's disposal. Furthermore, the political realists believe in the disharmony of national interests: the policy pursued by each country will therefore clash with the ambitions of all other states, and power will decide the outcome.

Since power is in this view a necessary means for the achievement of whatever goals a particular state may set for itself, it takes on the virtual status of an end. [37] In a system of independent states, where each state is the sole guardian and guarantor of all its own interests, it must possess power. It may not know when the power will be needed, nor for what national ends, since many of these are unforeseen, but the state tends to accumulate power as a reserve. [38] Moreover, power may be sought for the simple satisfaction of being powerful.

Many political realists regard the quest for power as rooted in a tendency inherent in individuals to dominate over others. Power owes its existence to the instinct to dominate. [39] The term power is defined as all that establishes and maintains one man's control over the minds and actions of other men. [40] This tendency is projected into the field of activity of states according to the classical assumption of political realism, statemen think and act, and thus states act, in terms of national interests defined as power. [41] Such words as dominance, control, or influence in political affairs are often used as synonyms. [42] A state has political power in relation to another state, therefore if it has influence over the minds of the leaders in the other country and thus control over the state's actions. [43]

The balance of power The main dynamic in the realist model is therefore that all states are engaged in an incessant struggle for power, each one aiming to achieve as much of the available power as possible in order to realise its interests; great powers may seek a dominant position. [44] The interplay between states is however, regulated by what some consider an automatic, others a semi-automatic, process. This can generally be described in the following way; [45] when a state, or a group of states, threatens to become powerful enough to upset the existing distribution of power (or the balance of power), some other state, or group of states, recognising a threat to their security and independence, respond to meet the challenge. [46] The incidental outcome of all individual or collective strives to enhance power is at each point of time a sort of equilibrium of power (or balance of power). The element of automaticity is supplemented, some claim, by the deliberate action of particular members, perhaps together, to preserve, often by diplomatic means, the system as a whole. [47]

As this brief survey shows, the notion of the balance of power has many different meanings, several of which are often implied at the same time. [48] It can be viewed in the sense of the automatic mechanism described above; it can be understood to mean any distribution of power from an equilibrium to a disequilibrium; it can be understood as a policy, which aims to create some equilibrium or even a favourable disequilibrium (hegemony); [49] and finally, it can be interpreted as a system in which one distribution of power is periodically transformed into some other. This last interpretation seems to include to some extent the other three.

The balance of power in the nuclear era The proponents of political realism admit that some features of the quest for power have changed owing to the new conditions arising from the division of the world into two antagonistic blocs, the collapse of the colonial system, and the advent of nuclear weapons.

In the first place, the simple picture of the balance of power between particular states has been partly complemented and partly replaced by the concept of a balance of power between the two antagonistic camps; in some variants it is is presented as a balance of power between the superpowers, or between the free world and the Communist system. It has also been proposed that the resolution of particular political problems depends not only on the world balance of power but also on the local or regional balance of power in the area concerned; this is also a new idea.

Secondly, the very essence of the balance of power has changed, since it has become a mixture of the traditional balance based on the ability to fight successfully, and of a balance of terror consisting in the ability of parties to deter war by threatening with severe punishment. [50]

Thirdly, the concepts of national interest has become more indefinite, especially for the United States, which, it is claimed, bears responsibility for the independence of all non-Communist countries and for world peace. [51]

Fourthly, the concept of quest for power is also more complicated since more power does not any longer mean a greater ability to fulfil national goals, and the problem is now to acquire more flexible power. [52]

Fifthly, military power would still be of ultimate importance in the defence of vital security interests, but would have less utility compared with other forms of power in bargaining over non-security issues. [53]

And finally, some scholars point out that one should distinguish between potential power and the actual power which can be quickly brought to bear in the regions of conflict. The latter depends not only on the material means at one's disposal, but on the ability to mobilise political support, both at home and from one's allies.

But, the modern political realists contend, the basic principles have not changed since the search for power remains the guiding principle of international relations, and the interest of the state (or the nation) remains closely connected to power. The search for power may now have taken the form of a search for flexible power, but the principle itself remains valid. [54]

War as a means of regulation in the balance of power system

The question of what role war plays in power politics and in the balance of power system is answered by the political realists somewhat ambiguously. On the one hand they present power politics as a search for stability and even peace. [55] The main historical argument they use to support this contention is the relative peace given by the balance of power system in Europe in the 19th century. [56] On the other hand, they assume that war is an indispensable instrument in the quest for power, and that it is one of the important means by which a balance of power system is maintained. [57]

This apparent contradiction may perhaps be resolved by the following interpretation. A power-seeking policy acts as a stabilising factor in international relations to the extent that the main stabilising powers thereby led to avoid a major war with each other, and that large changes in the territorial and legal *status quo* are prevented. War is one of several methods used from time to time to make adjustments in the basically stable balance, without destroying the system itself. The proponents of a power policy recognise the detrimental aspects of war, but they maintain that their position is based on an analysis of the realities: all independent states pursue such a power policy, and since the very fact means that war is a definite possibility, it must be prepared for. [58] The best way of preventing war, they maintain, is to establish some sort of stable balance of power. The school of political realism therefore contends that even in the nuclear era, war and the threat of war remains instruments of policy. In the society of sovereign states, vital interests can be defended only by an ability and willingness to employ force, [59] and conversely, a state's preparedness for war is the general criterion by which its ability to defend its interests and meet its needs is measured. The political realists have indicated, however, that the nature of power in the nuclear age has changed. The development of nuclear technology has led to weapons of such devastating efficiency that nations possessing such weapons hesitate to use them. That they might do so if hard pressed remains a possibility that acts as a damper on the use of military force in many situations. At the same time, non-military aspects of power, for example, psychological aspects have increased in importance. The necessity to tailor the use of power in accordance with the different circumstances and issues has led to a situation in which the power of nations must be measured relatively to all these particular circumstances. Nevertheless, the realists maintain that on the whole power has increased absolutely because nations ought to be prepared to meet all different challenges. In consequence to the change in the nature of power brought about by the changes in the structure of the international system, and by the advent of nuclear technology, the balance of power system has also changed. This is reflected in the research of the adherents of the school of political realism, inasmuch as they devote their efforts to discovering the best form of a balance of power for our time. Many different proposals of bi-polar, [60] or multi-polar, [61] stability have been presented and discussed. [62]

As to the question of morality, the political realists emphasise that the use of power is not in itself immoral. Statesmen act for their respective states according to the national interest and not according to any ethical principles. While the moralist may ask whether the policy adopted accords with moral principles or natural law, the political realist asks in what way the policy serves to increase the power of the state. [63] The realists defend their amoral position on the grounds that states do in fact act in accordance with their interests. A state that allowed some moral principle to guide its foreign policy would run the risk of being victimised by its own morality. For example, the policy of preserving peace at any price would lead to a situation in which the country could be at the mercy of more ruthless members of the international system.

Critique of political realism

The view that politics is simply a quest for power has often been criticised. Here these criticisms will be grouped under four headings.

Terminology In the first place, the critics have attacked the ambiguous nature of all the terms characteristic for the approach. The term 'national interest' can mean different things in different countries, it has been pointed out. Even within one state, there is certain to be a difference of opinion between political groups and movements about what lies in the interest of the nation.

The term power has similarly been given several different interpretations. [64] One source of confusion derives from its being used to mean both ends and means. Critics of the concept of international relations as an exchange in which power is the currency maintain that it is not historically true that states have exlusively used power in their conduct towards each other and that power always could be successfully used, unless the term power is given such a wide interpretation that it includes all the means by which states act. Given such a broad meaning, however, power becomes useless as a basic concept of a theory of international relations, for it lumps together phenomena which are incommensurable. With such a conceptual apparatus it is moreover impossible to answer such important questions of any theory of international relations such as why states make use of one means in preference to another in some given situation. Nor are matters improved by postulating increase of power as an intermediate goal, i.e. treating power only as a means, for unless one has some idea of the ultimate goal for which nations strive, one cannot explain why states use power, nor why they resort to war.[65]

The roots of power The second ground on which the realists' position is challenged is based on the connection they make between states and sub-national political levels. The contention that the search for power in which all states are engaged has its primary roots in human instinct is said to be unsatisfactory. The use of power is conditioned by many factors, some of which even force good men to act violently. Nor is it a convincing alternative to anchor a drive for power in human nature by postulating that it is the essence of 'political man' to search for power, dominance, and violence.

In spite of the fact that the state is said to act on the behalf of the whole nation, the realist model ignores the role domestic factors play in shaping the external policy of the state. The intentions, expectations, plans, and actions of individual states as well as the functioning of the international system as a whole are interpreted as if they were independent of domestic political forces. [66]

The dangers of realism The third group of criticisms admits that the realists' view of international politics is a reasonable representation of the actual behaviour of states. It denies, however, that the realist solution to the problem of how to prevent war is the right one. They maintain that war is an inevitable result of a system based on constant competition for increased power, for in such a system, short-run interests of a state, which are more directly connected with the power nexus, take the upper hand. [67] Long-run interests, which pre-suppose peace and stability, are sacrificed. The balance of power advocated by realists is therefore an unsatisfactory solution, for it can at best only prevent particular wars; it does not provide the basis for an effective preservation of peace.

To support their position, critics of the realist school also question whether the balance of power system in the 19th century really lived up to the claims the realists make for it. They point out that although it prevented larger wars, many smaller wars did take place; [68] and ultimately it proved incapable of stopping the First World War. Moreover, the system itself was based on a dis-equilibrium of power, since Great Britain acted as a balancer by threatening to join a coalition of states against any other state that attempted to upset the *status quo*. [69]

The critics do not accept the realists' claim to be neutral observers. On the contrary, say the critics, implied in the realist analysis is an acceptance or even approval of the power game, something which they believe to be a dangerous tendency, for power has a corrupting influence on men. [70] The danger is all the more pressing in modern times as violence has become increasingly identified with power as a kind of last resort to keep the power structure intact against foreign and domestic challenges. [71]

Finally, there is a danger involved in an analysis that, like the realist model, takes scepticism towards normative goals to the point of nihilism. The preoccupation of the realists with power draws their attention away from goals truly worthy of human endeavour and commitment. [72]

Invalid The last argument against the power theory is simply that it is invalid, especially under modern conditions. The whole concept of power derives from another epoch, when military force was the main component and military security the main test of the usefulness of power. In modern times, the ability of military forces to guarantee the security of state has greatly decreased.

The role of power has changed, because the state-goals have changed. For example, it may once have been the object of states to obtain a position of relative strength over neighbouring states and comparative economic advantages, but whether or not this is true of the past, states are now more interested in absolute economic gains that enable their citizens to enjoy a higher standard of living. Such economic progress requires a peaceful and stable world, in which international economic cooperation is possible. It is anomolous to view the international system as a collection of sovereign states, each of which is autonomous within its boundaries, and each of which must protect its autonomy by defending these boundaries. On the contrary, states in the modern world are highly interdependent. Many of the world's most difficult problems can only be resolved through joint efforts. It is this realisation that has led to the proliferation of international and transnational organisations.

The armed forces have become irrelevant to most of the goals pursued by modern states. [73] The tactics of blackmail, the threat of war, and ultimately the resort to war, which were the means by which the balance of power model operated, are becoming inapplicable because they may defeat the purposes of the state. The methods now used to resolve conflicts and differences of interests rely more on skillful diplomacy and trade-offs to achieve compromises acceptable to all.

To sum up, the critics maintain that the state has ceased to be the only actor in international relations, war and the threat of war are no longer the main means, and the accumulation of power is no longer the main direct goal for which states strive to achieve future goals.

The place of political realism among political approaches

Political realism has been presented in the previous sections as a set of ideas on international relations, with two main themes: (a) Power plays a decisive role as a means of promoting the interests of states and as a measure of their status and position in the international system; (b) war is a means of policy, the use of which is justified if it is necessary for the pursuit of the interests of a state. The general character of these ideas makes it possible to find political realists in many of the above mentioned political approaches. In particular, in the classification based on politico–military views they are included in all categories except the 'nuclear pacifist' and 'integral pacifist' schools. In the twofold division according to basic political values which divides scholars into adherents of 'war as a mistake' concept, and 'war as a natural instrument' concept, political realists belong naturally to the latter group, in accord with the realist tradition.

War as the continuation of policy

The process leading up to the decision to engage in war is one of the most important problems in any study of the nature of war, for it concerns the decisive moment when policy-makers decide to change the means by which they work for their objectives from peaceful to violent means, the moment when war is nascent.

A conscious political act?

Is the decision to start a war a conscious political act? Those who treat war as a fight for political goals, and in particular those who uphold the tradition of political realism would answer yes. For if war is the continuation of policy by other means, then states resort to war only when they believe that, all things considered, it is the most effective way of achieving their goal. Although many factors, some accidental or fortuitous, may influence the ultimate decision, it is nevertheless taken deliberately as the final link in the entire chain of political acitivity preceding war. [74] Moreover, the realists have traditionally held that war is a means for the continued conduct of foreign policy only.

Both of these assumptions – that the decision to go to war is deliberate, and that it is rooted in the foreign policy of the state concerned – have been challenged. To take the second of these points first, recent studies in the realist tradition have devoted more attention to the role played by domestic policy in the decision to engage in war. [75] This can be seen in four ways. First, such internal factors as the character of the socio-political system have been devoted more attention than ever before. Secondly, the activity of influential groups who for economic or social reasons may gain by war, has been the subject of some analysis. Thirdly, internal wars, which obviously are related to domestic politics, have also attracted realists. Fourthly, one aspect of such wars which has evoked particular interest is the link between internal wars and international ones.

As to the deliberateness of a decision to make war, the challenge on this point has come from various approaches that explain war as a function of some deep-seated cause (for example, the psychological or socio-psychological approach) or as the result of a set of 'explanatory variables' (systems analysis). The political realists argue, however, that although the final decision to unleash war may seem unavoidable in the context in which it is made, [76] there are always at least two options between which decision-makers must choose: to go to war or not to go, and these are political options. The so called deep causes of war do not inevitably lead to it. It can also be supposed, that some scholars consider the deep causes of contemporary wars to be already discovered; others, on the contrary, regard them as unknowable, or so remote from the resultant behaviour, that it is next to impossible to analyse the link between the cause and effect.

The Western realists as well as many scholars grouped with other political approaches, have therefore continued to concentrate their attention on the more immediate political causes of war and, in particular, on the decision-making process.[77] In part the direction they have taken can be explained by the development of new methods, particularly quantitative ones, which are more suitable to the examination of particular elements in the decision-making process than of deep causes of war. In part it is the result of interest in a number of situations that developed after the Second World War in which wars seemingly could be avoided, but some, perhaps over hasty, decisions contributed to their outbreak, and, on the other hand, in situations, in which war seemed about to break out, but in the end did not. 'Crisis management', in which a whole array of military and diplomatic instruments are used to avoid war, is a subject more easily analysed with the political realist approach than any other.

The central problems related to the decision-making process that nowadays interest scholars in the realist school are:

(a)   The definition of a 'crisis-situation', its nature and origin.

(b)   Key factors influencing the decision, as perceptions regarding threats and opportunities held by decision-makers, and the sources from which their information derives. [78]

(c)   The interplay of rational and irrational motives and considerations.

(d)   The formulation of alternative lines of action (initiating negotiations, appealing to the UN, setting up an economic blockade, declaring war, etc.) and the calculation of their relative costs and benefits. [79]

(e)   The conditions most conducive to escalation in a crisis situation.

(f)   Decisions as to the outcome of trade-offs between bureaucratic agencies, each of which tries to defend its interests and prestige.

(g)   The role of individuals charged with responsibility in the decision-making process.[80]

Interpretations

The realists thus tend to assume that war is a continuation of policy in the sense that to go to war is a deliberate step, a political choice. This assumption can be interpreted in different ways, however.

The ideal case In the classical interpretation, war is regarded as the continuation of long-term policy. It is assumed that in some cases non-military means do not suffice for achievement of important political goals, and the use of armed violence is deliberately planned and well prepared long before war actually breaks out. All tools – diplomatic, economic, military, propagandist, etc. –

serve preparations for war. War is in such cases the instrument by which the aims of the state are clenched. War is a continuation of policy in the fullest sense.

Is such a model realistic? Sometimes, perhaps, an aggressive war may be planned and prepared for in advance, and its outbreak may be consciously timed; the unleashing of World War II as an outcome of the policy of Nazi Germany was perhaps such an example.

Short-term policy But usually the decision is taken shortly before the outbreak of war, and it is based on many factors, some unique or unforeseen, and often only indirectly related to the general political line and strategic considerations of the state. [81]

One conflict situation may resemble many others, but the historical context in which it arises is necessarily unique in some respects. The politico-military background may therefore vary in ways that affect the decision whether or not to engage in war. The significance of the issue at conflict in relation to the political and long-term strategy of the state may therefore be seen by the leaders of the state in a new light. [82]

Moreover, the decision involves a choice between alternatives, and as such is dependent on both the actual resources that can be mobilised and, more importantly, the information about these resources that in fact reaches the decision-makers. [83] They may then attempt to analyse rationally the risks, the costs, and benefits of the alternatives that seem most realistic. In this process, the views of each decision-maker will, however, be influenced not only by his perception of national interests and by interests of groups, institutions, branches of armed forces which are represented by particular decision-makers, but also by psychological factors, irrational whims, and emotional impulses. Moreover, his appraisal of the situation and of the best solution can vary as events unfold in contact with the other decision-makers. [84] There is, in other words, a socio-political dynamic to the decision-making process. Time is also an important factor, for it may limit the possibilities of obtaining accurate information, of deliberating over existing alternatives, and of creating new ones.

Thus, according to this model, war would also be a continuation of policy in the sense that it would follow from the political events preceding its outbreak; but it would not necessarily be a continuation of policy according to the classical Clausewitzian doctrine, nor to the traditional interpretation of 'political realism'.

War as the result of a specific issue Another interpretation of the term 'continuation of policy' is to treat the outbreak of war as the result of specific issues. Luard states, [85] for example, that nearly two-thirds of the external wars that occured between 1865 and 1965 did not result from the long-term policy of the states involved, but were fought over some specific issue; he maintains that if these incidents had not occurred, the wars would probably not have occurred either. [86] Seen in this perspective, the decision to engage in war may be the result of a sudden change in the *status quo*, for example, the adoption by a potential opponent of a crash programme of rearmament, [87] or the appearance of a power vacuum, which seems to threaten the security of the state. It is also possible that the course of events presents the state with an opportunity to achieve a goal that may have been part of its long-term policy, but it may also have developed as a new goal out of the changed circumstances.

Combination War may also be the result of some combination of these three factors. The decision to engage in war may be taken in a crisis situation over some particular issue which is also related to the long-term goals of the state.

The need for a re-examination?

The development of weapons technology in the nuclear era has led Western scholars to study situations in which the traditional interpretation of the term 'continuation' can be put in question.

Basic to these analyses are two observations. The destructive power of thermonuclear weapons is of such magnitude that no constructive gains can be made by using them, but everything could be lost if, despite all reason, the enemy were to strike first. At the same time, the speed with which a blow can be delivered has increased so much that there is very little time in a nuclear war to deliberate over moves and counter-moves: devastation comes quickly. In a situation with this combination of parameters – high stakes and little time for deliberation – it may be difficult to regard decisions taken by the leaders of a country as deliberate acts whose purpose is to achieve definite political goals.

The problem becomes most acute in a situation of international crisis. Here it is very difficult for the participants to keep events under their control. [88] They may misinterpret the intentions of the enemy. [89] They may mistake some event for a signal that the enemy has launched an attack (it could for example be a military accident) [90] and begin hostilities themselves.

Unintended war More generally, situations may arise in which war becomes unavoidable, even though none of the participants have actually intended it. [91] In such situations, such socio-psychological factors as distrust and hostility between states and peoples, or the stereotype notions people have about one another, can make it more believable that the enemy has attacked, and thus add to the the risks of war. [92]

'Brinkmanship policy' which some scholars attribute to the great powers may increase the risk. [93] Bert Röling contends that wars stemming from erroneous, emotional, and irrational decisions are now the main danger, while the probability of wars in the Clausewitzian sense is small. [94] Terms unintended war, unwanted war, and inadvertent war are here used. [95] Raymond Aron presents more general ideas on the sources of unintended wars. Since
(a)   all states permanently increase their military potential;
(b)   they are always ready to resort to war if it promises some gain;

(c) the strategic-diplomatic behaviour of many states is essentially indeterminate, and

(d) psychological and irrational forces play a great role in the decisions of state leaders — for these reasons there is an everpresent risk that war will break out. [96]

Corresponding to the opinion that an unintended war is now the main danger, some scholars state that nowadays any decision on war will be taken with more reflection than ever before, because of the major risk which war presents: only extremely grave political tensions and only probability of achievement of the desired political objective without unacceptable catastrophe can justify such decision. [97]

Preemptive or preventive war Another possibility is of somewhat different character. Many scholars have postulated that it may be rational, given the stakes involved in a nuclear war, for one side to launch a surgical attack against the other if it believes it likely that the other side may attack first. Such a war would be in some sense intentional inasmuch as defensive goals would be pursued, but it could hardly be named a deliberate political act in order to achieve some positive political goal. Two kinds of such war have been described: pre-emptive war and preventive war. The former refers to hostilities begun in the conviction that an attack from the enemy is imminent. [98] Preventive war is based on the assumption that the other side will start a war in the near future, and that it would be favourable to be the first side to undertake military operations. [99] In both cases it is supposed on the basis of an analysis of political developments that the expected attack is inevitable. This supposition may be false, in which case a war unwanted by either side would break out through misunderstanding.

The continuation by other means

Armed and organised The use of armed violence is generally accepted to be one of the distinctive features of war [100] and the main means to attain its aims. [101] A second is the organised character of combat, [102] which emphasises the social aspect of war; that which distinguishes it from fighting in the animal world. [103] Men first began to fight wars when their social life had advanced to the point that they could create an organisation to divide the tasks and elaborate a common plan. [104] In order to underline the distinction between war, as a social phenomenon, and other kinds of use of armed violence (e.g. duels), the participation of closed groups in hostilities is said to be the indispensable attribute of war, [105] and the term armed forces in the sense of regular troops is often included in definition of war. [106]

When a clash occurs between regular troops that receive their orders from a central command and use conventional military tactics, there is no problem in describing their activity as organised combat. Guerilla warfare, which has become increasingly common in modern times, is much more loosely organised. [107] Leadership is usually given by a movement or political party instead of by a state, over irregular troops. Each military unit may be compelled to plan its activities by and large on its own. In guerilla warfare, political means may even play a larger role than military means; nevertheless, it is usually regarded as one kind of war.

The use of non-military means All scholars agree that non-military means are also used during war. This sometimes makes it difficult to determine when a war begins according to Clausewitzian maxim. [108]

Some scholars points out that since total wars with unrestrained efforts are unthinkable in the nuclear era, the role of non-military means in war has increased profoundly, while the relative role of military ones has decreased. The classical distinction between war and non-war, based on a difference of means, thus loses some of its clarity. [109]

Traces of these ideas can be found in the advice offered by some strategicians as well as in the analyses presented by experts on the socialist countries. For example, the proponents of the strategy of escalation underline the importance of increasing pressures at each stage by using many different means, for the use of military means alone may have the undesirable effect of strengthening the enemy's will to resist.

Theories which diminish the role of armed fighting in war may be traced to the well known strategy of indirect approach advanced by the prominent military theorist, Liddell-Hart.[110]

During the first period of a war, it holds, and before any decisive military battle is waged, the strategy must be to upset the opponents' military, political, and psychological balance by exerting indirect pressures with a whole range of diplomatic and military moves. Conditions are thereby prepared for a decisive blow. What is implied in this view is that combat is not the main method of war, at least in the first period, which may however last a long time. The only occasion on which military activities predominate would be at the short conclusion. It is even conceivable that the enemy would capitulate without any decisive battle.

In one recent version, it is stated that the world is moving from an age dominated by the 'overt, fighting use of military forces' for achieving victory by battles, into one in which manoeuvre and strategem, avoiding direct fighting wherever possible are dominant.[111]

Politics and the conduct of warfare

The influence of politics on war

Goals and strategy For those scholars who have adopted the theses that war is a continuation of politics, political considerations are of primary importance for the conduct of war. War is fought to obtain political goals. The first act taken in any war – to define these goals – is thus a political one. The choice of goals decisively influences the whole course of war, for on it hinge all subsequent decisions. From this follows, that politics determines the main outline of the military strategy to be applied: the strategic aims of war, the main methods of warfare used to achieve these aims, the size of the armed forces, their armament, and so on. Alternately, the decision whether to wage a general or limited war, a protracted or a short war, whether to adopt offensive or defensive tactics, whether to deliver a single crushing blow or successive ones and if so, where and how – all these decision are ultimately based on political aims. If war is fought by an alliance, additional forms of political influence and decision arise concerning the way in which each member of the alliance will participate.

The over-riding influence of policy on the choice of strategy has, if anything, increased in the nuclear era. Several reasons are given to support this view. [112]

In the first place, the central issue in international politics is the struggle between two global blocs antagonistic to each other. In this confrontation, military strategy is only one part of a coordinated political activity: military goals in peacetime and wartime have become integral parts of general political goals.

Moreover, since war has become a much more risky instrument than it was, a decision to declare war and a definition of its aims have become acts of enormous importance. They must be made on the basis of a careful evaluation of the global political situation. The military-strategic perspective is too dangerously narrow.

In the nuclear era war would more than ever require a total commitment from a society. The choice of military strategy for both offensive and defensive goals is therefore much more dependent on the economic and technological resources that society can muster. In a nuclear attack, enormous destruction can be inflicted in lightning-short time. Owing to this danger it is necessary that all military activities be conducted with great prudence under careful political control.

Finally, modern developments in mass communications have enhanced the importance of political factors by making possible more sophisticated techniques for politico-psychological warfare.

Military objectives According to the classical definition, the military objective of war is victory through which the victor would be able to impose his will on the defeated or, put another way, the vanquished would be compelled to meet the political demands of the victor. Victory in the Western analysis is presented from two points of view, with emphasis either on the achievement of political goals or on the purely military sense of victory.

By the political criteria, victory in a total war means the complete capitualation of the enemy and the fulfilment of all demands of the victor; in a limited war, victory would be defined in relation to some limited gains, for example, some conquests on the frontier. [113] One can also distinguish between achieving the immediate goals of a war, and improving the victor's general politico-military position, [114] which would enable him to achieve more long-term political goals in the future.

In the military sense, victory means inflicting such damage on the enemy's forces through superiority in military power that the enemy agrees to submit to the will of his opponent. The notion of defeating the enemy is thus relative: it may mean that the enemy is completely destroyed and disarmed (that is, absolute military domination is achieved), or it may mean that the enemy suffers greater or more strategic losses than the victor, who thus wins relative military superiority. Victory in the military sense thus corresponds to victory in the political sense. [115]

It has been suggested that different kinds of victory can be ranged on a scale containing all possible gradations of asymmetrical outcomes, with the symmetrical one – no winners, stalemate – in the middle, and the extreme outcome – absolute victory – at both ends. The extremes would thus constitute the ideal military objectives of war, victory in the sense of the complete political subjugation of the loser to the winner. [116]

The classical notion of victory is not so readily applicable in the nuclear era. It is now impossible to obtain absolute victory through total war (except perhaps in theory), for both the winner and loser would suffer losses that far outweigh any possible gains. [117] In thinking the unthinkable, Herman Kahn has proposed that in a nuclear war, the objective may simply be 'to prevail'. Victory in this sense would mean no more than to emerge from war in a better position than the opponent, be it with regard to the extent of devastation or to the strategic base from which future politico-military engagements can be planned.

What victory means in limited conventional wars has also become more complex because the political aims in such wars have become more intricate. In such situations, the Western powers, especially the United States, claim to pursue two sets of goals. Firstly, there are the goals directly related to the particular goal; these may be to re-establish the *status quo ante*, to prevent the victory of a Communist revolution, and so on. The second set includes long-range goals, such as the maintenance of world peace and stability, and the prevention of the success of international Communism. Both sets of goals are presented in the political sense as defensive, but to fulfil them may require in military terms an offensive strategy.

Moreover, it is difficult to achieve the same degree of success in both sets of goals in any given war. Even if local victory is total, the effects of it on the world-wide confrontation may be marginal, because the restoration of the *status quo* in one corner of the world does not eliminate the sources of international crisis and war. [118]

Political control over military operations The second main aspect of the primacy of politics, its control over the conduct of war, was largely neglected in classical analyses of war. In the nuclear era it has become a question of vital importance, however. Two main questions have been examined: the development of military strategies susceptible to political control, and the study of factors that impede or facilitate such political control.

In the American debate on the choice of military strategy, great weight has been given to the degree of political control that could be maintained. One of the main arguments Herman Kahn uses against a 'spasmodic' nuclear missile attack is that a war carried on in this way would quickly run out of control. By contrast, the 'escalation ladder' he proposes would have the merit of preserving political control at each step.

The strategy of flexible response, which has become officially adopted in the United States, is designed to secure political control over the use of armed forces in all kinds of war. Amongst the advantages of this strategy, which allows time for deliberation and a choice of military moves, the guarantee of control by the political leaders is always emphasised. To make this strategy possible, top priority has been given to building up command, control, and communications facilities to provide the American executive with information about the course hostilities may take, and to allow them to direct American forces. The American government has also maintained that political control through the strategy of flexible response is only possible under certain military conditions: namely, that the United States maintain a stable superiority over the Soviet Union in strategic nuclear forces and that the American 'sword' remain invulnerable; the term superiority has been avoided in the recent expositions of military doctrine, however.

Western scholars have also devoted a good deal of attention to the influence various factors may have upon the course of war, its possible escalation and termination. [119] Particular interest has been given to factors that may impede and limit the political leadership from exerting conscious control over war in accordance with some long-term plan.

(a) political differences between political decision-makers;
(b) the increase in the role of the military during war-time;
(c) the influence of public opinion in demanding extreme goals (e.g. victory);
(d) the conviction amongst decision-makers when facing defeat that if war continues, a better military position as a base for negotiations may be achieved;
(e) a simultaneous increase in demands and losses;
(f) differences of opinion amongst political and military decision-makers about the desired conditions for peace;
(g) deficiences in communication with the enemy. [120]

Factors which give stimulus to an increase of the control have also been presented: amongst them are the high priority given domestic political objectives, the pressure of world opinion, and the fear of immense losses.

Many military scholars point out that while the principle of the political control over military operations should be observed, another principle also remains valid: such a control cannot violate the essence of war, cannot make victory unattainable. Thus principles of cooperation between the political and military leaderships are presented. [121]

The influence of war on politics

During peace The relation between peace and warfare does not run in one direction alone; war also has an influence upon policy. Just as the craftsman makes adjustments to his tool so that it better fits his needs, so too the politician decides on improvements in his military fitness, and does it long before he is to make use of it. In the short run, however, the craftsman may be forced to make do with the tools he has and to adjust his work to them. Similarly, since political leaders must always take into account the risk of becoming involved in war, they must base their policy upon their actual military potential in relation to other countries in a short-run perspective. Clausewitz is quoted as having said that although war is fought to achieve political goals, these goals must in turn be set in relation to the military means available. The specific situation in the nuclear missile age consists in having tools of warfare so effective that they are too dangerous to use. It is therefore necessary to adjust policy to what is in theory possible and in practice advisable.

During war Once war has actually broken out, military issues will naturally have much greater political significance and may in some circumstances even get the upper hand over all other issues. In the first place, the course of war may force the combatants to clarify their political objectives, especially if these were vaguely formulated to begin with. It has been suggested, for example, that the war aims of the Allies were unclear at the outbreak of war. The decision to go for unconditional surrender was based according to the critics, less on the long-range interests of the Allies than on the fact that they had geared up their countries for total committment to the war effort.

There is thus a thin line between a clarification of political goals and their redefinition. The military staffs were convinced that World War I would be a short affair. The politicians defined their aims accordingly. In the event, it dragged on for four years and had far reaching repercussions in the social life of all participant countries. As losses mounted and emotions become inflamed with each month of fighting, peace through compromise became more and more unacceptable. In the end, nothing short of a decisive change in the *status quo ante* could be considered.

To take a modern example, many critics of the American intervention in Vietnam have contended that the United States has been forced to change its objectives to allow for an unforeseen course of events in the war.

The course of war may also give military leaders such an influential voice in the vital decisions about war goals that the broader political perspective becomes obscured. [122] Many who have analysed American policy in World War II maintain that the goal of unconditional surrender was primarily a military objective. In terms of political realism it was indefensible since by closing the door to any compromise it left the Germans and Japanese no choice but to fight. Moreover, it completely ignored the question of what political configuration would be established after the war, with the result that the Soviet Union was able to make large political gains. [123]

Many scholars have therefore reached the conclusion that war is now a crude instrument with which to pursue political ends. The violence and destructiveness of modern warfare set off a chain of consequences that can be neither perfectly controlled nor perfectly anticipated. In practice, the subordination of certain political considerations to military requirements may be the necessary condition for avoiding defeat. A nuclear war may become the master instead of being the slave. Moreover, some add, even the winning side in a nuclear missile war would undergo such a radical transformation that the effects on the whole social-political structure would be profound, and unpredictable.

The influence of minor wars on politics Owing to the involvement of several of the Western powers in minor wars at various occasions since the Second World War, there has been a special interest amongst Western scholars in the effect such wars have on the domestic and foreign policies of the powers concerned. Less attention is given in this regard to short successful wars in which the use of military action proved to be the right choice to achieve viable political goals. Such studies have rather concentrated on those instances in which the Western power has entangled itself in a prolonged and costly war from which it has great difficulty in extricating itself on favourable terms. Such unsuccessful operations are bound to put in question both the aims and the methods of the political leaders and to have far-reaching effects on both the domestic and foreign policy of the power concerned. They have usually led to a reappraisal of the global role of the country, a reduction of its military commitments abroad, a replacement of direct military support by military and financial aid to local forces, a re-examination of alliances to encourage greater military participation from allied countries, etc. [124]

Domestically, protracted and unsuccessful wars have generated a lack of confidence in the political and military effectiveness of the government, polarised opinion between those who call for more democratic and those who appeal for a more resolute government, and compelled politicians to give lower priority to social reforms due to lack of funds. The costs of a prolonged war also have a purely economic impact that can only add to domestic dissatisfaction. [125]

Notes

[1] Deutsch, 1968, p.77; Karl J. Friedrich, *Man and his Government*, McGraw-Hill Co., New York 1963; Laswell (1936) defined politics as the theory

about who gets what, when and how; and together with Kaplan (1950), as 'a projected program of goal values and practices' (p.71). Salisbury noted three meanings of the term: broad guidelines for the whole action, general frame of authoritative rules, and political behaviour (1972, p.65); cf. Frankel, 1972; Midlarsky, 1975; Levi, 1974; Barbera, 1973.

[2] In many definitions of policy, power is considered to be the main content and aim of political activity (for a review, see: Ruge, 1967, pp.7–14); Barbera writes that nation is concerned first with power and secondly with development, and search for power is the essence of politics (1973, pp.1–2 a.o.). Ruge analyses several definitions and concludes: '. . . policy is a managing activity, aiming at grounding the state and ensuring its security by means of legally controlled power (rechtlich geregelte Macht) and motivated by established values (under Rechtfertigung durch bestimmte Werte) of the social order' (Ruge, 1967, p.8).

[3] 'Politics consists in the more or less incomplete control of human behaviour through voluntary habits of *compliance* in combination with threats of probable *enforcement*. In its essence politics is based on the interplay of habits of cooperation as modified by threats' (Deutsch, 1968, p.17); policy needs force to compel the members of the society, and legitimacy to evoke compliance without coercion (Andrew Janos, 'Authority and Violence: The Political Framework of Internal Wars', in Eckstein, 1964, pp.131–3).

[4] '. . the state is the organization of the power of the society for its maintenance, protection, and expansion' (Withey, Katz, 1965, p.66); '. . . this purpose (of the state) is stated as the promotion of the power of the common good or general warfare' (*Encyclopaedia Britannica*, art. 'State', vol.21, p.336). In Röling's presentation, the state has the following functions: internal security, external security, welfare, peace (1970, p.159). Cf. functions of the state in Frankel, 1972. For an analysis, see Wright, 1965, 'The Analysis of War by Political Scientists', App.XXVII, pp.1376–81.

[5] Nation-states in the course of evolution became entities with a really decisive influence of the nation upon the policy, the citizens tended to see the nation-state as the creative element protecting and extending social gains, and advancing the interests of the nation against those of other states (Rosecrance, 1973, ch. II); 'Most of the time we tend to think of war only in relation to the nation-state and as an instrument of diplomacy' (Lang, 1972, p.133). Röling writes that 'human thinking is group-thinking, which in our time means national thinking' (1970, p.132–3). Another view is that notions of state and nation are not identical. 'The nation is a unit of society with a common language, tradition and culture which may not coincide with state boundaries' (*Encyclopaedia Britannica*, art. 'State', Vol.21, p.336, 1964); 'The state is not identical with the nation, although nowadays we are accustomed to regard the nation-state (that is, a country, in which one nation forms a state) as being normal' (White, 1947); 'Not only are states and nations not identical but the links between them vary greatly in time and place', sometimes states precede nations, sometimes nations precede states, some states are multi-national, others have large national

minorities, and some nations are divided into separate states (Frankel, 1972, p.13). Midlarsky observes that 'this preoccupation with the state as actor leads to a number of analytic difficulties' (1975, p.4).

[6] Aron, 1966, p.7

[7] 'This is clearly the traditional focus among Western students, and this is one which dominates almost all of the texts employed in English-speaking colleges and universities' (Singer, 1966).

[8] Nicholson, 1970, pp.3, 16. He regards the study of war as a part of the study of international relations. The primary interest of students of international relations is 'the hostile behaviour of states' (Dina A. Zinnes, 'The Expression and Perception of Hostility in Prewar Crisis', in Singer, 1968, p.85); 'We have grown accustomed to apply to wars the adjective which serves to characterize political units. Since the latter were national, wars were national too' (Aron, 1966, p.296).

[9] One of the earlier definitions: Bernard defines war as an organised conflict between communities, carried on by armed forces, for national goals; he calls them 'public goals' (1946, p.28). The threat of violence, and the 'constructive use of it is an indispensable condition of progress and stability. War breaks out when instead of threatening violence, which is a conscious instrument of national policy, the states take risk of using it. War is therefore a continuation of national policy' (Nieburg, 1963, pp.43 ff.)

War is 'a conflict between two states carried on in general under the recognized rules for such conflicts, and including both armed hostilities and the intention to carry on such a conflict' (White, 1947, p.309).

'War may be strictly defined as organized fighting between at least two politically independent nations in pursuit of goals . . . It is the nation that contends for world goals (security, autonomy, territory, prestige, allies, and ideology) not individuals or races or continents or subunits of the nation. War is made and unmade by the nation, and only the nation has this prerogative' (Barbera, 1973, p.1).

In Wright's work, while many various approaches are analysed, the focus is on war as an interstate conflict (1942, 1965).

[10] The decision on war is taken, when no alternative course of action is seen, and it permits the nation to reach a desired goal (Stagnar, 1965, pp.45–6); 'war is just an extension of national policy in other forms', an extreme form of an external conflict, classes and class struggle are not taken into account (Withey, Katz, in McNeil 1965, pp.64–88).

[11] For example, John Slessor, André Beaufre, and others (their views have been described in Lider, 1969). '. . . War, or the application of armed force in international relations, is a part of the measures or instruments which a nation can use in the pursuit of its policies and for achievement of its objectives . . . Only national objectives of paramount importance can excuse the unleashing of violence and the sacrifices of national assets which any war creates' (Sokol, 1961, pp.13–14).

[12] For example, writings of George Kennan, Henry A. Kissinger, Robert E. Osgood, Robert W. Tucker, Alastair Buchan, and Michael Howard.

[13] Holsti, Brody, North, 1965, pp.444–5; 'In international systems, units actors, whether empires, city-states, articulate goals based on such general considerations as 'the national interest' (Haas, 1968, p.215). A typical definition: 'Interests are highly generalized abstractions that reflect each state's basic wants and needs ... The only vital national security interest is survival – survival of the state, with an "acceptable" degree of independence, territorial integrity, traditional life styles, fundamental institutions, values, and honor intact' (Collins, 1973, p.1).

[14] But other conceivable actors, e.g. individuals embodying the state (king, prime minister, dictator), or governing bodies, or influential groups behind the government, were usually considered less appropriate as levels of analysis (Holsti et al., 1965). For an analysis of the discussion about whether nation-states ought to be chosen as the primary level of analysis see Holsti et al., ibid.

Midlarsky proposes 'three levels of analysis for the study of internationally related political violence': the international system itself, the boundary between the system and the nation-state, and the normal 'state as actor' (1975, pp.2–8).

[15] James N. Rosenau, 'Toward the Study of National-International Linkages', in Rosenau, 1969: Klaus Faupel, 'Internationale Politik und Aussenpolitik', in *Die anachronistische Souveränität*, Ernst Otto-Czempiel, ed. Köln-Opladen 1969. For a critical review of the above mentioned studies, see Gantzel, 1972, pp.56–62.

[16] Some scholars contend, that in small and underdevloped countries domestic pressures exerted by 'interest groups' on decisions that involve war and peace as a consequence, are not determining, as compared with systemic properties and the international situation (Holsti, 1973).

[17] Bernard, 1946, chapters: 'Personal Diplomacy', 'Secret Diplomacy', 'Obscure Motives in Diplomacy' (pp. 385–92).

[18] The literature on the subject is too vast to be presented here. Some items can, however, be listed: C. Wright Mills, *The Power Elite*, New York 1956; Fritz Vilmar, Rüstung und Abrüstung' in *Spätkapitalismus,* Europäische Verlangsanstalt, Frankfurt M.1965; Vagts, 1967; Deutsch, 1968; Kolkowicz, 1972; Egbert Jahn, 'The Role of the Armament Complex in Soviet Society (Is There a Soviet Military Industrial Complex?', *Journal of Peace Research*, 1975:3.

[19] Of the classical studies on the role of the military in society, see: Lasswell, 1941, 1950, 1962; Huntington, 1957, 1961, 1962; Jannowitz, 1960; Wright-Mills, 1956; J. Planchais, *Le malais de l'armíee*, Paris 1959.

[20] Karl Deutsch, *Nationalism and Social Communication*, MIT Press 1966; and W. J. Foltz, ed, *Nation Building*, Atherton 1963; 1967 ch.8–10; 'A group ... is a collection of persons who are linked by two things: they share some relevant common characteristic, and they fulfil some ... interlocking roles' (1967,pp.50–1).

[21] Dahrendorf, 1961; H. Lasswell and D. Lerner, *World Revolutionary Elites*, MIT Press 1966; T. B. Bottomore, *Classes in Modern Society*, Pantheon

1966; T. H. Marshall, *Class, Citizenship, and Social Development,* Doubleday 1964; S. Ossowski, *Class Structure in the Social Consciousness,* Free Press of Glencoe 1963; John Strachey, *Contemporary Capitalism,* London 1956.

[22] The main theme of Barnett, 1971, 1972.

[23] Senghaas, 1969, Vilmar, 1967, Pietacker, 1972. The main contentions are: (a) The majority of the democratic society has neither real political conscience and will nor real power, the powerful circles (Machtclique, Senghaas, p.147) indoctrinate the society to accept their political program. (b) Public opinion has no possibility to recognize the true national interests and the mechanism of policy-making. (c) Society has no interest in wars, powerful circles which are interested in war determine the government policy. (d) Powerful circles increase the nuclear weapons potential as a 'deterring power', more weapons produce, however, more suspicion and tension and a vicious circle arises. (e) Without changing the structure of power in the society (Herrschaftstruktur, Senghaas p.85) the 'deterrence system' cannot be abolished.

[24] For example, K. P. Misra, 'Intrastate Imperialism: The Case of Pakistan', *Journal of Peace Research,* 1972:1; 'Intrastate Imperialism as a Factor in Conflicts Within and Between States', Paper presented on the IXth World Congress IPSA in Montreal, 19–25 August 1973. Pakistan and Sudan served as examples of sharp internal contradictions and of use of internal armed violence.

[25] Levine, 1963. He applied a combination of two categorisations: (a) 'anti-war', 'anti-Communist' and 'middle' groups of scholars with (b) 'marginalists' and 'systemists'. The marginalists proposed small changes in policy and its mechanism, maximalists postulated changes affecting entire systems.

[26] The two schools of thought are called by some scholars 'utopians' and 'realists' (e.g. Carr, 1964, Pfaltzgraff, Dougherty, 1971, pp.6–8,65).

[27] The division into four approaches, the Rationalists, The Economocrats, the Traditionalists, and the Kremlinologists may be an example of a mixed classification based on the fields of interests and on the method of research (Roman Kolkowicz, 'Strategic Élites and Politics of Superpowers', *Journal of International Affairs,* 1972:1).

[28] Bruce M. Russett presented a mixed classification including twelve schools characterised by their methodological or theoretical approach and additionally, by the university to which the group of scholars had been attached, e.g. Yale – International Integration, Bargaining and Negotiation; National Interest; Simulation-Northwestern; and the like. ('Methodological and Theoretical Schools in International Relations', in Palmer, 1970).

[29] Lider, 1963, 1967, 1969.

[30] Orvil A. Anderson, Louis Johnson, Francis Matthews, Robert C. Richardson, George C. Reinhardt, W. D. Puleston, Alexander P. De Seversky (1950, 1961).

[31] Thomas S. Power (1964–65), Dale O. Smith (1955), Curtis LeMay, Thomas D. White.

[32] Robert Strausz-Hupé, William R. Kintner, James E. Dougherty, Alvin J. Cottrell (comp. bibliography).

[33] Hans J. Morgenthau, Henry A. Kissinger, William W. Kaufmann, Robert E. Osgood, Malcolm Hoag, Morton Halperin, Herman Kahn, Thomas C. Schelling, Maxwell D. Taylor, Roger Hilsman, Klaus Knorr, Morton A. Kaplan, Thornton Read, Glenn H. Snyder, Samuel P. Huntington, Kenneth Waltz, Warner R. Schilling, Bernard Brodie, Walter W. Rostow, J. David Singer, W.T.R. Vox (comp. bibliography).

[34] Seymour Melman (1961, 1963), Jerome Wiesner, Herbert York, Ralph Lapp.

[35] Charles Osgood ('Reciprocal Initiatives', in James Roosevelt, ed., *The Liberal Papers*, Garden City 1962; Amitai Etzioni (*The Hard Way to Peace: A New Strategy*, New York 1962; *Winning Without War*, New York 1964); Kenneth Boulding, 1962.

[36] The main theme of Hans J. Morgenthau's writings (1951, 1967); international politics is 'dominated by the quest for power' (Strausz-Hupé, Possony 1954, p.5); 'all civilized life rests in the last instance on power' (Spykman, 1942, p.11); 'power is the common denominator of the most general nations' goals, the search for a strong position in the balance of power dictates the foreign policy' (Aron, 1966, ch.III).

[37] Aron, 1966, ch.II; Deutsch, 1967, pp.44–7; Dougherty, Pfaltzgraff, 1971, p.67

> One generalization about war aims can be offered with confidence. The aims are simply varieties of power. The vanity of nationalism, the will to spread an ideology, the protection of kinsmen in an adjacent land, the desire for more territory or commerce, the avenging of a defeat or insult, the craving for greater national strength or independence, the wish to impress or cement alliances – all these represent power in different wrappings. The conflict aims of rival nations are always conflicts of power (Blainey, 1973, pp.149–50).

[38] All politics, by definition, revolve around the exercise and pursuit of power; power plays the central role in international politics, as an instrument for the achievement of national values (Frankel, 1972, p.75); he contends, however, that he doesn't approve the 'extreme approach', because the concept of national interest, which governs state behaviour, is not limited to power considerations alone.

> While one nation attempts to achieve its goals, it must watch and resist another nation competing with it, and it may occasionally need to block the achievements of still another nation. Thus a context of rivalry surrounds nations as a normal and perennial condition of power politics . . . If this context is the general condition of power politics, then war may be seen as a

particular aspect of that process . . . From the perspective of any government, then, the nation is concerned first with power and secondarily with development (Barbera, 1973, pp.1–2).

All relations between nations are governed by the 'power as capacity to act' (Puchala, 1972, ch.8); the ever-changing distribution of power within the international system is its dominant feature (Spanier, 1972, the main theme); cf. Levi, 1960, 1964, 1974.

[39] '. . . a man feels himself more a man when he is imposing himself and making others the instrument of his will'; this gives him 'incomparable pleasure' (Jouvenel 1952, pp.93, 100); Hannah Arendt states that an element of dominance and obedience plays a significant role in many theories and definitions of power (1969).

[40] 'Thus power covers all social relationship which serve that end from physical violence to the most subtle psychological ties by which one mind controls another' (Morgenthau, 1967, p.9).

Power is 'the ability to move men in some desired fashion, through persuasion, purchase, barter, and coercion' (Spykman, 1942, p.11); it is 'the ability to move others or to get them to do what one wants (Wolfers, 1962, p.103); power is 'the capacity of individuals or groups to control and organize their environment to conform with their physical requirements or their code of moral values' and 'the desire for, acquisition, and exercise of power is the raw material of politics, national and international, and violence may sometimes prove an effective means to secure or retain it' (Michael Howard, 'Military Power and International Order', in Howard, 1970, p.203).

[41] Morgenthau, 1951, 1967.

[42] Carroll, 1972, pp.585–616; Christian Bay, *The Structure of Freedom*, Atheneum, New York 1968; Bell, Edwards, Wagner, 1969; James G. March, 'The Power of Power', in D. Eastin, ed., *Varieties of Political Theory*. Englewood Cliffs, New Jersey 1966; Michael Parenti, 'Power and Pluralism: a view from the bottom', *The Journal of Politics,* 1970.

[43] 'Power might be said to be the capacity to induce others to behave according to patterns in one's own mind. A state has power when it has the ability to compel other states to pattern their behavior after its own ideas' (Arleigh Burke, 'Power and Peace,' *Orbis*, Summer 1962, p.188); power is used for influencing the opponent's behavior in four ways: by gradual transformation of his intentions, by stimulating a revolution from above, or from below, or by war (Strausz-Hupé et al., 1959; his 'The Protracted Conflict', in Hahn, Neff, 1960); power is 'the general capacity of a state to control the behavior of others' (Holsti, 1967, p.193); 'Power, put most crudely and simply, is the ability to prevail in conflict and to overcome obstacles . . .' (Deutsch 1968, p.19); 'The power of a system is its influence potential, i.e. its potential to induce its will on other systems' (Johan Galtung, 'The Balance of Power and the Problem of Perception', *Inquiry*, 1964:3). In an encyclopaedic presentation, power terms

refer to subsets of relations among social units such as the behaviour of one or more units depend in some circumstances on the behavior of other units (*International Encyclopedia of the Social Sciences*, vol.12, p.406).

[44] '. . . all nations actively engaged in the struggle for power must actually aim not at a balance – that is, equality – of power, but at superiority of power on their own behalf. And since no nation can foresee how large its miscalculations will turn out to be, all nations must ultimately seek the maximum of power obtainable under the circumstances' (Morgenthau, 1967, ch.14).

[45] For a detailed review of concepts about the essence and mechanism of balance of power system see e.g. Claude, 1962, ch.2; Frankel, 1972, ch.6; Sterling, 1974, ch.2. See also Schulte, 1975.

[46] Some scholars contend that the prevention of the domination of one state over all others is the essence of balance of power. 'It has, until recently, worked tolerably well; it has not prevented all wars but it has almost certainly prevented some wars, and since the rise of the nation-state system in the seventeenth and eighteenth centuries, it has prevented any one nation from gaining permanent domination over all others' (J. William Fulbright, *Introduction to Frank*, 1967, – viii). Of the 278 wars for the period 1480–1941, 135 'are considered to be balance of power wars, that is, wars fought to prevent any nation, or group of nations, in the community, from becoming disproportionately strong through conquest or elimination of rival states' (Coats, 1966, p.18).

[47] The process of interaction was described in a historical study by Kissinger, 1957.

[48] Ernst Haas (1953) discusses eight meanings of the term: (a) Distribution of power. (b) Equilibrium. (c) Hegemony. (d) Stability and peace. (e) Instability and war. (f) Power politics. (g) A universal law of history. (h) A system and guide to policy . . . making (pp.459–74); Alastair Buchan (1973) characterises four meanings of the term: (a) Even distribution of power. (b) Equal shares for the great powers at the expense of the lesser. (c) Holding a special role in maintaining an equilibrium or even distribution of power. (d) Predominance (pp.63–70); Zimmermann enumerates following meanings of the expression: any distribution of power; system in which any attempt to dominate is opposed; power politics generally; and also an interpretation of the politics as an automatic process according to historical and natural law (1969, p.158).

[49] Balance of power 'consists in an attempt on the part of one nation to counteract the power of another nation by increasing its strength to a point where it is at least equal, if not superior, to the other nation's strength' (Thompson, in Morgenthau and Thompson, eds, 1950, p.103).

[50] Glenn H. Snyder calls the two interacting systems: balance of terror and tactical balance of power (1960).

[51] Because of the bipolarity of the contemporary world, the leadership of the US, and the growth of the power of all Western countries under the leadership of the US, is the condition for the preservation of their independence. The

Western world has to become a system of states under the American leadership (Acheson, 1958). 'In principle, the United States is the one nation in the world which can afford to take a detached view of events . . . It does mean that United States is the natural arbiter of the world's problems, within realistic limits' (Walters, 1974, p.191).

[52] The following division into categories has been proposed: offensive power – the ability of a political unit to impose its will on another; defensive power – the ability to avoid coercion by another unit; and the deterrent power – the ability to prevent certain threats or actions by posing an equivalent or greater threat (Henry A Kissinger, 'American Strategic Doctrine and Diplomacy', in Howard, ed., 1965). One consequence of the lack of direct proportionality between the amount of military power and the ability to influence world events was formulated as follows: '. . . in terms of international order, the authority of the superpowers was by no means commensurate with their physical strength' (*Strategic Survey 1968*, The International Institute for Strategic Studies, London 1969, p.3).

[53] Sometimes it would have even a negative effect since the threat to apply it carries a high risk of devaluing the other bargaining chips in one's possession (Seyom Brown, 'The Changing Essence of Power', *Foreign Affairs*, Jan. 1973, p.289).

[54] For example Michael Howard asserts, that 'the strategic approach' consisting in taking account of the part which is played by force, or the threat of force, in the international system, is still valid, what is more, it is inevitable and necessary. A failure to adopt such an approach may place the Western countries at the disposition of their enemies. But one should not equate power with military power, as the political realists did in their earlier writings, and one should carefully control all measures based on power, so that other states would not see them as direct threats (1976, pp.67–75).

[55] To his study on power politics, Morgenthau gave the subtitle: 'The Struggle for Power and Peace' and presented the balance of power policy as one of the main methods developed throughout history to maintain international order and peace.

[56] '. . . the "Concept of Europe" describes a method and a reality. The empires of Europe, controlling hundreds of millions of people in Europe, Asia and Africa, and decisively affecting the conduct of hundreds of millions more, did keep the international conflict limited in scope and minimal in destruction. They did provide an economic system, accepted without protest, and they did establish political coherence unequaled in extent since the Roman Empire' (Acheson, 1958, p.4).

[57] 'As long as war lurks in the backbround of international politics, the question is not whether states require military power. Instead, the relevant issue is how much and what kinds of military power are most appropriate in trying to achieve a given state's objectives' ('The Military Dimension.' Introduction to ch.9, in Tanter, Ullman, 1972, p.188); 'There are inherent dangers of war in any

international system in which the nation state remains the arbiter of its own interest and the judge of the means by which its security is best assured' (Buchan, 1968, p.126); 'War, far from being excluded or banished by the application of balance of power rules, is in reality enthroned as the final method for preserving the balance as a system' (Ernst B. Haas, 'The Balance of Power as a Guide to Policy-Making', *The Journal of Politics*, August 1953, p.377); two contrary assumptions may be put together: 'War may be required for equilibrium; war may be prevented by equilibrium' (Claude, 1962, p.54).

[58] War 'is simply the use of violence by states for the enforcement, the protection or the extension of their political power'. (Michael Howard, 'Military Power and International Order', in Howard 1970, p.203); some scholars observe that any structure of relationships between states relies on the ability to execute armed violence, and balance of power system doesn't 'produce' more war danger than any other system (Buchan, 1973, pp.60–1).

[59] The main theme of Kissinger's, Robert Osgood's and Kahn's writings (for an analysis, see Lider, 1969).

[60] Buchan contends, that all concepts of balance of power other than bipolarity are unreal (1968, pp.30–3), neither an equilibrium of a multiplicity of states, nor a universal collective security, nor the superiority of one power. Only a military balance between antagonistic powers and blocs may have positive consequences exemplum 'peaceful coexistence', Bull contends that although a bipolar balance of power doesn't guarantee peace, it helps to preserve independence of states and to prevent their subjugation to a dominance of one power (1961, p.39). Waltz argues that the bi-polar world system has greatly contributed to the maintenance of peace since 1945 ('The Stability of a Bipolar World,' *Daedalus*, Summer 1964, pp.881–909). Cf. the analysis of bipolar, tripolar and quadripolar systems in Rosecrance, 1973, Part 5; cf. Buchan, 1973.

Kissinger distinguishes between the military bi-polarity and the political multipolarity, and presents concept of international order based on both above premises. (1969).

[61] Frankel contends that five powers are the minimum number facilitating a steady equilibrium of power (1972, pp.120–1). Two would end in a clash, four would tend to form two rigid combinations, balance based upon three units would be inherently unstable since each state would have little opportunity to re-alignment and therefore would have a strong incentive for striking first.

[62] Not only the number of parties to balance of power is important, however, but also the relationship of forces: 'for the international system of the 19th century, peace is generally associated with power polarity, but the 20th century peace generally has been maintained when the leading power or coalition had a clear preponderance of power over the challengers' (J. David Singer, Stuart Bremer, and John Stuckey, 'Capability Distribution, Uncertainty, and Major Power War, 1820–1965', in Hanson, Russett 1972, as summarised by editors, p.14).

[63] 'Men have long struggled to replace power by law based on such

principles as the right for self-determination: but in fact they have succeeded only in so far as they have induced decisive powers to support these principles. If a structure of world laws grows up and we achieve the peaceful settlement of disputes, it will be because those who possess power have decided to make it the servant of these things' (Beaton, 1966, p.10). For an analysis of the moral indifferentism of political realists, see Carr, 1946.

[64] Two examples: Luard observes, that power means armed strength, status, influence, capacity to coerce, control over territory or resources (1968, p.134); Hoffmann writes that power means both a condition of policy and its criterion, a sum of resources and a set of processes, it is applied as a potential and as power in use (1960, p.31).

[65] If the purposes are not named, and the growth of power is said to be the aim of war, it means that the real motives of war are not explained; instead, the present-day conflicts are explained by the search for capacity to win future conflicts – economic, strategic, ideological, and the like. (Pruitt, Snyder, 1969, pp.16–17).

[66] Some scholars contend, that the sharp distinction between domestic and external politics is one of three basic assumptions of Morgenthau's 'paradigm' of international relations, the two others being the treatment of nation-state as the basic unit of analysis, and of the search for power and peace as the essence of international relations. (Johan R. Handleman, John A. Vasquez, Michael K. O'Leary, William D. Coplin, 'Color it Morgenthau: A Data Based Assessment of Quantitative International Relations Research', Prince, Research Studies Paper Number 11. International Relations Program, Maxwell School, Syracuse University, Syracuse–New York 13210, p.11).

[67] Writings of C. Wright Mills, Hannah Arendt, Werner Levi, Anatol Rapoport, Alfred Vagts, Ernst Haas, Stanley Hoffmann and others. The quest for power arises from the will to satisfy one's interests which must lead to conflicts and wars, since the others will also satisfy their interests: 'The search for it (power-JL) becomes necessary to guard against the consequences of this search' (Levi, 1960, p.147).

[68] Even scholars who consider the period of balance of power system in nineteenth century to be a time of peace, in fact accept that peace in Europe was achieved to a great extent thanks to a possibility for colonial expansion ('a useful safety-valve') which allowed state-power to increase without territorial expansion in Europe (Frankel, 1972, pp.120–21)

[69] Ernst Haas, 1953; Alfred Vagts, 'The Balance of Power: Growth of an Idea,' *World Politics* October 1948; Paul Seabury, ed., *Balance of Power*, Chandler Publishing Company, San Francisco 1965.

[70] The main theme of Rapoport, 1964.

[71] Arendt, 1969, pp.11–17; 'All politics is a struggle for power; the ultimate kind of power is violence' (Mills, 1956, p.171).

[72] Carr writes that scepticism about any purposive action, aiming at perfection of the society and elimination of wars deprives humanity of high

ends, and of emotional appeal to any perfecting activity (*The Limitations of Political Realism*, ch.6, 1964).

[73] 'In the traditional theory of the balances of power, the kind of power involved was the physical capability of taking and holding territory. Equilibrium was said to be obtained when military capabilities and war potential on each side were roughly equal. This is not the case with nuclear dialectics which is essentially "political power" rather than physical capability' (D. K. Palit, 'Strategic Concepts: Old and New', *The Institute of the Defence Studies and Analyses Journal*, New Delhi, January 1972, p.322); because of the peculiarities of the nuclear era 'power cannot be considered now a homogeneous, highly interchangeable commodity on the analogy of money' (Robert O. Keohane, Joseph S. Nye Jr., "Power and Interdependence,' *Survival*, July/August 1973, pp.168–9). Stanley Hoffmann in his many sided critique of that approach to international relations in which power is the central concept, argues that 'the post-war era has witnessed radical transformations in the elements, the uses, and the achievements of power'. The main effect is that exercise of power has become uncertain and complicated, and that there exists a disproportion between ingredients and uses of power and their outcomes. The conclusion is that 'the old concept of the national interest, tied to a notion of power, as reasonably calculable and stable is being dissolved' ('Notes on the elusiveness of modern power', *International Journal*, Spring 1975, pp.183–206).

[74] Osgood, Tucker, 1967, p.9; Abel concludes from the analysis of twenty-five great wars, that they didn't break out as a result of emotional or sentimental tensions, from beliefs or other irrational motives, but were based on rational, cold, deliberate decisions ('The element of decision in the pattern of war', *American Sociological Research Review*, 6:6 1941, pp.853–9); Jessie Bernard observes that wars are viewed as 'often highly rational, purposive, deliberate, used coldly, even without hatred, a calculated choice based on policy or strategy' ('The Sociological Study of Conflict', in: Bernard, et al., 1957, p.40); modern wars 'were preceded by long and cool-headed preparations and finally started after carefully calculated decisions . . . the decision to go to war has usually been made, as history shows, upon careful deliberation of the usefulness of war as an instrument and can largely be understood as such' (Levi, 1960, pp.152–3).

[75] For example Michael Howard contends that World War II was rooted in the whole process of socio-political internal development preceding its outbreak; the societies were educated in the nationalist spirit and in social Darwinism ideas, the upper classes sought in army and war profits and prestige, and liquidation of the sharpened armed struggle; armed forces, rearmed and trained for war, were expecting it impatiently. At the same time, political events preceding war might be considered as fortuitous, and the search of great nations for revision of the balance of power could be settled without war ('Reflections on the First World War', in Howard 1970, pp. 99–109).

[76] A. L. Burns criticised the applications (by Richard Rosecrance) of the formula 'x (war) is a function or partial function of y(explanatory variable)' as a

tool of analysis, because it shifted the emphasis from the political decision on war, and wrote: 'Any war, despite all unintended aspects, is a deliberately continued conflict and is amenable to particular historical explanation only, not to general functional explanation' (review of Rosecrance's study, 1973, in *Survival*, March/April 1974, p.104).

[77] For example, Pruitt, 1969, esp. Part Three: 'Movement Toward War: From Motives and Perceptions to Actions'; Mack, Snyder, 1957; Snyder, Bruck, Sapir 1962.

[78] '. . . threat perception arises out of a situation of armed hostility in which each body of policy-makers assumes that the other entertains aggressive designs' (J. David Singer, 'Threat Perception and National Decision Makers', in Pruitt, 1969, p.41). Also Pruitt examined the so-called 'determinants of the perception of threat' and pointed out the role of predispositions to perceive threat, as of the general tendency to perceive threats, a distrust based on the past experiences etc. ('Definition of the situation as a determinant of international action', in Kelman, 1965).

[79] The following so-called determinants of expected gains are presented: importance of the goals, perceived threat, own military strength; the so-called determinants of expected costs are: military strength of the opponent, inherent cost of waging war, strength of the international norms against war, effectiveness of world public opinion, dependence on other states which do not want war (Pruitt, 1969, pp. 235–6).

[80] It seems that some students of the decision-making process are inclined to exaggerate the influence of individuals on the decisions and actions of states and to neglect the influence of groups (influential circles, classes) e.g. 'Nation-State action is determined by the way in which the situation is defined subjectively by those charged with the responsibility of making choices' (Snyder, Bruck, Sapir, 1962, p.212). Similar contention in J. G. March, H. A. Simon, *Organizations*, John Wiley and Sons, New York 1958.

[81] A. L. Burns, *Of Powers and Their Politics: A Critique of Theoretical Approaches*, Prentice-Hall, Englewood Cliffs, New York 1968; H. Speier, 1969; A. Wohlstetter, 'Theory and Opposed Systems Design' in *Journal of Conflict Resolution*, September 1968; John W. Chapman, 'Political Forecasting and Strategic Planning', *International Studies Quarterly* September 1971; Graham T. Allison, 'Military Capabilities and American Foreign Policy', *The Annals of the American Academy of Political and Social Science*, March 1973; 'The interdependence of national intentions and calculations, the presence of both shared and disputed aspirations, the environment containing both steady trends and the possibility of dramatic changes: all of these proclaim immediately that forecasting and planning are to be conceived less as sciences than as prudential arts' (Chapman, 1971, p. 319).

[82] 'Decisions about the use of military force are, and should be, "situational" (Allison, 1973,p.23); 'relations between policy and strategy have no solutions in terms of principles of general applicability', 'situational

considerations may be decisive', and 'our posture and our behaviour must be tailored to the situation and not tied to principles in abstraction from the emerging strategic environment' (Chapman, 1971, pp.345,347)

[83] From another perspective, one scholar states that the political value of arms is more elusive than their military worth, since it is constituted in the minds of the men (Speier, 1969, p. 34). Stoessinger states that the most important precipitating factor in the outbreak of war is misperception; such distortion may manifest itself variously: in a misleading leader's image of himself; in a wrong view of his adversary's character and intentions; finally, in a wrong assessment of adversary's capabilities and power (1974, pp.222 ff.).

[84] The American Government may be regarded as an 'alliance' of people with various views (Thomas C. Schelling, 'Nuclears NATO, and the "New Strategy", in: *Problems of National Security*,p.170).

[85] Luard, 1968, p.65.

[86] Richardson stated that in the period 1820-1929 there was a great number of occasions on which wars could occur, but only in a few of them a war in fact started (1960); some researchers quote or repeat this contention (Nicholson, 1970, pp.43 ff.)

[87] 'Under some circumstances military preparations in themselves can increase the probability of war' (Pruitt, 1969, p.23).

[88] Aron, 1966.

[89] Schelling calls such a policy a conscious manipulation of the shared risk of war (1966, pp.99-105, 'Brinkmanship: The Manipulation of Risk'). 'It means exploiting the danger that somebody may inadvertently go over the brink, dragging the other with him . . .' (p.99).

[90] Schelling, Halperin, 1961; Kahn, 1961; Singer, 1962; Strachey, 1962; Buchan, 1968.

[91] Gantzel sets the 'misperception' – theories against the 'calculation' – theories, the latter claim that war is a conscious and calculated action which aims at realisation of definite interests (1972, p.68).

[92] Examples of accidents listed in some studies: error in warning system; misinterpretation of tactical evidence; accidental detonation of a nuclear weapon, unauthorised provocative action by a pilot or a bomber or a missile commander; sabotage; catalytic war; mistakes in 'brinkmanship' (Schelling, Halperin, 1961); other conceivable cases: Buchan, 1968.

[93] Respectively Strachey, 1962, Röling, 1970, Schelling, 1966.

[94] Op. cit., pp.39-42.

[95] Misperceptions may lead to a situation where the outbreak of war seems to be unavoidable ('point of no escape') (Bruce M. Russett, 'Cause, Surprise, and No Escape', *Journal of Politics*, 1962).

[96] Irrational motives are also to be taken into account: R. L. Giddings Jr, 'On Understanding War', *US Naval Institute Proceedings*, 1968:7; Röling, 1970.

[97] Beaufre, 1974, p.3.

[98] Such an unintended war, distinct from the accidental one, may occur

'because of a series of unpremeditated events which produce in one, or both, sides the belief that they are in acute danger, and that the only safety lies in immediate action' (Strachey, 1962, p.78); such a category of war is 'our greatest danger in the nuclear age' (p. 80).

[99] Schelling, Halperin apply the term 'premeditated attack'.

[100] 'In modern war the propaganda, economic, and diplomatic front may be more important than the military front; but, if the technique of armed violence is not used or threatened, the situation is not war' (Wright, 1965, p.700).

[101] Statesmen were exposed to 'Clausewitz's doctrine that decision could be achieved only by victory in battle' (Buchan, 1968, p.83), and it was 'the view of Clausewitz and of his master Napoleon that it is defeat in battle that leads to the enemy's moral and political disintegration' (p.95). Gen. Beaufre characterised Clausewitz's doctrine as 'destruction of the enemy forces' (1965, p.25).

[102] If we define war as an armed combat between two or more sovereign units, using organised armed forces to fulfil specific goals the word 'organised' is here the most important (H. M. Kallen, 'On War and Peace', *Social Research*, September 1939, pp.37–8).

[103] Montague, 1942; 1969; G. Nicolai observes that only ants and bees fight in an organised way, but they do it against other species (*The Biology of War*, New York 1918, pp. 18–19).

[104] Turney-High, 1971, pp.23, 25, 30, 39, 52–3, 61, 123.

[105] (from about 1720) . . . 'the violent conflict assumes (setzt voraus) the existence of closed (geschlossene) groups to be recognized as "war" ' (Krieg, in *Sowjetsystem und demokratische Gesellschaft*, 1965 p.1027).

[106] 'War is the hostile conflict waged by nations, states, governors, or between groups in the same nation, or in the same state, by armed forces' (*The Shorter Oxford English Dictionary on Historical Principles*, 3rd rev. ed., vol. 2, Oxford 1965, pp. 2383 ff.); 'war is a conflict waged "by armed forces" ' (Wright, 1965, p.8).

From another perspective, Janowitz writes that 'war is differentiated from other forms of social conflict because war making relies on a highly professional and specialized occupation, the professional soldier' (1964, p.338).

[107] Aron quotes a definition of war as armed conflict carried on by organised forces and observes that in some cases it is difficult to say, whether the notion 'organized military forces' applies exactly to one of the fighting sides (1957, pp.180–1); he also writes about 'the refusal to allow the regular armies a monopoly on war' (1959, p.66).

[108] See note 101.

[109] Consequently, 'The distinction between diplomacy and strategy is an entirely relative one. These two terms are complementary aspects of the single art of politics . . .' (Aron, 1966, p.24).

[110] For some of Liddell-Hart's writings, see bibliography. I have described his views in Lider, 1971.

[111] Barber, 1975, p.7; Aron observes, however, that the obliteration of this

distinction may also be a confirmation of the Clausewitzian idea of the continuation of use of political means during the war (1969).

[112] The dependence of military strategy on policy is caused not only by the appearance of nuclear weapons, but also by other factors: 'One of these is the Communist reliance on political warfare in adroit combination with other types of Soviet power. Another is the American policy of collective security. The fact that the United States has entangling alliances with over forty nations dispersed throughout the globe has placed severe political restrictions on both the deployment and potential employment of US armed forces' ('The Politicalization of Strategy', in: Abshire, Allen, 1963, pp.388–9).

[113] American military historians often underline, that in earlier wars fought by the United States limited victories were pursued. The War of Independence offered a forestaste of a change of the concept, the Civil War and Indian wars were fought for a complete military victory, and since the United States rose to world power, especially in two world wars, total victory and achievement of unlimited political goals became the aim of war (Weigley, 1973, pp. xx-xxiii).

[114] In some 'search for power theories' war is considered as a means of increasing power for future victorious wars.

[115] Gen. Hoyt S. Vandenberg defined two types of victory: (a) a complete one, which emasculates the enemy, and (b) a combination of diplomatic and military victory allowing the enemy to remain a viable force and not changing the balance of power in the world (quoted in William I. Gordon, 'What Do We Mean By "Win"', *Military Review*, June 1966, p.5).

[116] Paul Kecskemeti, 'Political Rationality in Ending War', in *How Wars End*, 1970, p.107.

[117] 'Now if someone thinks we should have a nuclear war in order to win, I can inform him that there will not be winners in the next nuclear war, if there is one, and this country and other countries would suffer very heavy losses' (McNamara, After George Lowe, 'Neither Humiliation nor Holocaust', *US Naval Institute Proceedings*, June 1963, p.63).

Liddell-Hart observes that: 'History shows that gaining military victory is not in itself equivalent to gaining the object of policy. But as most of the thinking about war has been done by men of the military profession there has been a very natural tendency to lose sight of the basic national object, and identify it with the military aim. In consequence, whenever war has broken out, policy has too often been governed by the military aim – and this has been regarded as an end in itself, instead of merely a means to an end' (1967, p.351).

[118] Young analyses the war in Korea, the interventions in Lebanon 1958 and in Thailand 1958, and the Caribbean crisis, and observes: 'When attained, the announced limited objectives of each crisis restored the *status quo ante*, but failed to resolve the basic reason for the crisis – the threat of intervention of internal communism.'

[119] *How Wars End*, 1970, esp. essays by William T. R. Fox, George H. Quester, Quincy Wright, Morton H. Halperin and Paul Kecskemeti.

[120] For example, peace signals may be misinterpreted (Quester, ibid).

[121] The literature on the subject is too vast to be quoted. In several articles in the West German central military periodical a critique of the wrong interpretation of the Clausewitzian idea of primacy of policy is presented: instead of emphasising the control of the military leadership by the political, one should point out the necessity of cooperation between them (e.g. Münter, 1975). Haniotis lists four principles of cooperation between politics and the military direction of war: 'First: Mutual understanding of the respective views; Second: Acceptance by the military of the principle of the political primacy in the direction of war. Third: Non-interference of politicians within the purely military area of responsibility. Fourth: Politicians while issuing their directive and imposing restrictions of a political nature on the war should avoid, unless it is absolutely necessary, the violation of the fundamental principles and rules which govern the rapid and effective conduct of war' (1970, p.54).

[122] 'It remains a mistake, in the theoretical level, that the American decision was dictated by the exclusive concern to destroy the major part of the German army and that consideration of the political consequences of one method or another was regarded by Roosevelt and his advisers as an unwarranted intrusion of politics into the realm of strategy' (Aron 1966, pp. 27-9); '. . . the conduct of [the] Second World War was essentially political, i.e. dictated by consideration of the consequences remote from the scene of hostilities and victory – on the Soviet side' (p. 29).

[123] '. . . The Eisenhower strategy opened all Eastern Europe and much of Central Europe to seizure by Soviet armies. The great defeats suffered by the democracies in the first two years of cold war were ordained by that pivotal decision of strategy, made in the shooting war that preceded. And that decision emerged from the failure of American leaders to accept war as a tool of policy' (Hessler, 1949, pp.158-60); the same view Hanson W. Baldwin, *The Great Mistakes of the War*, Harper and Brothers, New York, 1950, pp.1 ff. Osgood wrote about the 'political blindness of American military strategy in World War II' (1957, p.115); cf. Haniotis, 1970, p.52, who quotes some critical comments on the subject.

Clairborbe Pell contends that the American leadership forgot that military aims have to correspond not only to the national current interest, but also to the far-reached national policy (*Power and Policy: America's Role in World Affairs*, W. W. Norton and Co., New York 1972, pp. 16-20).

[124] For example, some essays in *International Affairs*, 1967 (Editor's Foreword); Eqbal Ahmad, *War and Counter-Insurgency*; Jean Baechler, 'Revolutionary and Counter-Revolutionary War; Walter Goldstein, 'The American Political System and the Next Vietnam'.

[125] The failure of the intervention in Vietnam has thrown into question 'the capacity of our system to act effectively' (Ithiel de Sola Pool, in Richard M. Pfeffer, *No More Vietnams?*, Harper and Row, New York 1968, p.142).

4   Internal war as a continuation of policy

The study of war was traditionally confined to conflict between sovereign powers and ignored revolution or other conflicts involving a combatant without sovereign status. Clausewitz exemplifies, indeed sums up, this tradition in assuming that politics is tied up with sovereign powers and is an expression of the will of the sovereign. The liberal tradition put the concept of politics, and war, in a much broader perspective by showing that the basis of political authority lay in the nature of man's social relations and that it could in principle be questioned by any citizen and overthrown by a majority of them. However, only the magnitude of contemporary revolutions and civil armed conflicts (Russia, Spain, China) made distinctions between the nature of interstate and intrastate armed conflicts seem unnecessarily artificial, for these revolutions were as war-like in their political aims and violent means as the traditional wars.

Since the Second World War a growing amount of scholarly attention has been devoted to internal armed conflicts as a kind of war, owing to the increase in the number of revolutionary uprisings and wars of national-liberation, especially within the last two decades. [1] One result of the multitude of case-studies and analytical works that have been published, [2] is that many different definitions [3] and schemes of classification have been presented. Terms frequently used by American scholars include internal war, intrastate war, domestic war, guerilla warfare, intrasocietal war, subversive war, revolutionary war, sublimited war, irregular warfare, and partisan war. To these the British and French have added small wars and brush-fire wars, and the West Germans concealed wars. This large number of terms reflects a wide variety of approaches, which emphasise either the socio-political essence of such wars or their territorial, military, or other aspects. All this leaves a complicated and perhaps even confused picture of what in non-Marxian theory is understood as internal armed conflict.

The notion of internal war seems to have gained currency in more recent studies, [4] but an increasing number of studies are also devoted to civil wars, [5] a notion similar to that used in Soviet theory, but in the Western literature interpreted differently, and variously by various scholars. [6]

It should be mentioned however, that the focus remains on international wars and internal wars are still treated in many studies as an abnormal kind of conflict. [7]

The political framework

Internal war as a continuation of policy

Despite this diversification, attempts have been made in the Western study of

86

war to develop, if not a general theory of internal war, then at least a general theoretical framework in which to analyse it. The point of departure for such work is the general definition of war as a continuation and instrument of policy by violent means. [8] Whereas the aim of international wars is usually protection of some vital national interest, the political goals of internal wars are usually regarded as either socio-political, national or ethnic; for example: (a) to remove certain individuals from high political office (e.g. *coup d'état*), (b) to change the system of political authority and thus the process of policy-making, [9] (c) to change the basic socio-economic structure of society (e.g. socialist revolution), [10] (d) to achieve local autonomy for a national (or religious etc.) minority, to win national self-determination from colonial rule, or to enter union with another state. [11]

Internal war is thus generally understood as a violent conflict between social forces within a given state as the most acute form of civil strife. [12] This may be regarded as its broad definition, because it encompasses wars with all of the political goals mentioned above.

The ideal case becomes more complicated when support is given to one or another of the social forces by a foreign power, many examples of which can be found in modern history. Is such a war an internal or international one? There is no obvious answer to this question, so most scholars judge each instance according to the nature of the main issues and the tactics adopted; the involvement of foreign powers is not in itself a decisive criterion. [13]

It has been suggested however, that the use of the term internal war should be confined to purely domestic clashes; this may be regarded as the narrower concept. Wars waged for national liberation are here grouped with international wars. This may solve the problem inherent in the more general approach of deciding how much foreign involvement there can be before an internal war becomes an international one, but it has the disadvantage of ignoring military and political similarities between many wars fought for national goals and wars fought for socio-political reasons.

No generally accepted typology has been constructed; some scholars observe that it is difficult to present a typology based on the socio-political content, since such content is elusive, changing, and additionally, variously assessed by different authors.

One of the characteristic features of internal wars in consequence of the kind of aim for which they are fought is an asymmetrical relation between combatants. One side usually controls the state apparatus, including the armed forces, and can make claims on the support of the citizenry on the basis of legitimacy and authority. The opposition is much less well equipped technically, and is therefore spurred to find unconventional ways of waging war. [14] Internal wars have thereby provided scholars interested in the military-tactical aspects of war with a wealth of new material, which explains the number of terms used to describe internal wars according to the means used: guerilla warfare, unconventional warfare, *coup d'état,* etc. Some scholars point out, however, that the use of the notion guerilla warfare interchangeably with revolutionary war (or insurgency) as though they were synonymous is misleading, since the former refers only to a particular kind of military operation performed by irregular forces, while the latter is a war, a fusion of political and military activity. [15]

Wars of socialist revolution

The so-called wars of socialist revolution or revolutionary wars have been one of main themes in the Western politico-military analysis [16] because of their frequency, extreme goals, and supposed relation to the goals of international Communism. [17] They have been usually presented as having double aims; one internal, to bring about a Communist transformation of the whole social structure, and the other international, to strengthen the socialist coalition and its allies, and speed the world revolution. [18]

A great deal of attention has been given to the extent to which internal wars are related to the world Communist movement. Although it has usually been recognised in this approach that civil wars (as they are often called) may be rooted in economic distress or political oppression, [19] the main point of interest is the extent to which these conditions are exploited by domestic Communist forces or by world Communism to weaken the global strength of the defenders of Western values. [20] This focus of attention has led to a four fold taxonomy: war instigated and supported by international Communism (e.g. the Greek Civil War, the Korean War, the war in Vietnam), wars that the Communists apparently did not initiate but attempted to gain control over and exploit (e.g. Cuba and Algeria), wars supported politically by Communist powers but with very little direct participation from world Communism (e.g. Congo 1960 and 1964, Guatemala 1946–54), and finally, wars lacking Communist support.

Simpler classifications have also been presented, in which revolutionary wars have been divided into wars internally rooted, although exploited by international Communism, and wars directly instigated by the latter. [21] This approach should be perhaps regarded as a child of a time in which the Cold War coincided with the growing movement for self-determination within the Third World, for it was then that it had its heyday. British and French analysts linked most of the wars in their respective empires to revolutionary movements; [22] American scholars believed anti-Western tendencies in the Third World to be largely supported by external Communist forces; [23] and West German analysts warned of the dangers of hidden Communist subversion within Europe. [24] In recent times the assumptions on which such interpretations have been based have been challenged in the West by many scholars and politicians alike. [25] Others observe that Communists instigated revolutionary wars (as well as so-called national-liberation wars) up to 1972–73, but the era of international involvement in the revolutionary struggle of other peoples then came to an end. [26]

Roots and preconditions

The structural roots of war, whether internal or international, is a subject treated more fully in the next chapter on war as a kind of conflict. In this section therefore we will review the debate about the weight that should be given to deep or, on the contrary, more immediate causes of internal war in explaining their occurrence and about their necessity or, on the contrary, their accidental character.

The fact that internal wars may be fought for so many different reasons and in a variety of ways makes it difficult to make any generalisations about the character of their roots or the conditions under which they occur. Naturally, wars of national liberation spring from the insufferable yoke of a colonial power. If a national minority fights for autonomy, the reason is usually that it considers itself oppressed. The source of civil war between classes usually lies in intolerable socio-economic inequalities. In societies in which there is no recognised procedure for replacing political leaders, questions of succession may have to be solved by war. [27]

However, while such basic conditions may be necessary for internal war to break out, they are not in themselves enough to give a full account of the matter. It is unlikely, for example, that internal war will break out in a country beset by serious economic problems if the government in power is effective in acting to alleviate the situation. [28] Nor is it enough to attribute wars of national liberation to the existence of colonialism, for not all colonies have gained independence that way. In some countries there are national minorities that do not strive for autonomy, and some of these do not find their living conditions oppressive even though there is no great difference between these conditions and those of a minority in some other country that does feel oppressed. [29]

Considerations such as these have led some scholars to seek an explanation to internal war elsewhere. Internal wars may spring from many sources, and there may be a great number of necessary conditions to be fulfilled before an insurrection will succeed, but if these necessary conditions are not also sufficient conditions, they cannot explain internal wars. Scholars who have reasoned thus have consequently searched for the immediate cause that catalysed a situation of potential violence into one of civil war.

Some scholars argue that both the outbreak and the success of any internal war is dependent upon certain special circumstances that set the country in a state of crisis. They point out that it was just such extraordinary conditions that made possible the victory of the historical revolutions: Tsarist Russia had been broken by World War I at the outbreak of the October Revolution; the Japanese invasions of China stimulated the development of the Communist revolutionary movement; [30] and the course and outcome of World War II gave impetus to the wave of national and social uprising in many countries both in Eastern Europe and, later, the Third World. [31]

A second reaction is to deny any essential, and, what is more, any determining causal ties between particular types of internal war and socio-political or socio-economic conditions. Such a view is implied, for example, in some studies of revolutionary movements in developing countries. It is suggested in these that the Communist-inspired goal of social revolution is not endemic with those living in such societies but is a foreign ingredient. The natural cause of these internal wars is a desire for national liberation. The Communists, being better trained and organised and having support from abroad, are often in a good position to seize control of a struggle for national liberation. In support of this view it is maintained that all successful Communist revolutions in Third World countries have followed this pattern. [32] Some add as evidence that in both the Soviet doctrine of revolutionary war and the Chinese doctrine of 'people's war', the main condition for the success of a communist struggle is the exploitation of national liberation movements. [33]

A similar stress on immediate causes can be found amongst those students of *coups d'état* who deny the relevance of deeper social problems to the explanation of this type of internal violence. This should be regarded as an extreme position, however. More usually some distinction based on political aims is made: for example between governmental *coups* ('palace revolutions') which introduce no significant changes in the social structure and even in the structure of the authority, revolutionary *coups,* which aim at radical social and economic changes, and reform *coups,* which aim at some reform at least of the political, economic, or social structure. [34]

To sum up, although there seems to be general agreement in the West that internal wars, or at least organic ones, are rooted in some fundamental conflicts, be they economic, political, or social, and that such conflicts are a potential from which may develop a division of antagonistic social forces, it is also held that to point out such roots of internal wars does not suffice to give a full account of the causes of their outbreak. Situations in which a resort to violence is imminent can be described as a set of conditions. The search for an explanation to how such situations become transformed into violent ones has led to an emphasis on immediate causes.

The military picture

The political nature of internal wars is, as we have seen, impossible to sum up in one general formula, but the main political goals of particular wars can usually be identified relatively easily. It is often more difficult to characterise the military aspect, for no two internal wars are conducted with the same strategy. Since surprise is a major weapon of insurgents, and since they are in general less well equipped and trained militarily, they usually have a great incentive to innovate and improvise. As a result, most internal wars are from a military viewpoint a mixture of old and new.

Nevertheless, it might be suggested that there are four main ways in which armed force has historically been used in internal wars of the present century. [35]

1 The rebels gain the support of some part of the existing armed forces (military, police, or security forces) and seize the offices of government through a *coup d'état.* [36] *Coups* may differ substantially in the amount of popular support they receive; they may be ranked from 'palace revolutions' to socio-political actions supported by popular masses. The October Revolution was recognised by some Western scholars as a *coup* supported by workers' movement in Petersburg and Moscow; the Portuguese revolution 1974 may be a recent example of another kind of *coup.*

2 The rebels build an armed force of their own and launch an armed uprising at the centre to defeat government forces (e.g. Cuba).

3 Again, the rebels base their military strength on their own irregular troops, but avoid major clashes with government forces; instead, guerilla tactics are employed (attack and disperse), and a prolonged war of attrition is waged (e.g. China).

4 Terrorism is a fourth tactic, which, although not itself a sufficient means of overthrowing an established government, may be used in an initial or transitory phase of an internal war (e.g. Ireland and Palestine). [37]

Faced with such a variety of political goals and military means, scholars have either tried to develop some general ideas about the strategy of internal war or concentrated on some particular type (defined in terms of goals and means) of internal war. The type that has received by far the most attention is the guerilla war fought for socialist revolution.[38] The great success of the Chinese Communists has inspired revolutionaries in underdeveloped countries, particularly in South–East Asia and Latin America, where social upheaval is generally regarded as most probable.

Revolutionary guerilla war

It is generally recognised in the West that revolutionary guerilla war is distinct from international war not only with respect to the political goals for which it is fought, but also with respect to its military conduct. [39] The main reason for this is the great initial disparity in the resources available to the combatants. The government authorities have great superiority in tangible assets, military and economic, and their political resources to command support both domestically and from abroad are normally considerable. The revolutionary party must rely on its superiority in such intangible assets as ideology and devotion in the hope that these can win the support of dissatisfied masses, especially the peasants, and be exchanged for more tangible assets. [40]

Because of their military inferiority, the partisans move in small groups and avoid big battles. Their strategy is to deliver swift damaging blows and then to retreat quickly into the countryside. The success of their operations is therefore largely dependent on their ability to gain popular support amongst rural inhabitants. They must rely on the peasants not only for aid and shelter and for secure caches of ammunition and stores, but also for reinforcements of manpower. Only after the revolutionary forces have succeeded in increasing their numbers and enlarging their bases can they undertake regular military operations and fight conventional battles. [41]

Many scholars emphasise the extreme importance of the utilisation of space by guerillas. Since the enemy can move its conventional land forces great distances, guerillas must base their tactics on an optimal exploitation of space; dispersion and mobility being the main tactical principles. They must make it impossible for the enemy to control the whole territory militarily or politically by continually menacing the adversary over the largest possible area and by waging a protracted war. In contrast to the strategy of traditional forms of fighting, [42] which aims at the destruction of the enemy forces by a direct attack from without, the basic assumption of guerilla war strategy is that the capacity to fight of the armed forces of the enemy can be destroyed from within. [43] The emphasis is therefore on the protracted nature of the struggle; on transforming the war into a war of attrition. [44]

In such circumstances, political activity plays a much greater role than in international war. Not only are the general aims of war and the strategic goals defined in political terms but also the tactical objectives: to win the active support of the population is regarded as the aim of operations at all levels. Each military move is therefore planned to achieve direct political outcomes. Some theorists of guerilla warfare even maintain that military operations should not be undertaken until the final stage when the political struggle has been substantially won. Compared with international wars, the change of method from the period of peace is therefore much less radical. [45]

The view has even been expressed that political action replaces military action as the main method and principle instrument in guerilla warfare. [46] The question arises, of course, whether such a struggle can then be called war.

On the other hand, some individual scholars express a completely contrary view. They contend that many guerilla wars are fought without any clear political doctrine, are dominated by military considerations, and are wholly directed by military leaders. The bands of partisans are said to have full political autonomy; the decisions they take subordinate political aims to military realities. [47]

One result of the growing attention devoted to guerilla warfare has been that the necessity to study and to elaborate a theory of counterguerilla warfare has been emphasised. [48]

The relation between internal and international war

Up to now, the focus in the Western research has been on the differences between international and internal war. [49] One important kind of relation between them has been discussed however, namely the effects internal war may have on the incidence of international war and vice versa.

The effect of internal war on international war

Seen from outside, a country torn by internal strife constitutes an element of uncertainty. Since the outcome is unknown, foreign powers may be tempted to intervene in order to ensure a result most favourable to themselves. The occurrence of internal war may therefore increase the likelihood of international war. [50]

Even when internal war has not in fact broken out but is in the air, the latent tensions may have an international impact. A government faced with a difficult political situation at home may try to save its position by redirecting attention to an external threat. [51] The enemy may be more fictive than real, or at least less of a demon that he is made out to be, but by playing on the fears of its citizens a government may stave off threats to the internal stability of the country. It may even find an artificial cause to launch war against some other state. [52]

Some scholars of former colonial states analyse the history of these countries in terms of the redirective mechanism. They describe the colonial policies of the imperialist countries as one of 'externalised civil war'. The foreign adventures of these states allowed them to ease domestic conflicts by giving their citizens a sense of world mission, by providing territories in which dissatisfied groups could settle, and by obtaining resources with which to raise the material standard of living in the home country. The loss of colonies has led to an aggravation of internal contradictions in the former imperial countries. At the same time, the success of the struggle for national liberation has marked the end of an external factor that generated national cohesion in the former colonies; in particular, ethnic cleavages have once again become important divisive forces with the result that internal wars have increased. [53]

Another topic of analysis is the impact the outbreak of internal war may have on a state's participation in international war. Obviously, this leads to a weakening of the external military effort and in some cases may cause its termination. The October Revolution may serve as an example.

The effect of international war on internal war

Participation in an international war may have the effect of unifying a country so deeply divided that internal war would otherwise have been a likely development. It is also possible, however, that the sacrifices required by war create social dissatisfactions or sharpen existing ones. International war may therefore foster internal war as well as the other way round. [54]

Some scholars argue more generally that external variables are the primary determinants of the outbreak, course, and termination of all internal wars. [55] Encouragement and support for both sides from foreign powers is always present since even apparent inaction can be interpreted as a form of passive intervention or support for the stronger party. [56]

In a more general approach, internal strife is seen as a part of a larger conflict between opposing socio-political camps, between great powers and colonies (or former colonies), or between races. [57]

Special attention has been given to the effect the Cold War has had on the incidence of internal wars. [58] Since the rivalry between the two global camps cannot be manifested in the form of a great international war because of the capabilities each has for enormous destruction (the so called 'nuclear paralysis'), it takes on the less dangerous form of ideological conflicts and internal struggles that may break out in war. Sometimes one camp may become directly involved in an essentially internal struggle to improve or defend its strategic position in relation to the other. Other internal conflicts unintended by either of the superpowers may also arise if competing domestic groups believe they can exploit the limited manoeuvrability of the superpowers.

The most far-reaching contention is that internal and international wars are now interchangeable as complementary forms of the great global struggle. In this view, the basic international cleavage has generated four kinds of internal war: (a) wars by proxy, fought by the great powers in a roundabout way; [59] (b) internal wars combined with some external intervention (Vietnam); (c) internal wars fought by forces participating directly in the world-wide struggle (a possible war between the two Germanies); [60] and (d) internal wars influenced by the superpowers although without their open participation.

Denial of any regularities

Against these various views should be set the contention that there does not exist any relation between internal and international war, [61] or that there is at least no regular relationship between them. [62] According to the more extreme position, there is an essential difference between the behaviour of groups acting within the framework of a state and the behaviour of states acting in the international context. Tempting though analogies may be, wars that occur in one setting must be basically different in their origins and nature from wars occurring in the other. [63] The more moderate view allows that one kind of war may be dependent on the other, but argues that the relation is so contingent on the particular nature of each case that no general statements can be regarded as valid. [64]

Notes

[1] Luard, 1972; Lincoln P. Bloomfield, Amelia C. Lewis, *Controlling Small*

Wars, Allen Lane, The Penguin Press, London 1970; 'the vast majority of local wars after the Second World War have been civil wars or have started as internal conflicts or insurrections that have led to the intervention of other states.'
Adrian Guelke, 'Force, Intervention, and Internal Conflict', in Northedge, 1974, p.99; McNamara estimated in 1966 that within the previous decade '149 serious internal insurgencies' occurred.

[2] See bibliography, esp. studies by Huntington, Galula, Thompson, Paret and Shy, Campbell, Paget, Delmas, Heilbrunn, Bloomfield and Leiss, Luard; cf. in articles *Military Review, Revue Militaire Générale, Journal of Conflict Resolution. Journal of Peace Research* and *Orbis.*

[3] Internal wars are 'attempts to change by violence, or threat of violence, a government's policies rulers, or organization' (Eckstein, 1964, p.1), and in another definition he calls internal war a resort to violence within a political order aiming at changing its constitution, government or policies. Internal war is 'a violent conflict between parties subject to a common authority and of such dimensions that its incidence will affect the exercise of the structure of authority in society.'
(Andrew C. Janos, 'Unconventional Warfare: Framework and Analysis', *World Politics,* July 1963, also in Eckstein, 1964, p.130). Internal war is distinguished by confinement of the fighting to one country, of the causes to internal ones, and of the fighting forces to local ones (Morton A. Kaplan, 'Intervention in Internal War: Some Systemic Sources', in Rosenau, 1964, p.92).
Civil war is a breakdown of a legitimate order of government aiming at over-throwing a government and capturing it (Osgood, Tucker, 1967, p.31).

[4] Eckstein, 1964, 1965; R. Tanter, 'Dimensions of Conflict Behaviour Within Nations, 1955–60: Turmoil and Internal War', *Peace Research Society Papers* 3 (1964); J. K. Zawodny, F. M. Osanka: 'Internal Warfare', in: *International Encyclopedia of the Social Sciences,* D. Sills, ed, Macmillan, New York, 1968, vol.7; Lang, 1972, Guelke, 1974 (comp. n.1). Tefft writes that two types of war are identified: (a) Internal war, which is warfare between political communities within the same culture unit, and (b) External war, which is warfare between political communities that belong to different culture units (1975).

[5] For a detailed analysis of the notion of civil war as distinct from a more general term internal war, see: Martin Edmonds, 'Civil War, Internal War, and Intrasocietal Conflict: A Taxonomy and Typology', in: Higham, 1972, ch.I; cf. Luard, 1972; cf. Definitions of civil war in dictionaries: C. W. Bain, 'Civil War' in *Dictionary of Political Science,* J. Dunner, ed., Vision Press, London 1966; W. White 'Civil War', in *White's Political Dictionary* World Publishing Co., New York 1947; cf. J. K. Zawodny, 'Internal Warfare: Civil War', in: *International Encyclopedia of Social Sciences,* vol.7. Martin Edmonds writes that while in any definition of internal war the context within which violence takes place (an autonomous political system, with a legal government) is stressed, the distinctive feature of civil wars is that the opposing forces are

politically organised, and that the war is a direct challenge to the authority of the government.

J. K. Zawodny distinguishes, however, two types of civil war, of which only one is a planned one and occurring either in an absence of effective formal and informal channels for settling political grievances or based on the assumption that there is no recourse other than violence for securing redress. The other is the spontaneous type, without any planning or even actual leadership (ibid., p.499).

[6] The lack of an agreed view what are the most typical features of civil war can be illustrated by the assumption of Robert Higham who in the introduction to a collective work on civil wars, of which he is the editor, denotes such war as 'the work of right-wing reactionaries or conservatives seeking to keep arbitrary freedoms they enjoy or think they are entitled to exercise' (1972, p.1). And he adds:

> Some of the features of a civil war are: 1) that it is an overt internal conflict with international overtones; 2) that it is organized by a socially cohesive class seeking to protect arbitrary freedoms, which makes the conflict an esentially negative response; 3) that an extralegal government is created which possesses, or believes it possesses, not only economic self-sufficiency but also the loyalty of a sufficient part of the regular part of the regular armed forces to defy the rest of the country; 4) that this government possesses a contiguous territory; and 5) that this government's actions will be a mixture of professional governance, aristocratic or gentlemanly standards, and cruelty mixed with great humanity.
>
> (Editor's Introduction, in Higham, 1972, p.2)

In his review of the modern writings on war, Lang concludes that 'most of the time we tend to think of war only in relation to the nation-state and as an instrument of diplomacy' (1972, p.133) and states that 'these struggles [civil wars, national-liberation wars, etc.] becomes regular wars when the former subject group attains some legitimacy in the eyes of the other' (p.134). Therefore e.g. Hedley Bull postulates to nullify the 'artificial distinction' between international and internal war in law and diplomacy ('Civil Violence and International Order' in 'Civil Violence and the International System', Part II, *Adelphi Papers,* no.83 1971, p.32).

[7] In the most outstanding modern study on war, civil strife is recognised as war only if there is an equality in status of combatants, which sets extreme limits to the number of cases included in the term: insurrections, colonial revolts, etc. are not assigned the status of war (Wright, 1965, p.695 a.o.). Many scholars observe that civil war is even now seldom dealt with in the major theoretical works which discuss war. M. A. Nettleship notes that since civil war is thought to occur within rather than between political units, it is treated as a different order of phenomenon than war ('Definitions', in Nettleship, et al., 1975, p.85).

[8] Therefore many definitions or descriptions of internal war are related to

definitions of international war, either (a) treating internal war as a class in the general classification of wars, or (b) as a species of international war, or (c) contrasting it with the latter. In the first approach, Wright proposes four classes: balance of power war, civil war, defensive war, and imperial war (1942, p.641), or three classes: civil wars, imperial wars, and international wars (1965, p.695). Singer, Small (1972) characterise three types of war, two international – intrasystemic i.e. interstate wars, and extra-systemic, the latter are divided into imperial and colonial wars – and one type of civil war. But they add one type of extra-systemic war, namely an internationalised civil war, i.e. civil war accompanied by an intervention from outside (1972, pp.31–2). In the Western military literature in the 1940s and 1950s civil war was often treated as a kind of local war, e.g. Colonel Nemo presents three types of war: generalised war, limited war and cold war, the latter encompassing civil wars ('The Place of Guerilla Action in War', *Revue Militaire Générale,* Jan 1957); Montgomery makes a similar proposition and divides limited wars into interstate wars and wars 'provoked by a rebellion within a state' (after Nemo, ibid.).

[9] Osgood, Tucker, 1967, p.31; most of civil wars aim at the overthrow of a legally established government, and in this sense they have been revolutionary as contrasted with interstate wars which aim at a territorial aggrandisement (Andrew J. Kaufmann, 'On Wars of National Liberation', *Military Review,* October 1968, p.36); struggle for authority distinguishes internal wars from international (Andrew Janos in 'Authority and Violence: The Political Framework of Internal Wars', Eckstein, 1964, p.133); internal war with rebellion, is a 'way of making demands on authority, whether for the change of specific acts or rulers or of structure of authority' (William Kornhauser, 'Rebellion and Political Development', in Eckstein, 1964, p.142); George Modelski, who calls internal war a 'national one', describes it as a 'conflict over national authority and the definition of the state' (1972, pp.304, 310).

[10] The diversity of the socio-political goals of various internal insurgencies has led to a presentation of various classifications. Jack P. Greene: revolutions may be directed toward a change of either/or (a) a government leadership; (b) a regime (form of government and distribution of political power), (c) the whole society (its structure, system of property control and class relations), (d) governmental policy ('Theories of Revolution in Contemporary Historiography', *Political Science Quarterly,* March, 1973); he presented the first three types after Chalmer Johnson and added the fourth one; Eckstein: internal wars are fought for either/or (a) a change of authority/with a possible radical change of its structure/, (b) a change of some political decisions or even of the whole direction of government's policy, (c) obtainment of an influence on the policy and a participation in its shaping (1964); Harold Lasswell and Abraham Kaplan present three types of internal wars (revolutions): palace revolutions, political revolutions (change of the structure of the political authority, e.g. replacing democracy by dictatorship, or vice versa), and social revolutions (*Power and Society,* Yale University Press, New Haven 1950); the same three types, but the

first called *caudilissimo* are presented by Edwin Lieuwen, *Arms and Politics in Latin America,* Praeger, New York 1960; Huntington presents various classifications, the most detailed consists of four types: the palace revolution, the reform *coup,* the revolutionary *coup,* the internal war; although only the last term is related to armed violence *sensu stricto,* the former may also be connected with some form of it. ('Patterns of Violence in World Politics', in Huntington, 1952).

An interesting typology based formally on external features but closely related to the political content is presented by D. E. H. Russell, who distinguishes between three types of 'rebellion' (defined as 'a form of violent power struggle in which overthrow of the regime is threatened by means that include violence'); mass rebellion, military rebellion, and anticolonial rebellion (*Rebellion, Revolution, and Armed Force:* A comparative study of fifteen countries with special emphasis on Cuba and South Africa, Academic Press, New York, 1974).

Some scholars presented a two-items typology, distinguishing between civil violence aiming at moderate changes, in principle performed as a *coup d'état.* and revolutions aiming at radical political, economic, and social changes (Crane Brinton, *The Anatomy of Revolution,* Vintage, New York 1952; George Blanksten, 'Revolutions', in H. E. Davis, ed., *Government and Politics in Latin America,* Ronald Press, New York 1958).

[11] Accordingly, various classifications were proposed. Evan Luard: (a) conflicts reflecting the world-wide divisions in the cold war, (b) conflicts connected with other ideological struggles (e.g. radical Arab national forces against the traditional and conservative ones), (c) conflicts of a post-colonial type (new established regimes contra opposing forces), (d) conflicts arisen as a form of protest against dictatorial or oppressive government (1972, pp.11–14). Osgood, Tucker: revolutions (modern internal wars) aiming at: (a) a change of the political system without a radical change of the social system, (b) a change of the whole political system, (c) expulsion of a colonial power and achievement of national independence, (d) achievement of a local autonomy or independence for a minority. An additional kind: wars for determination of the succession after the expected liberation (Dougherty, Pfaltzgraff, 1971, p.249). Leo Heimann: (a) people's war against the enemy occupation, (b) revolutionary civil war against corrupt and reactionary regime, (c) national liberation war, (d) ideological war for political or religious reasons ('Guerilla Warfare: An Analysis', *Military Review*, July 1963); Francois Duchéne: (a) revolutions for the achievement of better social and economic positions, (b) revolution of the underprivileged groups in extreme cases aiming at seizure of power, (c) attempt of national secession during the national-state building, (d) manifestation of expansionist nationalism ('Introduction', in 'Civil Violence and the International System', p.I, 1971, p.6).

[12] 'Civil violence ... covers any private political violence which aims to influence or overthrow government, rising at the upper end as far as guerilla wars but excluding conflicts in which the regular forces of states are engaged on

both sides' (Francois Duchéne, 'Introduction' in 'Civil Violence and the International System', p.I, *Adelphi Papers*, no.82, 1971. Aron observes that the definition of civil war, as each definition, describes the 'perfect' phenomena; civil wars may be treated as marginal cases in the graduation of civil strife, and therefore: 'On the borderline, civil war and international merge together, as do the clash of armies and guerilla warfare.' (1957, p.181).

The character of internal war as a social strife has led to the contention, that such a war, in contrast to classical conventional wars, is a wholly social phenomenon (Jan Baechler, 'Revolutionary and Counter-Revolutionary War: Some Political and Strategic Lessons from the First Indochina War and Algeria', *Journal of International Affairs* 1971:1, p.71).

[13] Lincoln Bloomfield takes, however, the degree of foreign intervention as the basis for ranking various types of internal conflicts on a scale from 'basically internal disorders', through 'externally abetted internal instability' to 'externally created or controlled internal instability' (*International Military Forces*, Little Brown and Co, Boston 1964, pp.28–30). For a critique of his proposal, as not exhaustive, too rigid to deal with elusive and changing phenomena, and too dependent upon subjective judgments, see Miller, 1967, pp.5–6. The latter proposes an informal classification of 'intrastate violence' into colonial wars, post-colonial civil strife, aggressions, and subversion from outsiders (pp.11 ff.).

[14] A. J. Mack contends that the asymmetrical character of war can also be noted in internal wars in their broader sense, e.g. in civil wars accompanied by interventions of external powers (Vietnam) or in national-liberation wars (Algeria) ('Why Big Nations Lose Small Wars: The Politics of Asymmetric Conflict', *World Politics* January 1975).

[15] John Baylis, 'Revolutionary Warfare', in Baylis et al., 1975, pp.134–5. Cf. M. Osanka, *Modern Guerilla Warfare*, Free Press, Glencoe 1962 and esp. Huntington's introduction; Johnson, 1964: M. Rejai, *The Strategy of Political Revolution*, Anchor Press, New York 1973.

There are, however classifications based on the distinction between the use of guerilla warfare as an instrument of subversion, of insurrection, and of defence (Paret, Shy, 1962).

[16] 'The term [revolutionary war] connotes conscious efforts to seize political power by illegitimate and coercive means, destroying existing systems of government and social structures in the process' (Collins, 1973, p.47); cf. Cyril Edwin Black, T. Thornton, *Communism and Revolution: The Strategic Uses of Political Violence*, Princeton University Press, Princeton 1964; Thompson, 1970.

[17] E.g. C. Johnson, 1964, 1966, 1973; W. Laqueur, 'Revolution', in *The International Encyclopedia of Social Sciences*, vol.13, pp.501–7; Carl J. Friedrich ed., *Revolution*, Atherton Press, New York 1966; Leiden, Schnitt, 1968.

[18] 'Revolutionary war is the whole of non-military and military means and actions directed against a non-communist authority, i.e. performed according to a plan, without or with use of armed violence by the leading centre of the world communism'. (Arnold, 1961, p.197); revolutionary war is 'a form of warfare which enables a small ruthless minority to gain control by force over the people

of a country and thereby to seize power by violent and unconstitutional means' (Thompson, 1970, p.4). Revolutionary war has a single objective: 'the overthrow of the established order and the seizure of power' (Ximenes, 'Revolutionary war', *Revue de Défence Nationale,* February–March, 1957); cf. Ruge, 1967; Hahlweg, 1967, 1968a, 1968b; Willy Rothe, 'Der revolutionäre Krieg', *Wehrkunde,* 1960:5; Udo Eulig, 'Bürgerkrieg im Atomzeitalter?' *Wehrkunde,* 1963:2; Erich Vorwerck, 'Der revolutionäre Krieg, I–II, *Wehrkunde,* 1964:4–5; Claude Delmas, *La Guerre Revolutionaire,* Paris 1959; Pater, Shy, 1962; George Tanham, *Communist Revolutionary Warfare,* Fr. Praeger, New York 1961; Geoffrey Fairbairn, 1974. Revolutionary war results from the action of the insurgent to seize power, or at splitting off from the existing country, and from the reaction of the counter-insurgent aiming to keep the power (Galula, 1964, p.3).

[19] Ted Robert Gurr, *Why Men Rebel,* Princeton University Press 1970; Huntington, 1968; J. C. Davies, (ed.) *When Men Rebel and Why,* New York 1971; among the listed reasons are: poverty, sharp reversal after a period of economic growth, relative deprivation.

[20] Revolutionary war is seen by the Communist camp as 'the best way of using force to expand the Communist empire with the least risk', and it 'enables Moscow and Peking to manipulate for their own purposes the political, economic, and social revolutionary fervor which is now sweeping much of the underdeveloped world' (Roger Hilsman, 'Internal War – The New Communist Tactics', *Military Review,* April 1962); '. . . revolutionary war, as preached by well known communists profiting originally from the experience in World War II, has been used subsequently as an instrument of Russian and Chinese foreign policy in order to spread communist influence and control and to encourage world revolution' (Thompson, 1970, p.3).

In a more general form: some imperial powers are said to attempt to exploit the very frequent domestic conditions of instability in many countries to gain control over them by establishing allied or puppet-regimes; thereby interstate war takes the form of fighting within single states, and internal war with multi-national participation may be treated as a species of international war. (Richard A. Falk, Janus Tormenred 'The International Law of Internal War', in Rosenau, 1964, pp.216–9).

[21] For a simplified classification, containing only two kinds of war, with and without Communist participation, see Speier, 1969. In another kind of classification, internal wars are divided into wars inspired by the international Communism (by the Soviet Union or China), wars which the Communist powers attempt to get control over, wars supported but not controlled by them, and wars without any external help. A variant uniting the two intermediate classes in the above classification: Donn A. Starry, 'La Guerre Revolutionaire', *Military Review,* February 1967.

[22] Writings of John Slessor, see bibliography.

[23] The main theme of Thompson, 1966: start and timing of each insurgency

depend on an 'order from Moscow'; Communists sponsor and exploit radical nationalism, they push it to armed social revolutions which result from international conspiracy (Roger Hilsman, in T. N. Greene, ed., *The Guerilla and How to Fight Him*, Fr. Praeger, New York 1962, p.24).

[24] Described by Lider in 1966, 1971 (the co-called covert war).

[25] William Raymond Corson, 'In Search of New Wine and New Bottles', *Journal of International Affairs*, 1967:1; '... the assumption that a guerilla force, like a conventional army, can be controlled and commanded by a foreign or externally based government ignores the organizational, psychological, and political facts of revolutionary warfare... The conditions leading to revolutionary wars are not created by conspiracy. They are partly inherent in a situation of rapid social change, but the outbreak normally results mainly from the failure of a ruling élite to respond to the challenge of modernization' (Eqbal Ahmad, Revolutionary War and Counter-Insurgency, ibid. pp.14–15).

[26] Johnson, 1973, pp.1–4.

[27] Gurr, 1970; Huntington, 1968.

[28] Thus some scholars call civil war 'a case of pathology in politics'; it occurs only if there are no effective channels for settling political grievances and there is a conviction that this can be done only by using violence (J. K. Zawodny, 'Civil War', in *International Encylopedia of Social Sciences*, vol.7, p.499). Social conflicts were regarded as amenable to peaceful adjudication and the scholars have been taught to 'abhor' internal violent cleavages (Kelly, Miller, 1969, pp.2–4).

[29] For example, Samuel P. Huntington contends, that poverty and the desire to liquidate economic exploitation are not the causes of civil wars in the underdeveloped countries. ('Civil Violence and the Process of Development', in: Civil Violence and International System, p.II, *Adelphi Paper* no.83, 1971, p.1).

[30] Calvert adds to the 'favoring conditions' the fact that China relapsed into a large collection of provinces led by warlords (1970, p.103).

[31] Fairbairn, 1974; Johnson, 1973; Ulam, 1968; Bell, 1971; C. P. Fitzgerald, *Revolution in China*, Cresset Press 1952. R. G. Wesson points out, that each Communist revolution, including those in Russia and China, has regularly been associated with war, especially an unsuccessful or prolonged and costly one, which has discredited the socio-political order, weakened the state apparatus, and made the country receptive of an extremist movement, led by a strong organisation ('War and Communism', *Survey,* Winter 1974). Edgar Snow paraphrasing Mao Tse-tung's replies to his questions wrote: 'They [the Japanese] created conditions which made it possible for Communist-led guerillas to increase their troops and expand their territory. Today when Japanese came to see Mao, and apologised, he thanked them for their help.' (Interview with Mao Tse-tung, 9 January 1965 in Snow, *Long Revolution*, Random House, New York 1972, p.199).

[32] Communism has succeeded only when it has been able to co-opt a national liberation struggle. [This is] 'the characteristic pattern (China, Cuba) of a

communist party coming to power through a program of national liberation, anti-imperialism, antifascism, or anticolonialism, and then launching a 'dictatorship of development' justified in the name of an advance toward socialism and communism'. National liberation revolutions give way to communist-dominated nation-states which build communism (Johnson, 1973, pp.11–12).

[33] 'Lenin became the most important communist to perceive that the lure of "self-determination" for colonized peoples, or other internationally or ethnically dependent groups, provided the only basis for authentic revolutionary situations' (Johnson, ibid.). And: 'The Chinese doctrine of "people's war" arose from the historical effort to make the desire for self-determination part of communism's appeal' (p.13).

Revolutionary wars are 'nationalistic in nature'. The cause is that 'the strongest appeals since World War II have been patriotic and nationalistic' (Collins, 1973, pp.48, 50).

[34] Huntington, 1968.

[35] Many classifications based on external features of the revolutionary actions have been presented: personnel wars (palace revolutions which aim at seizure of power), authority wars (revolutions aiming at the change of the system of authority), and structural wars (aiming at economic and social changes) (Rosenau, 1964, pp.63 ff.); civil wars, military coups, guerilla, insurgencies, riots, acts of terrorism (*Strategic Survey, 1970*, Institute for Strategic Studies, London 1971, pp.75–97). Johnson, 1973, presents a classification based on four criteria: the aims of the revolution, the actors, ideology, and character of preparations. These criteria result in six types of revolution, and two of them are said to be the most typical in our century, namely the militarised armed insurrection for nationalist or communist goals (1964, 1966). Tanter and Midlarsky propose classifications based on the intensity of social activity: the degree of mass participation, the duration, and the degree of violence are used as indicators of the intensity. Four types of revolution are ordered on this basis: palace revolutions, reform coups, revolutionary coups, and mass revolutions. Intensity of revolution is increasing in all three indicators, from the 'absence of mass participation, very short duration, and no violence' in palace revolutions, to the 'high mass participation, long duration and high violence' in mass revolutions. A fourth indicator, the intentions of the insurgents, increasing from 'no desire for change' to 'postulates of fundamental changes' has been added, seemingly related, however, to the socio-political characteristic of the revolutions not to their intensity ('A Theory of Revolution', *Journal of Conflict Resolution,* Sept., 1967); cf. definitions presented by Lasswell and Kaplan, Lieuwen and Huntington in note 9. Luttwak divides 'illegal seizures of power' into: revolutions (actions conducted by uncoordinated popular masses aiming at changing the social and political structure), civil wars (warfare between elements of the national armed forces leading to the displacement of a government), *pronunciamento* (the South American version of military *coup d'état*), *putsches* (attempts by a formal body within the armed forces to seize power), liberation

(overthrow of the government by foreign military or diplomatic intervention), wars of national liberation and *coups d'etat* (see note 29 a) (1964, pp.23–7).

[36] For a detailed analysis of *coup d'état,* see Luttwak, 1969; cf. Adam Roberts, Civil Resistance to Military Coups, *Journal of Peace Research,* 1975:1. 'A coup consists of the infiltration of a small but critical segment of the state apparatus, which is then used to displace the government from its control of the remainder' (Luttwak, 1969, p.27). The author posits that while in the revolutionary war armed forces, policy, and security agencies are the prime target of attack, the *coup d'état* uses them for seizure of power (p.16).

[37] For an analysis of the political terrorism as used mainly by revolutionary movements, but also by state (repressive terrorism), see Paul Wilkinson, *Political Terrorism,* Macmillan, London and Basingstoke, 1974; cf. Calvert, 1970, pp.84 ff. Brian M. Jenkins, 'International Terrorism. A New Mode of Conflict', in Carlton, Schaerf, 1975; Gaston Bouthoul, 'Definitions of Terrorism', ibid.

[38] Galula, 1964; Fairbairn, 1974; Johnson, 1973: Wallach, 1972; Frank Kitson, *Law Intensity Operations,* Faber and Faber, London 1971; Thompson, 1967; Hahlweg, 1968; Hyde, 1968; Paret, Shy, 1962; Heimann, 1963; Lewis Gann, *Guerillas in History,* Hoover Institution on War, Revolution and Peace, Stanford, Calif. 1971; Andrew Scott, et al., *Insurgency,* The University of North Carolina Press, Chapel Hill, N.C., 1970; Edward F. Downey, 'Theory of Guerilla Warfare', *Military Review,* May 1959; George B. Jordan, 'Objectives and Methods of Communist Guerilla Warfare', *Military Review,* May 1959; Trager, 1974; Franklin Mark Osanka, 'Guerilla Warfare' in *International Encyclopedia of Social Sciences,* vol. 7, pp.503–6.

[39] Guerilla warfare (called here insurgency) 'is a protracted struggle conducted methodically, step by step, in order to attain specific intermediate objectives leading finally to the overthrow of the existing order' (Galula, 1964, p.4); the concepts of peasant mobilisation, anti-imperialist national front, clandestine munitions and possible sanctuary support from allies, and protracted war, are the central concepts in guerilla warfare (Johnson, 1973, p.47); 'Guerilla war is the ensemble of the actions supported by violence and conducted by an adversary on the territory whose political administration, economic management and military occupation is officially in charge of the other adversary' (Colonel Nemo, 'The Place of Guerilla Action in War', *Revue Militaire Générale,* January 1957); he calls guerilla war a part of a broader political-military activity intermediate between cold war and limited war. For an analysis of the military side of guerilla war, see Wallach, 1974, 13. Kapitel: 'Der subversive Volkskrieg: Mao Tse-Tung, Giap. Guevara'.

[40] Eckstein, 1964; cf. 'The socio-political characteristics of guerilla warfare', in Osanka, *Guerilla Warfare,* loc. cit. pp.504–5.

[41] Theorists of guerilla war contend that such a war cannot ensure the final victory, it prepares the conditions for decisive conventional regular battles. The examples are: Mao's third phase of people revolutionary war (the strategic

counter-offensive), Giap's second phase, when the mobile conventional war develops, and third phase, when it becomes the main form of warfare, and Che Guevara's second decisive phase, when the guerilla army becomes a regular army, and when it wages conventional warfare.

Cf. Wallach, 1972, ch:13; Thompson, 1970, pp.16–17.

Cf. the military-political characteristics of guerilla warfare and review of guerilla methods in Osanka, *Guerilla Warfare*, loc.cit., pp.503–5.

[42] Beaufre, 1972 (transl. 1974, Ch.III, 'An analysis of conventional warfare against guerillas').

[43] Michael Elliott-Bateman, 'The Age of the Guerilla', in Elliott-Bateman, Ellis, and Bowden, 1974, p.5.

[44] Battle fronts in a revolutionary war include the political, socio-economic, cultural and ideological, psychological and international planes, among which the political one is the most important, since the population is seen as the key to entire war. Conflict on the political front, especially the struggle for public support, dictates the military strategy and tactics (John Baylis, 'Revolutionary Warfare', in Baylis et al., 1975, pp.134–9); cf. M. Elliott-Bateman, *The Fourth Dimension of Warfare,* Manchester Univ. Press, Manchester 1970.

[45] Millis, 1961, pp.200 ff.; Osgood, Tucker, 1967, p.164; 'Revolutionary warfare means first of all political warfare, psychological, ideological, economic and social; military operations are employed only to score political, psychological, ideological, economic, and social points;' it is a war, because the aim is to employ various means, with a mixture with military ones, but military operations are not the main ones (Reuben S. Nathan, 'Planning for Likely Wars', *Orbis,* Winter 1971, pp.851, 861); revolutionary war is the sum of a partisan war and a psychological war ('La Guerre Revolutionaire', *Military Review,* February 1967, p.62): all direct strategic goals are political (Galula, 1964, pp.7–9).

[46] 'These wars are not simply military conflicts with a complex political background. They are, rather, political conflicts which involve an unusually high level of violence' (Michael Howard, 'The Demand for Military History', *Military Review*, May 1971, p.42); Revolutionary war is a 'political war' (Oscar Bettschart, 'The Strategy of Political Wars', *Military Review,* April 1966): Galula concludes: 'Paraphrasing Clausewitz, we might say, that "Insurgency is the pursuit of the policy of a party, inside a country, by every means"' (1964, p.3).

[47] Wallach characterises the Cuban variant of guerilla war as waged exclusively by military forces: 'The Cuban guerilla movement was completely independent of any political party and provided itself after the victory with its political institutions'. This was its peculiarity, and it attempted to export this feature to other countries on the sub-American continent (and to other continents) (1972, p.290). Che Guevara was said to contend, that revolution should not wait until all conditions for it would ripen; military action would itself create such conditions. Regis Debray was also said to replace the idea of

the necessity of creating political conditions for revolutionary uprising by the creation of a revolutionary force, and to replace the political leadership of the party by the autonomy of military forces; revolutionary fighting would create revolutionary conditions (Luis Mercier Vega, *Guerillas in Latin America*, Fr. A. Praeger, New York–London 1969). 'It is the guerilla, and not the traditional, political parties, no matter how leftist they may be, who dictates the politics of the revolution' (Enrique Martinez Codo, 'Insurgency: Latin–American Style', *Military Review,* November 1967, p.7).

[48] The main theme of Galula, 1964; many definitions of a counter-guerilla warfare have been presented, e.g. as military, paramilitary, political, psychological, and civic actions taken by a constituted government to defeat subversive conditions falling short of civil war which result from a revolt or insurrection against the government ('Some Reflections on Counter-insurgency', *Military Review,* October 1964, p.73); cf. Gustav J. Gillert, 'Counterinsurgency', *Military Review,* April 1965; Josiah A. Wallace, 'The Principles of War in Counter-insurgency', *Military Review,* December 1966.

The main theme of Trager, 1974. He postulates 'effective planning for wars of national liberation' (i.e. for countering them) which 'begin in a political context and have as their goal a political revolution.' 'The familiar patterns of pre-combat psychological warfare and subversion violence, and finally armed conflict are their means. Planning therefore should be done by a multidisciplinary team' (pp.104–5).

[49] The difference between international and internal war is one theme in Raymond Aron's book on peace and war (1962). War is implicit in interstate relations; but on the contrary, it cannot be recognised as compatible with the continuity of a domestic political order, or it means a radical break in it. Thus civil war can seldom be conceived of as the continuation of domestic policy by other means. Cf. 'armed conflict between states has been looked on as inherent in the very nature of the international political system' (Guelke, 1974, comp. for 1).

[50] What is more, some scholars observe that internal wars influence the whole structure of the international system, as the revolutionary war in Russia 1917, the guerilla war in Greece 1946–50, the 1948 *coup d'état* in Czechoslovakia, the civil war in China, the revolt in Cuba (Rosenau, 1964, pp.49–50).

[51] One of the themes of Rosecrance, 1973; also analysed in Dougherty, Pfaltzgraff, 1971; leaders of some countries 'will sometimes use foreign adventures to strengthen their position at home; history is filled, for example, with instances of wars which had their origin in some leader's desire to blame foreigners for his own obvious failure to deliver on the premises he had made to his own people' (Sen. Claiborne Pell, *Power and Policy,* W. W. Norton and Co., New York, 1972, p.30); '. . . aggression release within a society is inversely proportional to outlets outside' (Clyde Kluckhorn, *Mirror for Man,* Fawcett Books, Greenwich, Conn., 1960, p.213).

[52] Western authors usually charge the Soviet Union with such exaggeration and demonisation of the 'imperialist states', but some ascribe such policy to all

great powers. The image of the arch-enemy is probably created in order to mobilise against the danger of internal disruption which exists even within the highly integrated society (Herbert Marcuse, in 'Internal conflict and Overt Aggression, Discussion', in Anthony de Reuck, Julie Knight, eds, London 1966, p.206). 'Mobilized' nationalism and aggressiveness are then redirected against an external enemy (Karl W. Deutsch, ibid. pp.206–7).

[53] Ali. A. Mazrui, 'The Contemporary Case for Violence', in 'Civil Violence and the International System', p.I, *Adelphi Papers,* no.81, 1971; he called colonial policy 'a permanent aggression', a 'frozen warfare' (pp.18–19).

[54] Wilkenfeld, 1968, 1969; Coser, 1945; John Collins, 'Foreign Conflict Behavior and Domestic Disorder in Africa', APSA Annual Meeting, New York 1969; Michael Stohl, 'Linkages between War and Domestic Violence: A Quasi-Experimental Analysis', IX World Congress of the IPSA, Montreal 1973. Stohl hypothesises that external war stimulates activisation of new groups in the productive process, increase of the status positions of underdog social groups, and other economic and social changes, which all generate demands for a new allocation of political power and rewards, and consequently an intensification of internal conflicts and violence.

[55] Modelski, 1964; the outcome of internal wars is said to be always dependent on external factors.

[56] And more generally, in a bi-polar international system the probability of an active intervention in internal affairs is always greater than in a multi-polar, or balance of power system (Kaplan, 1964).

[57] Miller, 1967, p.35. And: 'Internal disorders . . . reflect major influences in world politics: the cold war in its military, political, ideological, and psychological aspects; the transition from colonial administration to new regimes; and the uncertain balance induced by nuclear technology' (p.10).

[58] It is sometimes called 'the modern war', manipulated by the superpowers for their own purposes (Hans Rechenberg, 'Necessity of a New Definition of War – Real or Fictitious Problem?' *International Relations,* The Journal of David Davies Memorial Institute of International Studies, May 1973, p.249): Enrique Martinez Codo writes, that the destruction of the Organisation of American States was a regional, strategic aim of internal uprisings in Latin American States ('Communist Revolutionary War in Latin America', *Military Review,* August 1963); revolutionary violence very often breaks out in areas, where vital or important American interests are involved (Thompson, 1970); cf. the aforementioned characteristics of revolutionary war (Thompson, 1970, Hilsman, 1962 and others); or 'revolutionary national-liberation war' which has a double function, being an instrument of antigovernmental subversion, and serving the world-wide struggle of the Communist bloc against the Western powers (Klaus Hornung, 'Warten auf Scharnhorst, Die Budeswehr vor neuen Problemen', *Die politische Meinung,* 1969:1).

[59] These are defined as 'an international conflict between two foreign powers, fought on the soil of a third country, disguised as conflict over an

internal issue of that country, and using some or all of that country's manpower, resources and territory as means for achieving preponderantly foreign goals and foreign strategies' (Karl W. Deutsch, 'External Involvement in Internal War', in Eckstein, 1964, p.102).

[60] Such a war would be a war of the world capitalism against the world socialism, because in Germany, main forces of both camps are confronting each other directly (*Einheit,* 1966:3, East Berlin).

[61] Gerald Maxwell, 'Conflict over proposed group action: a typology of cleavage', *Journal of Conflict Resolution,* December 1966; Tanter, 1966; for a review of some quantitative studies on the subject see Converse, 1968.

[62] John Farrell, Asa P. Smith, eds, *Theory and Reality in International Relations,* Columbia University Press, New York–London 1968, p.42.

[63] In one view, the relation depends on the character of state system: there is a tight connection between internal and international conflicts in the polyarchic nations, a somewhat looser connection in the personalist nations, and the loosest in the highly centralised ones, where external conflict decisions are made in a relative isolation from the possible reactions of the people (Wilkenfeld, 1968).

[64] Connections between civil violence and international relations are 'unpredictable and uncontrollable', partially because of the 'diversity and elusive character of civil violence' (Pierre Hassner, 'Civil Violence and International Relations', in 'Civil Violence and International Relations', Part II *Adelphi Papers,* no.83, 1971, p.16). Related to this theme are studies devoted to the impact of the degree of the development of society on war; some scholars have found that no relationship between societal development and war exists, and domestic institutions were least successful in explaining the occurrence of wars (Manus I. Midlarsky, Stattford T. Thomas, 'Domestic Social Structure and International Warfare', in Nettleship et al., eds, 1975, pp.531–48). And in another study, Midlarsky observes that domestic social structure is found to have no important effect on the frequency, duration, or intensity of war (1975).

5　War as a kind of conflict

To increase their understanding of the nature of war and to research for the possibilities of maintaining peace, many social scientists have attempted to relate war to other kinds of social conflicts, to find similar roots or other features, by examining the process leading up to the outbreak of war, and especially to the decision to wage war. Although most of these researchers seem to assume that the act of going to war is primarily a political one, by treating war as a kind of conflict they try to determine how the dynamic properties of the social system and the interplay of economic, ideological, cultural, psychological, and other factors influence this act. Some have even attempted to develop a general theory of conflict, which, however, is still at a very early stage of development. [1] Others have objected to such a general theory on the grounds that there can be little point in trying to encompass so many disparate phenomena under one set of covering laws. [2]

The various merits of these two positions will not be discussed here. Some of the elements of this embryonic theory will be reviewed, not as part of the general theory, but as independent contributions to a better understanding of the concepts of the nature of war from the sociological point of view. [3] These are: definitional elements of the concepts of war, hypotheses about the social function of war and its role in social development, and finally, some problems of conflict resolution.

Elements of definition

In order to create a general theory of conflict, it would seem that a certain amount of agreement must be reached at least about the basic social unit or set of units regarded as actors in conflict, and about what the idea of conflict itself includes.

Actors

As was pointed out in the previous chapters, the study of war has usually been limited to those situations in which sovereign states have resorted to armed violence; the delimitation of actors according to a legal criterion was thus an intrinsic feature of most older ideas of war. A general theory of conflict would require that some general criterion could be applied to distinguish actors at the various levels of conflict. [4] The recognition of such a need can be seen in the following definition, which seems to be a representative one: 'conflict refers to a condition in which one identifiable group of human beings (whether tribal, ethnic, linguistic, cultural, religious, socio-economic, political or other) is engaged in conscious opposition to one or more other identifiable human group because these groups are pursuing what are, or appear to be, incompatible goals'.[5] The choice of 'social group' to denote the actor does not resolve the problem, however, for there is no generally accepted taxonomy of social groups,[6] and additional criteria must always be applied. Consequently, although the general term was intended to obviate the competition between the many proposals that have been put forward, it cannot do so.

Conflict

The search for a general theory has also brought to the fore the question of the scope that should be given to the notion of conflict. In one view, conflict denotes interaction or behaviour alone: [7] antagonistic interests, tensions, competition, and hostile attitudes are regarded as predispositions that may accompany or intensify conflict, but they are not part of conflict itself.

This view is very close to the traditional treatment of war as a kind of behaviour characterised by the use of organised armed violence. For example, in 'The Military Balance' and some other publications of the International Institute of Strategic Studies, the terms war, conflict behaviour are used interchangeably.[8] The disadvantage of this point of view is that attention is directed to overt clashes, while other situations that may share many other important features with these openly hostile ones are ignored.

This has led those who wish to develop an all-encompassing theory of conflict to plead for a broader definition of conflict. It has been argued that all relations involving differences of objectives have to be included into the term conflict. [9] The study of conflict should therefore properly include the whole life-cycle of an issue, from its emergence to its resolution. [10] An even more radical view holds that overt forms of struggle are only manifestations of conflict and attempts to resolve it, but that the conflict *per se* comprises the issues in dispute. [11]

In all general schemes of conflict, war is seen as an extreme form. There is no generally agreed method of classifying conflicts, however, and the range of proposals made reflects the variety of basic approaches to an understanding of the nature of war reviewed in chapter two.

One solution is to distinguish between conflicts in which organised armed violence is used and those in which it is not. This is a solution similar to the Clausewitzian position that political issues can be resolved either by peaceful means or by war.

More common are perhaps models in which several levels of conflict are proposed. Usually the stages in the graduation of conflict are defined by their external features without regard to the issues in dispute or to the actors involved. To cite only one such example, the stages defined run from contention, through rivalry, conflict, and crisis, to war.[12] An extremely extended model of this type is undoubtedly the famous 'escalation-ladder' of Herman Kahn. Even though he treats only international conflicts, and bases each progression solely on the intensity and magnitude of conflict, he nevertheless arrives at forty-four different levels of conflict and war.[13].

Such models have considerable appeal amongst official military strategists in many Western countries. For example, the defence preparations of both France and West Germany are based on a threefold division of the level of conflict: crisis (a state of increased tension), threat (whether of international or internal origin), and war.[14]

In these examples, conflict is subdivided into a number of discrete stages. [15] Some scholars prefer to conceive of conflict as something more or less continuous, ranging from a low level to a high one. For example, it has been suggested that conflicts increase in intensity the less they are governed by rules of conduct. The extremes of the continuum are thus defined to be fully institutionalised conflict (e.g. duels) at the low end, and conflicts of unlimited violence (i.e. absolute wars) at the high end. Actual wars would be located at the upper end of such a continuum, but the point at which conflict becomes war is left undefined. [16] This is not considered to be an important problem, however, for conflicts often come to be called wars not so much because of any objective criteria, but because some political leaders decide to regard them as wars.

Some researchers do not consider models based on the level of violence alone to be sufficiently sensitive to other theoretically important aspects of war. Several factors, including not only the level of violence in a physical sense but also the aims of the conflict and its private or public character, are combined to rank different conflicts. Those who use such a multi-dimensional approach tend to favour the flexibility offered by the notion of a continuum.

Roots of conflict

Implied in the notion that all conflicts share some basic characteristics is the idea that they may spring from the same fundamental roots. One of the main ambitions of conflict theorists has accordingly been to trace the origins of conflict back to its roots in some inherent feature of social life. [17] Some believe they have found the roots of all conflicts; others make less broad claims, but maintain that in each epoch certain characteristic conditions will give rise to conflicts and wars. No consensus has yet been reached, however, and there are scholars who argue that so many conditions of each war are unique that any generalisations must represent a low level of theory.

The point of departure adopted by virtually all such general theories is that conflict springs from the fundamental division of human society into many entities. Since each of these usually has a will to survive, and to achieve better conditions of life, it must establish and preserve an identity and develop means of taking autonomous action in relation to its social and material environment in order to fulfil this and other essential goals. [18]

From this structural view of society various theorists have developed somewhat different hypotheses to account for the occurrence of conflict. In a socio-psychological perspective, the explanation lies in the fact that the members of each group come to identify themselves with each other and to develop hostile attitudes towards other groups and those who belong to them. [19]

A more general variant observes that the interests and values of the many social units are often incompatible with each other. It is from this inherent feature of society that conflicts arise. If the interests in conflict are vital for the survival of the groups that have them, it may be imposible to resolve the conflict peacefully. [20]

According to the 'structural theory of aggression', all social systems are stratified, and all aggressive behaviour and violent conflicts derive from this stratification. The social units at the top and the bottom of the system are in a state of equilibrium; that is to say that neither position gives any impulse for change, since those at the top have or can acquire what they want, while those at the bottom do not have anything but have given up trying to get it. The social units in between are in a state of disequilibrium, for they have enough values at their disposal to want to obtain more. It is therefore from this middle group that social conflicts originate. In a system of individuals (group), the struggle to achieve the lacking value may take the form of a crime, in a system of groups (nation), the form of a revolution, and in a system of nations (international system), the form of a war. [21].

Theories that trace 'revolutionary potential' to a high rate of social change reflect a similar set of ideas. The material and psychological situation of many people becomes changed in ways that may lead them to raise their expectations about future improvements. If these hopes and demands cannot be met, a growing number of people will become dissatisfied and frustrated (the group in disequilibrium becomes larger) with the existing state of politics and will press for more radical changes. [22]

Other structural theories are based on political concepts. The theory of dominance relations, for example, assumes that the primary cleavage giving rise to violent conflicts derives from the inequal distribution of power within a social system. [23] Sometimes the model is further elaborated with the inclusion of economic and social values (wealth and status) in addition to political power. Since complete concordance never exists between what the different social groups consider to be a just allocation of values, conflicts and wars to redistribute them must arise. [24]

In a similar model, the structural source of conflict lies in the hierarchy of authority required by any political system. There is always a basic conflict of interest between the governing and the governed. [25]

Although these theories may seem to be most obviously pertinent to domestic conflicts, they can be equally well applied to international wars if by society is meant the society of sovereign states. Indeed, the theory of dominance relations has been given a prominent place in the study of international relations as the structural theory of imperialism.

A similar but more sophisticated theory also takes what it terms the dependency structures between centre and periphery nations as its starting point in describing the hierarchical structure of world society, but adds that several sets of such structures exist. This leads to a more complicated set of possible relations, for example, between centre and periphery capitalist countries, between centre and periphery socialist countries, between developed and under-developed countries, between centre capitalist and centre socialist countries, between periphery capitalist and periphery socialist countries, and so on. [26] The traditional balance of power theory of political realism is a prime example. It is the anarchical nature of international society that leads states to seek power, if necessary by war.

Some researchers have entered into the nature of the international system from a somewhat different perspective. They argue that the stability of the system, which is to say the relative frequency of war, may be a function of the number of main actors. [27] Besides this general formulation of the hypotheses, two contradictory theories about the future likelihood of war as the current inter-national system develops have been put forward. According to one, a move from the situation that emerged after World War II with two main actors ('bi-polarity') to one with several actors ('multi-polarity') will cause a decrease in the number of wars. [28] Supporters of the second view believe that a bi-polar system is more stable than a multi-polar one. [29]

A variant which derives from all of the above theories postulates a relation between 'status inconsistency' and the occurrence of war. A condition in which a state is attributed less importance and in which it has less influence on the course of world affairs than might be expected on the basis of its power (capabilities) is considered to generate conflict and war. [30]

A second focus of those seeking in particular the roots of international wars has been on internal features of the states involved. One such theory postulates that with a totalitarian or semi-totalitarian political system (or ruled by professional soldiers) are more prone to become involved in wars than states with a liberal-democratic regime. [31] A variant of this approach offers four different 'social orders': Conservative, Fascist, Liberal, and Socialist. Of these, the first two are said to have an interest in expansion and to be more inclined to use military force to accomplish it. A Liberal order is considered to have good reasons to expand economically but to be less ready to use military means. A Socialist order has theoretically the least incentive to expand. [32]

Special attention has also been devoted to the role of the military and the armaments industry in the conduct of foreign policy, regardless of the type of political system. Wherever spokesmen of such interests gain influential positions at a high governmental level, it is held, that country will be more likely to adopt an offensive posture in its relations with other countries. Some scholars, putting their position even more strongly, maintain that war is the only

112

activity which justifies the existence of the military-industrial complex. [33]

To conclude this review of the roots of conflicts, mention should be made of theories that connect the occurrence of war with peculiarities in the cultural (in the broad sense of the term) development of particular states. The basic hypothesis of these theories is that as states become more highly developed and their internal structure becomes more advanced and complex, latent pressures are created within the society at the same time as the need to compete with other states increases and causes a more aggressive attitude towards them. Several variations of this basic theme could be mentioned. In one it is pointed out that endemic in the structure of industrial societies is the problem of converting ever growing amounts of energy to meet social needs. The process of industrialisation thus contains a source of conflict that becomes increasingly difficult to resolve. [34] In another variant, it is claimed that in industrialised countries, great 'cultural pressure', which cannot be completely eased through social and economic reforms, is exerted on the population. For such countries, war acts like a safety-valve. [35] Finally, a third approach links wars to a combination of population growth and technological development. Together these two factors cause demands upon given resources to increase rapidly and generate pressures for expansion in order to gain influence or control over new territory with raw materials, markets, or other needed resources. [36] The tendency to military expansion of this kind is believed to be greater when needs cannot be met through domestic extraction (because of high cost or inaccessibility, for example), or when acquisition of resources by non-military means is impossible. In this view, the source of international wars lies in the technological, economic, and ecological differentials between nations. [37]

Functions

The third contribution made by the theory of conflict to an understanding of the nature of war pertains to the social functions of conflict and war.

Conflict

Broadly speaking, views on this matter can be divided into two groups: those which emphasise the disruptive consequences of conflict, and those which emphasise its benefits for society.

Representatives of the latter category usually begin with the assumption that society is in a state of permanent change, and that conflicts and violence must be regarded as normal everyday phenomena because they are the driving forces of change. [38] The ubiquity of conflict is therefore a necessary condition for the process by which society has evolved. [39] Conflict is given a key role, for example, in social processes of adaptation, by preventing established patterns of behaviour from becoming inflexible.

Conflict stimulates innovation in all fields of human endeavour and is therefore a major factor in, for example, economic and technological development. Even if a society should stagnate for a time through suppression, conflict will eventually destroy the old forms and clear the way for new ones. Contradictions in society are always expressed as conflicts, the resolution of which may come about through the creation of new institutions and new norms. Ultimately, the resolution of such basic cleavages in a society will reduce hostility and strengthen the bonds between men. At the international level, conflicts clarify the identity and territorial boundaries of social groups that constitute states and thereby contribute to their social integration. At the very least, conflict can be taken as a warning signal and indicator that change is needed somewhere in the system.

Not all scholars consider conflict to be an exclusively beneficial force in social life. Some, like the 'structural functionalists', adhere to many of the above ideas in principle but note that conflicts may at times be dysfunctional. [40] Advocates of communications theory contend that failure to reach agreement is the result of a lack of proper communication flows. Conflict, as opposed to consensus, is thus attributed to some malfunction or shortcoming of interrelations.

Equilibrium theory in systems analysis probably goes furthest in conceiving of conflict as a disruptive force. Here the point of departure is the view that society is normally in a state of equilibrium, that is to say without conflicts. [41] When conflicts arise, the system moves into a state of disequilibrium. The élites in the system must therefore act to restore equilibrium. In doing so however, they may under certain conditions aggravate the conflict and move the system even further from the equilibrium point; to compound a conflict in this way would be dysfunctional for the system since it would ultimately lead to its breakdown. [42] In this view then, conflict is always a threat to the social system, even when adaptation is possible. [43]

Finally, it is frequently assumed, that no general assessment of the role of conflict in social history can be presented, and that each conflict ought to be estimated separately according to its own individual merits.

War

The lack of unanimity regarding the functional nature of conflicts is even more marked when the focus is narrowed to the role of war. Some maintain that war is a pathological condition, others that war has some definitely beneficial social functions. Between these a third view is also expressed that war may have had a beneficial role to play at earlier stages of man's history but that it is dysfunctional in the modern era.

Those who regard war as a social disease contend that it is never necessary to offer human lives in order to resolve a conflict. From the point of view of humanity as a whole, the total effect of war is always negative.

The second group, who see a functional side to all wars, is a much more diversified one. [44] As we have seen (ch. 2), some believe war to be a biological, others a psychological necessity; some anthropologists stress its role in cultural development; [45] to the high-priests of violence, war is the means to moral perfection. Perhaps the most numerous are the historians, political scientists, and others who, often implicitly, treat war as a natural and indispensable means of resolving conflicts that cannot be settled in any other way. Reflecting the belief that 'whatever is, is right', they assign to war its just place in the natural scheme of things.

Those who belong to the third group of scholars recognise that war has in various historical epochs had certain favourable consequences, but they do not claim that it is a necessary means of resolving conflicts, nor that the same benefits may accrue from future wars. Some scholars have, for example, argued that in primitive societies war has contributed to the regulation of key variables: economic (possession of or access to goods and resources), demographic (population size, sex ratio, generational balance, land distribution), and psychological (anxiety, tension, hostility). [46] Others have discussed the role war has played in the rise of the nation-state as a means of protecting territorial integrity, creating new sovereign states, eliminating those that were no longer viable, and adjusting boundaries. [47] Economic historians have indicated how wars have on occasion led to the redistribution of national resources and capital supplies in accordance with the requirements of greater efficiency. In a sociological perspective, war has acted as a stimulus to economic change and to technological and scientific innovation, and has increased social cohesion.

Function and the nature of conflict.

These differences of opinion about the functional role of conflict and war are perhaps more apparent than real since they stem to some extent from the different interpretations (reviewed at the beginning of this section) that can be given to the notion of conflict. Scholars who are mainly concerned with conflict behaviour (i.e. the use of violence) make a conceptual distinction between this method and the alternative, non-violent method. This frame of reference makes them sensitive to the problem of choosing between the two means on the basis of the costs and benefits of each, a problem which is of immense importance in the era of nuclear weapons. That conflicts are resolved may generally be a good thing with respect to the disputed value, but it may involve sacrifices with respect to other values, especially when indiscriminate destructive forces are let loose. Those who view conflict in this perspective consequently have reservations about whether it is in general functional or not. For those scholars who, in attempting to formulate general statements about all conflicts, do not put emphasis on the distinction between conflicts in which violent and non-violent means of resolution are used the problem of functionality in principle does not exist; it has been eliminated by their choice of concept. This may serve as one more example of the difficulties which block the path to a general theory of conflict.

Some problems of conflict resolution

The foregoing discussion has indicated that one of the major stumbling-blocks to a general theory of conflict has been the controversy about means – is violent conflict essentially the same thing as non-violent conflict, and are they equivalent according to some functional criterion? These problems suggest that the idea of conflict resolution must be the keystone that bears up any general theory of conflict. Although attempts have been made, and work is still going on, to develop generalisations about conflict resolution, no coherent theory has yet been formulated. Some of the most important questions related to the nature of war as a kind of conflict have been: what are the conceivable ways armed violence may be used (openly or covertly) to resolve war-generating conflicts? How can it be prevented? When and why do peaceful means fail so that non-peaceful means are resorted to?

To answer the first question efforts have been made to analyse situations in the last decades in which war either seemed imminent or actually broke out. [48] These may be divided into three groups: conflicts in which some temporary or more stable settlement was reached by peaceful means without any apparent resort to military might; conflicts in which an open threat to use armed violence was employed to force the opponent into submission (forced withdrawal, concessions, and the like); and finally, actual wars, usually terminated by conquest or annexation.

Of these three, much attention has been devoted to the first. Several basic types of peaceful solution have been studied: compromise, where both sides agree to a partial withdrawal of their initial demands or actions; avoidance, where both sides cease acting in ways that brought about hostile responses, awards, where a third party is endowed with the authority to announce the terms of settlement; passive settlement, where, even though the conflict is never formally settled, it gradually disappears and a new *status quo* becomes tacitly accepted as partially legitimate; frozen conflicts, where both sides remain committed to incompatible positions over a period of time, but in recognition of the futility of seeking a peaceful solution, agree to disagree. The study of these basic types and mixed forms through empirical and statistical research has cast some new light on the nature of peaceful settlements and suggested new ways in which conflicts might be resolved without the use of violence. [49]

The third problem – the step from peaceful to violent means of conflict resolution – has been tackled either very generally or with special attention being paid to the decision-making process. To the former group belong, for example, the systems approach, [50] with its very abstract description of conflicts in terms of equilibrium, disequilibrium and stability, as well as such empirical studies as the correlates of war project.

Studies of the decision-making process have taken one of two main directions. One strategy has been to try to discover the background conditions that have influenced historical decisions to wage war. This option might be exemplified by a model in which eight factors are proposed: four of them hinder the outbreak of violence (facilities, especially for coercion available to the government, previous successful repressions of disorder, adjustive concessions, and divisionary mechanisms), while four act in favour of it (facilities for violence available to insurgents, inefficiency of élites, disorienting social processes, and subversion). [51]

The other main direction taken in the analysis of decision-making has been to presume that all conflicts can be either peacefully resolved or kept under control, a resort to violence is unnecessary. The object is then to discover deficiencies in the decision-making process that account for particular instances in which violence actually occurred. This is a method that has recently been much exploited to explain the outbreak of internal armed violence.

In such studies as these, the boundary between peaceful and violent means of resolving conflict is of course crucial. The main research strategy adopted by some of those who wish to develop a general framework for the analysis of conflict resolution has been to turn to 'game theory'. [52] It is assumed that those in conflict will act on the basis of rationality, which thus comprises the 'rules of the game'. [53] The nature of the conflict is then determined by the objectives of the players. The outcome of the game (resolution of conflict) can be predicted if the objectives are known and the players follow the rules. What emerges are highly abstract models of behaviour, which, critics of the approach maintain, have little bearing on real-life situations. [54] Advocates of 'game theory' argue that in international politics, the assumption of rational behaviour is a legitimate simplification since the policy goals pursued by a country are usually stable over a long period, and decisions are strongly influenced by weighing the advantages and the disadvantages likely to result from alternative policies. In the nuclear era the restraints of rationality are all the stronger because of the much greater destructive power at the disposal of the superpowers. [55]

To sum up, it cannot be said that a coherent theory to explain the outbreak and resolution of armed conflict, still less the resolution of conflict in general, has been developed. Nevertheless, there is no doubt that these efforts have contributed to a better understanding of the nature of war, and that in itself is no small achievement.

Notes

[1] See 'Editorial', *Journal of Conflict Resolution,* March 1957, pp.1–2; Dougherty, Pfaltzgraff, 1971, p.138; Nicholson, 1970, p.20.

[2] Three kinds of objections, the 'specialist', 'idiographic' and 'gradualist' ones, are reviewed in Fink, 1968, pp.413–4.

[3] Three concepts of the notion conflict are recognised; conflict in the whole universe, in the whole living world, and in human society. Sociologists deal with the last concept.

[4] Some examples: (a) To divide conflicts into interpersonal, intergroup, interorganisational, and international; (b) To divide them into intrafamily, intra-community, intercommunity, intercultural conflicts (Robert A. Levine, 'Anthropology and the study of conflict: an introduction', *Journal of Conflict Resolution*, March 1961, pp.4–5). (c) To regard the 'system' as the main actor in conflict. System has been defined by some scholars as an aggregation of units which have some distinguishing features in common, characterised by a set of postulated interdependent fundamental variables, determining the behaviour of the system. Many classifications of systems have been presented, and the occurrence of conflicts and wars has been analysed as rooted in properties and basic functions of systems (studies of Morton A. Kaplan, Anatol Rapoport, D. Easton, C. A. McClelland, J. W. Burton and others, for some items: see bibliography).

[5] Dougherty, Pfaltzgraff, 1971, p.139; Wright characterises sociological approach as considering war to be an institutionalised intergroup conflict involving violence (1965, p.423).

[6] Fink, 1968; Sorokin (1928, 1947) enumerates about twenty kinds of groups which may participate in conflicts, as states, nations, nationalities, races, castes, classes, orders, families, religious, political, sex, economic, occupational, ethnic, ideological, ethical, artistic, scientific, philosophical, and territorial groups; Mack and Snyder (1957) regard, as parties in conflicts, individuals, social classes, nations, groups, and additionally cultures, coalitions, personalities, organisations, organisms, and systems. Fink writes: 'Diverse classifications of social units imply diverse classifications of conflicts, and each classification of conflicts implies a different set of special theories' (p.417).

[7] Coser, 1956, pp.37–8, 135; he states that the actual process of conflict has to be distinguished from the underlying economic contradictions and psychological patterns e.g. interests of labour and management may be said to be antagonistic, but real conflict occurs only occasionally. Cf. Mack, Snyder, 1957; Levine calls feelings and beliefs preceding and accompanying social conflict 'attitudinal concommitants of conflict' (1961, p.7). Morton Deutsch says that conflict is an action incompatible with other actions, preventing them, hemming, blocking, interfering with them, making them less effective. ('Conflicts: Productive and Destructive', *Journal of Social Issues*, 1969:1, pp.7–42); cf: Bengt Höglund, 'Concepts of Conflict', Department of Peace and Conflict Research, Uppsala Univ., Report No.2, Uppsala 1970, pp.11–14. The emphasis on conflict behaviour is reflected in the conditions of defining a relation between or among parties as a conflict, presented by Mack and Snyder

(1957, p.218): (a) Conflict is by definition an interaction relationship between two or more parties. (b) It arises from 'position scarcity' or 'resource scarcity'. (c) The parties can gain only at each other's expense; the conflict behaviour is one designed to destroy, injure, thwart or otherwise control another party. (d) It requires mutually opposed actions and counteractions. (e) It always involves the attempt to acquire or exercise power. Cf. Burton, 1968, p.80. Michael Nicholsson defines conflict as mutually inconsistent acts (1970).

[8] '. . . a conflict is a situation where the regular forces of a country or community are involved (either on both sides, or one one side only) and where weapons of war used by them with intent to kill or wound over a period of at least one hour'; the identification of the term conflict with war is strengthened by the classification based on the mentioned definition, where civil riots (only police or para-military security forces involved), mutinies and *coups d'état* (force is not used), as well as unopposed movement of military forces into the territory of a foreign country, are not included (Wood, 1968, pp.1 ff.).

[9] Some scholars emphasise the importance of motivations and point out that the confinement of conflict to conflict behaviour may exclude from the analysis the basic contradictions, namely the political, social, or psychological conflicts of interests: they treat conflict as a unity of motivations and actions (Herman Schmid, 'Peace Research and Politics', *Journal of Peace Research*, 1968:3; Carroll, 1970; Michael Banks defines conflict as a situation of incompatibility (unvereinbarkeit) of either interest (aims) or values leading to hostile mutual attitudes and hostile behaviour ('Zur Theorie des Konfliktes', *Österreichische militärische Zeitschrift*, 1971:1, p.13). Some scholars use the term conflict in two meanings, a broader and a narrower one (Wright, 1965).

[10] E.g. Dahrendorf has created a model of social conflict as developing from the first stage when the conflicting parties begin to organise themselves as groups of conflicting interests to the last stage when an open action leads to structural changes (1957, 1958, 1959, 1968). According to another presentation, social conflict is 'any social situation or process in which two or more social entities are linked by at least one form of antagonistic psychological relation or at least one form of antagonistic interaction' (Fink, 1968, p.456); the scholar calls the extreme opposite views of conflict as 'motive-centered' and 'action-centered'.

[11] Jessie Bernard holds that the real, deep, much more basic issues, which are disputed, constitute the conflict itself, and not the conflict behaviour. ('American Community Behavior', in *An Analysis of Problems Confronting American Community Today*, Dryden Press, New York 1949, p.105). Some scholars, although they interpret conflict as encompassing all stages, underline that for the identification of the essence of conflict, the underlying motivations be analysed (Boulding, 1962, Galtung, 1965, Stagner, 1967).

[12] Singer, 1972, pp.263–4.

[13] Kahn, 1965. Reviewed by the author of this study in *Ludzie i doktryny*, 'Iskry' Warsaw 1969, the analysis of Kahn's writings.

[14] Crise, cas de menace, guerre, Krisenfall, Bedrohungsfall, Verteidigungsfall.

[15] Some examples: Quincy Wright proposes the following stages of escalation from conflict to war: awareness of inconsistencies, rising tensions, pressures short of military force, military intervention or war (1965, pp.434–5), and in an earlier model: inconsistencies in the sentiments, purposes, claims and opinions, social tension, conflict, violence, war being its special case ('The Nature of Conflict', *Western Political Quarterly*, June 1951, pp.193–7); Holsti writes about esculation from incompatible positions, through conflict behaviour and crisis, to an overt use of armed forces (1967, 1972). Some scholars write about civil war as the highest degree of conflict. In one presentation the following stages are enumerated: civil conflict, civil strife, civil war, and in degree of means, individual violence, terrorism, insurrection, internal conventional warfare between a state and rebellion (Hassner, 1971). An indirect graduation is inherent in presentations of sets of internal violence, war being the highest form of it (Eckstein, 1964, p.3; Wood, 1968).

[16] Coser, 1967, pp.40 ff. Many scholars write about the elusiveness of the choice-point, caused by the gradual intensification of conflicts (e.g. Converse, 1968, Wright, 1965) because of the great variety of forms of conflict 'no definition can determine precisely at what point of the scale conflict becomes war, the latter is therefore only a matter of degree' (Osgood, 1957). Clausewitz is quoted: '. . . we have only to bear in mind the diversity of political objects . . . which may cause a war or to measure with a glance the distance that separates a death-struggle for political existence from a war which a forced or tottering ·alliance makes a matter of disagreeable duty. Between the two, innumerable graduations occur in practice' (p.36). (But Clausewitz spoke about the graduation of wars not conflicts). Another scholar put it in another way: 'the occurrence of war is independent of the amount of violence, when the parties declare that war begins, it begins' (Eccles, 1965, p.38). 'War is a part of a continuum ranging from individual antagonistic actions to its present theoretically maximum development in nuclear holocaust' (M. A. Nettleship, 'Definitions', in Nettleship, et al., eds, 1975 p.86); M. Melke treats war as part of a hostility continuum varying on axes of intensity (from threats to killing) and number of people involved (from two people to millions) ('Comment' to Nettleship, ibid, p.91).

[17] A great variety of terms has been used to point out the causal relationship between war and various factors, which adds to the difficulty to formulate general ideas. E.g. in one single study the following terms have been applied: 'societal conditions related to fluctuations of war and peace' or 'linked to the aggressive behavior', systems which have 'congeniality to decisions to enter wars', 'antecedant factors in accounting for wars', 'associated with wars', 'factors which correlate positively with foreign conflicts', 'systems which are "pre-requisites" for war', 'underlying conditions which prompt war', conditions which 'make war more probable', and additionally a set of other expressions

'psychological bases of war', 'related to the genesis' of war, 'domestic conditions related to warlike behavior', 'internal weakness invites war', 'war is a function of the type of government' it 'is rooted in. . .', it is 'a product of the very structure' (Haas, 1965).

[18] Quincy Wright, 'The value for conflict resolution of a general discipline of international relations', *Journal of Conflict Resolution* 1957:1, p.3; Converse, 1968.

[19] Coser, 1964.

[20] Nieburg, 1963. The scholar holds, that without violence the *status quo* would tend to remain frozen and the society would not develop.

[21] Galtung, 1964, pp.98–9. He presents the theory of 'structural violence' and holds, that such a violence exists 'when human beings are being influenced so that their actual somatic and mental realization are below their potential realization' (1969; cf. Tord Höivik, 'Structural Direct Violence: A Note on Operalization', *Journal of Peace Research*, 1971:1. For various theories relating the occurrence of wars to the rank disequilibrium in the international system, see R. B. Gray, *International Security Systems: Concepts and Models of World Order*, Peacock, Itasca, Ill., 1969; Wayne Ferris discovers that the greater the power disparities between states and the greater the change in such disparities, the higher the probability of an intense conflict between them (*The power capabilities of nation states: international conflict and war*, Lexington Books, 1973).

[22] Rosenau, 1974, 'Editor's Introduction'; Arnold Feldman, 'Violence and Volatility: The Likelihood of Revolution', in Eckstein, 1964.

[23] Dahrendorf, 1958. He calls the dominance relations 'imperatively coordinated associations'.

[24] Coser, 1967, pp.29–32.

[25] Nieburg observes, that in such a system, threat of violence and actual violence are always instruments of the authority, and threat of counterviolence and a periodical use of it are the instruments of the governed (1963).

[26] Klaus Jürgen Gantzel, 'Dependency Structures as the Dominant Pattern in World Society', *Journal of Peace Research*, 1973:3, Special Issue, pp.203–15.

[27] Cf. Kaplan, 1958.

[28] Karl W. Deutsch, J. David Singer, 'Multipolar Systems and International Stability', *World Politics*, April 1964, p.390.

[29] Kenneth N. Waltz, 'International Structure, National Force, and the Balance of World Power', *Journal of International Affairs*, 1967, p.229.

[30] Michael Wallace, 'Power, status, and International war', *Journal of Peace Research*, 1971:1; Midlarsky, 1975: Midlarsky combines the idea about the 'status inconsistency' with that of the 'environmental uncertainty reduction' which states that war is an attempt to reduce uncertainty (or to achieve certainty) in the international situations in which normal political relations have failed.

[31] E.g. Haas (1965) holds, that among the various types of systems. the

democratic system is the least, the authoritarian the most war-generating, and the totalitarian falls in between. Buchan contends, that the totalitarian (despotic) system, as opposed to the liberal one, generates wars, because it can mobilize great military power, is predisposed to unlimited action (as recognising no limits upon the internal authority), is ready for wars for economic reasons, and it exploits the situation that government can decide on war without any approval by the people (1968, pp.21–4). Some causal links between the type of the political system and the decision on war (external conflict) are presented by Jonathan Wilkenfeld in: 'Domestic and Foreign Behavior of Nations,' *Journal of Peace Research* 1968:1, pp.56–69; cf. his 'Some Further Findings Regarding the Domestic and Foreign Conflict Behavior of Nations', *JPR*, 1969:2, and his 'Models for the Analysis of Foreign Conflict Behavior of States', in Russett (1972). He asserts that the relationship between domestic and foreign conflict behaviour exists and that it takes on various forms, depending upon (a) the type of state under consideration (polyarchic, centrist, or personalist) (b) the type of conflict behaviour involved, and (c) the type of temporal relationship. Of these three factors, the first is perhaps the most crucial (1972, pp.275–6).

[32] Wallensteen, 1973, pp.41–52.

[33] 'Wars are planned, instigated, conducted, and justified by war-making institutions. These institutions serve no other purpose than to make wars. They should, therefore, be viewed as parasitic formations in the composite human organism' (1972, p.100).

[34] Gutorm Gjessing, Krigen og kulturene, Oslo 1950, 'Adherence to the Group Territory', *Folk*, 1964:1; 'Ecology and Peace Research', *Journal of Peace Research*, 1967:2.

[35] Klaus Jürgen Gantzel, 1972; cf. his: 'Internazionale Beziehungen als System', Sonderheft 5 der *Politischen Vierteljahresschrift*, Westdeutscher Verlag, Köln-Opladen 1973.

[36] Nazli Choucri, Robert C. North, 'Dynamics of International Conflicts: Some Policy Implications of Population, Pressures, and Technology', in Tanter, Ullman, eds., 1972. In some cases, however, the combination of a relatively low population level with relatively high levels of technology may diminish the probability of occurrence wars (Norway, Sweden). (Choucri, with the collab. of North, 'In Search for Peace Systems: Scandinavia and the Netherlands, 1870–1970', in Russett, 1972).

[37] Nazli Choucri, Michael Laird, Dennis L. Meadows, Resource Scarcity and Foreign Policy: A Simulation Model of International Conflict, paper prepared for the IPSA IX World Congress, Montreal 1973, pp.14–17. The authors call the sources of the kind of pressure which they analyse 'the dual imperative of growth and constraints'; cf. Nazli Choucri, Population Dynamics and International Violence, 1974; R. C. North, P. Lagerstrom, *War and Domination: A Theory of Lateral Pressure*, General Learning Press, New York 1971 (population, technology and resources interact with one another such that an increase in the first two create increased demands unless available resources

increase commensurately, and there is a tendency to expand state's control over resources in other states, which may result in war); Choucri, North, 1975 (the scholars trace the origin of war to the consequences of different paces of national growth – economic, demographic, and technological).

[38] 'Coercion, conflict, and change do seem, on balance, to be more basic societal attributes than consensus and equilibrium' (Bert N. Adams, 'Coercion and Consensus Theories: Some Unresolved Issues', *American Journal of Sociology,* May 1966; p.717).

[39] Coser, 1956, 1967 (pp.82–9); Bernard, 1957; Dahrendorf, 1958; Dougherty, Pfaltzgraff, 1971, pp.233 ff; North, Koch, Zinnes, 1960; George Simmel, 'Conflict', in *Conflict and the Web of Group Affiliations*, Free Press of Glencoe, New York, 1964.

[40] 'Normative functionalism' which considers society as functioning under relatively stable conditions, may serve as an example. Violent conflict and temporary disequilibrium are considered to be a special case of equilibrium.

[41] Arnold S. Feldman observes that the study of social violence is typically viewed as an area of social pathology, and violence is seen as incidental to the basic character of social structure and social life ('Violence and Volatility', loc. cit., p.111). Rosenau writes that political and social scientists are not interested in the study of internal conflict because they treat social disorder as an aberration of social order (1964, 'Introduction', p.2).

[42] The use of violence for suppressing the dissatisfaction, may be a failure. Some unforeseen triggering factors may also intervene, multiplying the dysfunction (Johnson, 1964, 1966).

[43] A third approach has appeared: conflict and order, disruption and integration are regarded not as rival processes, but as intertwined and complementary; both are fundamental social processes which in various proportions and admixtures are present in every society in every stage of its evolution (Lewis A. Coser, 'Introduction', in: Collective Violence and Civil Conflict, *The Journal of Social Issues*, 1972:1). Elise Boulding writes: 'Earlier battles over whether the "conflict model" or the "integrative model" was more basic to the workings of society have been superseded by increasing agreement that these are equally researchable complementary processes' ('The Study of Conflict and Community in the International System: Summary and Challenges to Research', in *The Journal of Social Issues*, Conflict and Community in the International System, XXIII:1, 1967, p.149). Frankel holds, that conflict and harmony are two recurrent modes of social interaction, and war and peace their extremes; the two modes may be applied simultaneously (1969, 1973).

[44] For a review of the merits of war, as stated by various scholars, see Walter Millis, 1961, pp.3–15.

[45] War 'may constitute a counteractive response made by a system when a variable or activity within the system has been disturbed in some way' (Andrew W. Wayda, 'Hypotheses about Functions of War', in: Fried, Harris, Murphy, 1968, p.85).

[46] Robert Bigelow, *The Dawn Warriors*, Atlantic-Little, Brown, Boston 1969; Turney-High, 1971, p.148. Many scholars may be quoted, who directly or indirectly point out that warfare may have been the principal engine of human evolution through the mechanism of selection in favour of the winners, i.e. of groups that better adapted themselves to the conditions.

[47] 'One simply cannot comprehend the rise, spread and decline of ancient civilizations and peoples, or the creation, unification, expansion, and protection of modern nation-states, except in relation to force' (armed forces) (Osgood, Tucker, 1967, p.5). The boundaries of states, their rights, strength, influence, all factors deciding their identity and sovereignty, were determined by the military power and fortunes of war.

[48] K. J. Holsti, 1966.

[49] In another study, a six-valued scale was devised which graded the types of solution along the dimension of 'co-operativeness' conquest, containment, mediation, arbitration, compromise, reconciliation (Alan Coddington, 'Policies Advocated in Conflict Situations by British Newspapers', *Journal of Peace Research*, 1965:4).

[50] Galtung regards the system as an aggregate of actors with a certain unity of ends and means ('action-system') and defines conflict as follows: 'An action-system is said to be in conflict if the system has two or more incompatible goal-states' ('Institutionalized Conflict Resolution', *Journal of Peace Research*, 1965:4, p.348). Cf. note 4).

[51] Eckstein, 1965.

[52] Neumann, Morgenstern, 1964; Rapoport, 1960, 1964, 1971; Boulding, 1962; Schelling, 1960, 1966; Richard Falk, *The Strategy of World Order*, 1966; McClelland, 1966.

[53] This is the application to war of the 'theory of a rational decision in situations involving conflicts of interest among two or more independent actors' (Rapoport, 1971, p.69).

[54] Other critical comments: (a) game theory has elaborated in detail only games with two participants, and only one kind of them, the 'zero-sum games' which are applicable only to some military situations (of operational and tactical character) and have no general value; (b) even those models which have been elaborated are lacking of accuracy (I. Horowitz, *The War Game*, New York 1963; J. Neumann, *The Rule of Folly*, New York 1962; Oran Young, in 'World Politics', XXIV, 1972, Supplement p.179; Gerd Junne, *Spieltheorie in der indernationalen Politik. Die beschränkte Rationalität strategischen Denkens*, Bertelsmann Universitätsverlag, Düsseldorf 1972); (c) the treatment of both cold war and thermonuclear war as zero-sum games has stimulated the elaboration of extremely uncompromising methods of struggle (Lowe, 1964, pp.95–6, books and articles by Ralph Lapp); (d) threat and intimidation are viewed as the best means of strategy, which leads to aggravation of conflicts and their perpetuation; (e) in all practical cases games theory has proved wrong, the most evident example being the failure of all predictions and projects connected

with war in Vietnam (Neil Sheehan et al., 'The Pentagon Papers' as published by The New York Times, New York, Bantam Books inc., 1971; Meghnad Desai, 'Social Science Goes to War: Economic Theory and the Pentagon Papers' *Survival*, March-April 1972; Desai and Sheehan hold that the US military strategy in Vietnam was influenced by the calculations and projects elaborated by the 'game theorists'.

[55] In some analyses of the conduct of war, the notions of fights and debates have been applied. The fight type of conflict is presented as a chain of escalating antagonistic actions and counteractions, which may take the form of escalation of arms race, threats and counterthreats, demonstrations and counter-demonstrations of force, strikes and retaliations; in sum a competition carried on by relatively peaceful means may become armed fighting (Deutsch, 1968, Ch.XI); debates consist of conflicts, in which adversaries are attempting to change the motives, values, views of each other by declarations, negotiations, unilateral moves, etc.

6 War – permanent or transitory?

The question whether war is a permanent or transitory phenomenon in the social relations of mankind is intimately interconnected with different ideas concerning the nature of war. An analyst convinced that war is part of the natural order of things will clearly take a different stand on evaluative or prescriptive aspects of war than a scholar who regards war as a dispensable social construct.

The problem may be divided into two questions: has war always been part of man's social history or, did it appear first at some particular stage? And regardless of the time of its first appearance, has war now become a permanent feature of social life, or is it more or less likely to disappear? The answers to these questions can in theory be combined into four different possibilities. In practice, however, there are two main answers which seem most representative of Western thought. According to the first, war has been and will always be a permanent feature of social existence. The second contends that whatever the origins of war it may in any case be eliminated in the future. This latter view may in turn be divided into an optimistic and a pessimistic interpretation. The optimists believe that what is required to eliminate war is a concerted effort from all peoples to do so. The pessimists agree in general with this position but are at the same time dubious about whether such a joint effort is a realistic possibility in the foreseeable future.

Let us now look at these two questions and some of the answers that have been given to them.

The origin of war

No scholar would deny that fighting between or among groups of men has always been part of human history. The point at issue is rather whether or not such fighting might properly be called war.

The historical permanence of war

By far the greater number of scholars of war treat it as something which has lasted as long as men have lived on the earth. For most of them the question of what criteria to use to distinguish war from other kinds of fighting does not arise. For example, adherents of the biological approach have no need to address themselves to such a question. Within their frame of reference, all fighting is the result of an aggressive drive that man has inherited as an animal. [1] Historians more concerned with description than analysis usually discuss primitive warfare as the first stage in the history of war. [2] Even when great fundamental differences are found to exist between primitive and modern war, it is nevertheless taken for granted that the same phenomenon is being studied. [3] Moreover, in social science literature there are innumerable general comments reflecting the assumption that war, as a product of social structures, has always coexisted with man, since man as a social animal has always lived in organised societies, be they tribes, city-states, nation-states, or empires. And political realists, who disdain discussions of the origins of war, take it as an article of faith that men have in fact always pursued their goals by waging war, and that many goals could not have been achieved otherwise. [4]

There are scholars who, cognizant of the problem, distinguish between various kinds of fighting in primitive societies. Wars are differentiated from reprisals, feuds, and punishment.[5] Some scholars contend that the differences between wars in primitive and modern societies are less important than they might seem, for they perform the same function of fulfilling social needs (e.g. attainment of political goals, territory, the enforcement of authority, material requirements) that could not be satisfied without using violence. [6]

War not contemporaneous with human existence

The view that wars did not appear until men had developed to a certain stage in their social relations is based on a more stringent definition of war. Two criteria in particular are usually regarded as vital: fighting may be termed war only if the hostilities are waged by organised forces for political goals.

Malinowski's analysis of war is perhaps the best known example of this point of view. Dividing the history of social fighting into six stages, he reserves the term war for hostilities fought in the pursuit of national policies by organised forces. [7]

Not all scholars are as restrictive as Malinowski with regard to goals. For some, for example, it is enough that a clear motive for fighting can be formulated. Many suggestions giving the idea of organised forces a more precise meaning have been put forward: [8] a division of labour with a chain of command that enables leaders to control their forces according to a plan of military operations; the ability to conduct military operations even if the first battle ends in defeat. Such a level of organisation and planning could not be realised at the earliest stages of social development, these scholars maintain. [9]

The future of war

The most widely held opinions about whether war will continue to be a feature of man's destiny can perhaps be divided into two groups: pessimistic and optimistic. Such a classification implies that these views all agree on the undesirability of war; they differ mainly in their estimate of the likelihood that war can be eliminated. Excluded from this discussion are approaches in which war is a means of achieving moral perfection (pessimistic and optimistic would have quite another meaning in that perspective) or a manifestation of some universal and timeless law (to eliminate war is therefore impossible) since they lie outside the mainstream of Western thought.

Pessimistic views

Political realism Viewpoints expressed by the political realists often seem to imply a belief in the everlastingness or war. In contrast to those who declare such a conviction openly, however, the political realists do not contend that the occurrence of war is a law-bound necessity, that it is an inescapable attribute of the nature of man as an individual or social being, nor that there are *a priori* reasons why it must always have existed as an integral part of man's social relations. What many of them do say is that war has in fact been a feature of all known societies and that it is very unlikely to disappear in the foreseeable future. 'War and the threat of war are endemic in international relations', [10] because there has been and still is no other way for states to pursue their interests except by using these two instruments. War has thus been indispensable for the functioning of the international system by providing a means for conflicts to be resolved and some legal order to be established and enforced. [11] Since there does not appear to be any realistic possibility of finding an alternative way of carrying out these essential functions, both international and internal wars are very unlikely to disappear. [12]

Moreover, there are no indications that there has been any weakening in conflicts of interest of the kind that traditionally have led to war. The advent of nuclear weapons has made the superpowers less anxious to settle their differences in a hurry, but the threat of war has assumed much greater importance as a political instrument and the use of conventional weapons has by no means disappeared. In addition, realists point to the goal of world domination held by the socialist bloc and the growing gap between the haves and have-nots as two explosive features of the present-day international system that are likely to remain in the foreseeable future.

Neither are many political realists more hopeful about the prospects for a reduction of internal wars. On the contrary, the political systems of many historically stable countries have shown signs of having difficulty coping with new strains put on them, e.g. connected with their industrial development. [13] At the same time, many of the new sovereign states have emerged from the age of colonialism with an imported political system unsuited to the indigenous political culture.

In sum, without explicitly asserting that war is eternal, political realism finds war to be an almost inseparable element of existing political systems, both domestic and international. Since no radical changes can be expected in these political systems, war will remain as a fact of social life at least in the foreseeable future. At best; what can be hoped for, is that some improvements in the political systems will lead to a reduction in the incidence of war. [14]

To digress, a special brand of scepticism usually characterises the attitude of historians towards the possibility of eliminating war. [15] Since the conclusions each historian comes to are his own, there is quite a wide range of views – that what is believed to be the main cause of war cannot be eliminated; that war has so many causes that it is impossible to understand and counteract all of them; that no two wars have the same causes, nor do they fall under any set of general laws; that nothing can be said about the future of war since its nature cannot be completely comprehended – these are only some examples. Like the realists, they generally regard the elimination of war as a utopian ideal: the goal may be worth striving for, and mankind may succeed in coming closer to it, but like all absolute goals, it is ultimately unattainable.

Likewise, scholars who focus on conflict and believe war to be understandable in terms of general laws of conflict behaviour often conclude that the elimination of war is highly unlikely. One of their basic assumptions is that conflicts are in any event unavoidable; the possibility that some conflicts may generate into wars cannot be excluded.

Quo vadimus? When the pessimists interpret recent historical trends, what they regard as significant are the number of unsettled conflicts, the occurrence of limited wars, and the increase in various form of armed intervention.

The balance sheet of measures and developments which many hoped would reduce the frequency of international war is even more discouraging. Both disarmament and the outlawing of war have proved to be ineffective. [16] The belief that democratic states (or socialist ones, depending on the political values of the analyst) are inherently peace-loving and that the road to peace lies consequently through domestic changes has been shown to be false. The growth of international organisations and the integration of sovereign states, two developments seen by many as steps that would preclude war and lead to world government, have come to be viewed with greater scepticism with the realisation that an increase in the size of political units may also magnify the differences between them. [17] Those who consider that no transformation of the international system is possible without a change in human nature have had little cause for optimism for in their view there has, on the whole, been no significant improvement in the material and spiritual conditions under which people live.

Optimistic views

The difference between the pessimists and optimists is often more a question of belief or attitude than one of theoretical perspective. For the most part, the optimists base their opinion upon what they regard as a radical change that has taken place in the role international war plays in man's social life. There was a time when war was rational in terms of political economy, and the idea of fighting for the conquest of a cause one believed in belonged to the moral

129

virtues. The technology of destruction has turned all this upside down: the costs of a potential major international war far outweigh any conceivable benefits; the notion that men can be won over to some system of values by force is now soundly denounced, especially with respect to international war. [18] On the other hand, industrialisation and the growth of trade has made it possible to increase man's productive capacity, to improve his material living conditions, and to give him a spiritually richer life; conquest has become obsolete. At the same time, the subsequent international division of labour has increased the interdependence of states and broken down barriers separating them. In consequence, there is increased incentive to settle disputes between states through compromise and increased disincentive to solve them by war.

Whither mankind? The optimists put a somewhat different interpretation on the trend of current events than do the pessimists. They emphasise the significance of the change that has taken place in relations between the superpowers: nuclear war is generally regarded as unthinkable, the Cold War has become much less virulent, and the superpowers have ceased quarrelling over irreconcilable principles and begun to concentrate their efforts upon immediate interests in which they may cooperate.

Improvements in relations between other states have also been achieved. There has been a growth of security communities consisting of countries which have virtually excluded the possibility of using armed violence as a means of settling their political differences. When violent clashes have occurred, great efforts have been made to confine and reduce the amount of warfare and to prevent the situation from developing into major war. [19] Moreover, the process of creating a suitable mental climate for peaceful transformation of the international power system has begun. One sign is a growing awareness that the use of armed violence seldom brings about a definitive resolution of a conflict yet costs much and risks even more. A second is the reformation that has occurred in the mythology surrounding the nation-state as the supreme bearer of welfare and security. The quality of a state is no longer measured by its ability to secure a high standard of living for its citizens through subjugation of other peoples: greater weight is attached to its efforts to make the best use of the resources at its disposal. This perspective has brought an awareness of the definite limitations to what each state can accomplish on its own and a new estimation of the value, indeed the neccessity of international cooperation. This new spirit has become manifest in the creation of international institutions (some of which are supranational), which successfully perform functions formerly believed to be inalienable components of sovereignty. [20]

Idealism The difference in faith between the optimists and pessimists can also be seen in the streak of idealism that very often marks the optimistic analyst. While the archetypical pessimist has difficulty seeing any possibility of major improvements, the optimist lends his support to some scheme of world reform. Some examples are the following:

(a)   The principle of the sovereignty of states should be discarded and a world-state with a world government created.

(b)   Major improvements in the existing international system should be made; amongst the concrete proposals are disarmament, a system of collective security, strengthening the authority of international law, the redistribution of global resources, the increase of flow of communication between nations, etc.

(c)   The internal structure of states must be bettered to eliminate the influence of those with a vested interest in war, to bring out the peace-loving side of all nations, and to redirect men's aggressive tendencies into productive activity.

The distinctions between optimism and pessimism, or between realism and idealism, are useful descriptive devices, but reality is less clear-cut. Hope does not preclude doubt: those who argue that the international functions historically performed by military means can and should now be carried out by non-military means sometimes state that internal war will likely remain because the possibility that violence may be used to settle an internal conflict cannot be eliminated. [21] Nor is idealism always devoid of *realpolitik*; there are optimists who do not exclude the possibility that war might be used for progressive change. [22]

Notes

[1] 'The instinct of pugnacity has played a part second to none in the evolution of social organisation, and in the present age it operates more power-fully than any other in producing demonstrations of collective emotion and action on a great scale' (William McDougall, *An Introduction to Social Psychology*, Luce, Boston, 1926 ch. 11); '. . . modern man inherits all the innate pugnacity and all love of glory of his ancestors' (William James, 'The Moral Equivalent of War', in Bramson, Goethals, 1964, p.22).

[2] Wright, 1964, p.5; *Encyclopaedia Britannica*, vol.23, London 1964, pp.311–12.

[3] Andrew Vayda, 'Hypotheses About the Functions of War' in Fried et al., 1968, pp.85–91; Napoleon A. Chagnon contends that primitive war differs from the modern one only by the small scale in military operations, short duration of active hostilities, poor development of command and discipline, and some other not too significant features ('Yanamamö Social Organization and Warfare', in Fried et al., 1968, p.110). M. A. Nettleship observes that the definitions of 'the social institution of War' by major thinkers on the subject 'do not routinely treat primitive fighting as different from war', although, as he thinks, this may be a profitable differentiation ('Definitions', in Nettleship, et al., eds, 1975, pp.80–1). He points out the differences: primitive societies are not sovereign states, they lack armed forces, and their fighting is of inconsiderable magnitude and especially brief duration.

[4] 'Of course, men may resolve their conflicts by methods other than war, but

the decisive issue is whether they can always secure the ends they desire by methods other than war. For the past, at any rate, the evidence is overwhelming that war, the organised use of armed force, has been a means for securing the variety of desired ends, ends that could not be secured other than through war' (Osgood, Tucker, 1967, p.330).

'The world would undoubtedly have been a more peaceful place if man were less resolute (or aggressive), but in that case we should probably also still be living in the stone age . . .' (Brodie, 1974, p.313). 'Since there have been organized societies, whether tribes, city-states, nation-states, or empires, there has been war among them, as the most extreme form of resolving conflicts . . .' (cf. Halle, 1973).

[5] Respectively, Wright, 1965, ch.VI; Leonard T. Hobhouse, Gerald C. Wheeler, and Morris Ginsberg, *The Material Culture and Social Institutions of the Simpler Peoples: An Essay in Correlation*, Chapman and Hall, London 1930, p.228; Joseph Schneider, 'Primitive Warfare: A Methodological Note', in Bramson, Goethals, 1964, pp.279–82. Some scholars state more generally, that fighting in the primitive society was between individuals, and the causes were more personal and familistic, than in modern wars waged for national interests, and characterised by essentially impersonal involvement and lack of personal motives (R. L. Bears, H. Hoijer, *An Introduction to Anthropolgy*, Macmillan, New York, 3rd ed. 1965; Alexander Lesser, 'War and the State', in: Fried et al. 1968, pp.94–5; Elman R. Service, 'War and Our Contemporary Ancestors', ibid., p.160). Wright writes that the view of the origin of war in history depends on the meaning of war. If by war is meant the use of firearms to promote the policy of a group (war in the technological sense) people of modern civilization have been its 'investors'. If by war is meant the reaction to certain situations by resort to violence (war in the psychological sense), it has always been fought by animals and men. War as a legitimate instrument of group policy (the legal, political and economic sense of war) probably originated among civilisations, while war as a social custom utilising regulated violence in connection with intergroup conflicts (war in the sociological sense) appears to have originated with permanent societies, that are found among the social insects and were probably characteristic of man from the beginning (1965, IV 2, pp.33 ff).

[6] The main theme of Osgood, 1957, Ch. 1, esp. p.16. Blainey contends that primitive war and civilised war seemingly have more causal similarities than differences; e.g. some anthropologists posit that primitive wars have a 'scapegoat' background, like modern wars (1973, p.74). Beaufre writes that the struggle of two tribes or two peoples (*Stämme oder Völkerschaften*) is the simplest, archaic form (*Erscheinungsform*) of war. All features of the 'more perfect' war can be found in that type of war (1975[1972], p.15). Coats describes primitive warfare as the first epoch in the history of war, characterised by a great variation of techniques and methods of fighting (1966, pp.60 ff.). M. Melko contends that war ought to be defined in terms of physical violence between

groups; the violence must be sufficient to cause serious physical injury to some of the participants. Civil war, revolution, gang warfare, and tribal raids can be included with this definition ('Comment' in Nettleship et al., eds. 1975). C. Richards stresses the similarities between early and modern fighting: only the decision-making process distinguishes the fighting of tribal peoples from that of nation-states ('Comment', ibid, p.91). S. Tefft writes that both civilised and tribal war involve armed conflict between sovereign political units no matter how small or large, structured or unstructured, decentralised or centralised, rather than stress the difference between them one should recognise the similarities. ('Comment', ibid., p.92).

[7] The presented states are: (a) Fighting within groups, being a breach of custom; (b) Collective and organised fighting between groups of a larger cultural entity, being a juridical mechanism for the adjustment of differences: (c) Armed raids for head-hunting, cannibalism, human sacrifices; (d) Warfare as the political expression of early nationalism, for making tribe-nation or tribe-state: (e) Military expeditions of organised pillage, slaveraiding, and collective robbery; (f) Wars among the culturally differentiated groups as an instrument of national policy, leading to conquest. Fightings classified in (d) and (f) are called wars (Malinowski, 1941).

[8] Turney-High, 1971. Five conditions necessary for a 'true' war are enumerated: (a) Tactical operations; (b) Definite command and control; (c) Ability to conduct campaigns; (d) Clear motive; (e) Adequate supply (pp.30 ff).

[9] To add, however, that even without the presentation of a 'law' many 'pessimists' among the Western scholars express the view, that the fact that war has been accompanying the entire history of mankind is a sufficient proof of its eternity (Baldwin, 1970, p.3).

[10] Osgood, in Osgood, Tucker, 1967, p.13. A great deal of West-German scholars continue to regard the traditional sources of war as still valid. Erich Weede contends that territorial disputes are among the most potent causes of war among nations in the 20th century (*Weltpolitik und die Kriegsursachen im 20. Jahrhundert*, R. Oldenbourg, Munich, 1975).

[11] Osgood, Tucker, 1967; Kissinger, 1957; Hedley Bull, 'Society and Anarchy in International Relations', in: Herbert Butterfield, Martin Wright, eds, *Diplomatic Investigations*, London 1966; Frankel, 1969, cart. 3.

[12] 'As long as humans are organized into state entities, as long as these entities are constructed as vehicles of power-accumulation, and as long as war remains an efficient means of power-accumulation, there will always be one party who will initiate war. If there is one party initiating war other parties will be forced to engage in it too. Thus an "anti-war system must prevent all states from using war all of the time in order to succeed", the invention of such a system does not appear probable, and there is no evidence to indicate that it is possible.' (Luttwak, the conclusion from the description of the term 'war' in a dictionary, 1972, p.214).

[13] The modern urbanised and collectivised life, with its greater than ever competitiveness, greed and struggle for material conditions, full of frictions and frustrations, magnifies the aggressive tendencies in human nature and in the nature of intergroup relations, and since the international system remains unchanged, the likelihood of war remains (Stanley Hoffmann, 'The Acceptability of Military Force', in: 'Force in Modern Societies: Its Place in International Relations', *Adelphi Papers*, no.102, 1973, p.3).

[14] Brodie's conclusion from the analysis of the causes of war: to stop thinking about curing war *per se* and to think more about avoiding particular wars. (1974, p.340).

[15] Aron, 1957, pp.177–203. It is also one of the main themes of his treatise on Clausewitz (1976) that in an unforeseeable future the world society will remain to consist of nations which don't understand each other, of states which want to preserve their independence, and of incompatible ideologies.

[16] Arguments for and against disarmament proposals were analysed by this author in 1966. Edward Tellers stated that even if realised, disarmament would not eliminate war; all its political causes would remain, and armament could always be rebuilt, and even developed in the course of non-military scientific effort as a by-product of many kinds of research ('Alternatives for Security', *Foreign Affairs,* January 1958, pp.201–8).

[17] One of the themes of Frankel, 1969.

[18] The main contention in Walter Millis' writings. '. . . the system of organized international war, as we have known it through ages, has reached a point of no return. Such useful or necessary social ends as it has served in the past it can serve no longer, or can serve only as an ultimate price which has become intolerably exorbitant.' (Millis, 1961, p.4).

[19] Frankel, 1969, 1972; Frank, 1967 observes, that the following changes in international relations could be viewed as slight movements toward the goal of abolishing war: militant enthusiasm is being gradually eroded; all wars since World War II have ended in stalemates; war has lost its glamour; international organisations that may become means for peaceful resolution of conflicts seem to become stronger despite setbacks (p.25).

[20] Quincy Wright points out, however, that the occurence of wars can be diminished within the contemporary system by establishing an effective international law outlawing war. If every state would be really subject to the international law, all conflicts would be soluble without war (1965, pp.1229–31).

[21] Arendt, 1963; Millis, 1961; Waskow, 1962; Osgood, Tucker, 1967.

[22] 'A new idea is that war is *intrinsically* evil, the general view is however that it may still occasionally prove necessary' (Brodie, 1974, p.3).

7 The assessment of war

Up to this point the review of the non-Marxist interpretation of the nature of war has dealt mainly with the way in which the general phenomenon of war has been analysed, either as a means of achieving certain political goals, or as a kind of behaviour distinguished by the application of armed violence in an organised fashion, or as both. An understanding of the nature of war may also be enhanced by putting it in a third perspective, namely by assessment of particular wars in relation to some system of values, political, socio-historical, and moral.

The point of departure adopted here is to ask what conditions must be fulfilled before a particular war, or particular kind of war, can be sanctioned by the dominant set of values. One way of answering this question is to examine the grounds on which concrete instances of war have been approved or condemned; another is to compare hypothetical instances of war to determine why one is justifiable and the other is not.

The assessment of the character of war touches on several ethical and other social problems – the interrelation between man's social relations, his perception of these relations, the ideal on which these relations should, he believes, be formed, and so on. To delve into all these issues goes beyond the scope of the present study. In the following discussion, the notion of the character of war will therefore be restricted to three features that are most relevant to the purpose of this study, namely, to compare Soviet, i.e. Marxist–Leninist and Western in the first line political realist ideas of war.

These three elements might be expressed in the form of three questions, which an analyst might ask when trying to characterise some war. Firstly, what are the political features of the war: who are the combatants and what are their immediate aims? Answers to this question will normally be descriptive, but underlying the terms in which the analyst touches his perception of reality will be a set of political values. Secondly, what role does the war play in the history of man's social progress? This question can only be answered with reference to some socio-political ideals. Thirdly, is the war just or unjust? Here the war can be confronted with fundamental principles of good and evil.

In a sense, these three questions constitute one inquiry at different levels of abstraction into the socio-political value to be attributed to some actual war or some type of war. The answers given may also be said to be reflections of the prevailing political ideology. Further discussion of this matter will be postponed, however, until the appraisals made by West and East are confronted with each other.

The political and socio-historical characteristics

The political evaluation

The political appraisal of wars in the United States has been closely linked with the broad assessment of the political situation on which its foreign policy has been based. [1]

In the immediate postwar period, the total struggle between the two super-powers completely dominated the American perspective. All armed struggles, actual or potential, were assessed in this context. As a result, three main types of war were usually discussed: total war between the superpowers and their allies, local limited wars, and anticolonial uprisings.

According to the premises on which American foreign policy was based in the immediate postwar period, the main threat to world peace came from the Soviet Union, which appeared bent on imposing the Communist system upon the rest of the world. The task of the United States was therefore seen to be to contain Communism by uniting the freedom-loving states of the world and by assuming the position of their leader. It was therefore assumed in the United States that a potential total war would take place between the two superpowers and their allies. The political goals of such a war would be ideological – to further or on the contrary, to prevent the socialist revolution – rather than, say, economical or territorial. Europe was judged to be the principal arena in which such a total war would be conducted, but a struggle in Western Europe would be only a prelude to total war of the socialist states against the USA. Such an assessment served as the theoretical and political basis for the creation of NATO. And the concept of the Cold War was based on the interpretation that some such total war was in fact being waged, even though the military forces were 'frozen'.

When assessing the political features of smaller wars, that broke out in various parts of the world, the Americans usually placed them in one of two categories: anticolonial wars or local interstate wars. The United States was generally well-disposed in principle to anti-colonial movements, partly because it sought to replace the influence of the colonial powers in the Third World. However, the generally anti-Western character inherent in many wars of national liberation was considered to be dangerous since it could be exploited by the Communists. The term local interstate war was used to denote wars which were limited in the scope of hostilities. Although the name itself says nothing about the political features of such wars, it was usually assumed in the United States that the aggressive party was being supported by international Communism in order to expand the socialist influence and to improve the strategic position of the Soviet Union in the Third World.

After the climax of the Cold War was past and the bi-polar view of world politics began to be replaced by a more complex one, a shift occurred in the description given to the political features of wars, especially those in the Third World. The global perspective was not abandoned, but somewhat more emphasis was put on the internal political, economic and social conditions and processes in the particular countries. The possibility that internal wars might break out in non-Communist countries owing to their unstable economic or political situation, or that local interstate wars might have causes other than the Communist conspiracy was entertained.

A corresponding change occurred in the foreign policy guidelines towards particular wars. The Dulles's doctrine of massive retaliation assumed that the United States would not participate directly in all wars even if they were considered to be important for their national interest. In the 1960s, the necessity of intervening in each conflict was put openly into question, and after the experiences of intervening in Vietnam, the policy of flexible response based on a careful political assessment of each individual case became the official political line.

In other Western countries the political assessment of wars was basically the same as the American approach, though perhaps with some change of emphasis. The British and French tended to stress the threat to Western interest posed in the colonial wars: the immoderate economic demands of the revolutionaries, the subversive support given them by Communist powers, the strategic implications of the withdrawal of the former colonial powers. A special interest of West German analysts was the possibility that a war in Central Europe had already been subversively begun with the active support of the Communist countries. This concealed war was believed to be an introduction to war on a global scale.

The socio-historical characteristics

In comparison with the Marxist–Leninist treatment of war, it is much less common in the West that scholars ask what role a particular war plays in the history of man's social progress. In part this can be explained by the fact that the end-state of social progress is much less clearly defined, if at all, in Western systems of values. In part it reflects the precept of traditional historical analysis that the ultimate consequences of a given event cannot be known before a considerable period of time has elapsed. There is therefore a general reluctance amongst Western scholars to deliberately try to assess a given war in relation to some ideal of social progress.

Nevertheless, contained in virtually all Western analyses of war are assumptions which together constitute an elementary and at the same time a very general concept of social progress. It is taken for granted that if Western powers were to become involved in some war (not necessarily against the socialist countries), although the immediate purpose of their participation would be to defend the concrete national interests, it would at the same time be to preserve peace, or to defend the independence of nations, or some such progressive aim; likewise, it was assumed that wars waged or supported by Communist powers would create a danger for world peace and hinder man's social progress. As was noted above, the favourable attitude of the United States towards movements struggling for national liberation in colonial countries was greatly tempered by its fear that they could be exploited by Communists for subversive anti-Western purposes.

The moral characteristics

The third question that might be asked when assessing the value of some particular war is whether it is just or unjust, what its relation is to some set of moral principles. Traditionally the aims of a war were taken as that aspect which determined its whole moral nature, [2] and three aims in particular were considered justifiable: to protect the innocent from an unjust attack, to restore rights wrongfully denied, or to re-establish an order necessary for decent human existence. [3] The question of means, and more generally, the way in which a war was conducted, was sometimes added for the purposes of evaluation. The question of the justness of the consequences of war was not usually considered separately since the answer to it would follow logically from the other two: just aims, if achieved, imply just consequences, and one of the conditions that must be fulfilled before the means could be regarded as just was that the beneficial results must outweight the costs involved in obtaining them.

Theories

On the basis of these considerations, five alternative opinions have been expressed. Two represent the extremes: that all wars are good, and that all wars are evil. A variant of them maintains that wars may be good in some historical epochs and evil in others. A fourth contention is that no moral assessment of war is possible, since wars themselves cannot be submitted to moral judgement; a political act can only be regarded a rational or irrational, purposeful or inneffective, favourable or harmful in relation to a given party. The fifth school of thought contends that each particular war, of each type of war, must be taken and evaluated on its own merits. This last approach has been a stimulus to a rich literature on the theory of just war.

The traditional theory of just war

Aims For a war to be considered just it was necessary and often sufficient in the traditional view for its aims to be just. The justification most widely accepted throughout history and in all belief systems has been defence against aggression. There is something fundamentally convincing about the idea that states should be allowed to protect their territory, their people, or more generally their rights against attack and thus ensure their survival. The basic notion is by no means unambiguous, however, and much of the literature on the justness of wars has been devoted to clarifying such controversial concepts as aggression and rights. The line between defensive and aggressive aims has often been very fine. Furthermore, each historical epoch has tended to weigh various political values differently: territory and political allegiance have been important since the Middle Ages, the freedom of commerce since the growth of commerce, national independence and freedom from feudal tyranny since the eighteenth century. [4]

A second powerful idea that has long been used to give moral sanction to a war is the just cause. From the religious wars of the Middle Ages, [5] to the Napoleonic battles against monarchical oppression, to the civilising mission of the imperialist countries, to the war to end all wars, to the war to give the superior race *Lebensraum*, or alternatively, to free the world from Nazi tyranny – the appeal in all cases to high ideals beyond the more immediate political objectives has been made. An interesting feature of this kind of justification is that by presenting a war as a conflict between good and evil, a leader may successfully imply that the moral distinction discussed above between aggressive and defensive is not applicable. For example, it was not wrong for the Crusaders to attack the infidels, for being un-Christian the latter were beyond the pale of the Christian ethic.

The above distinction between moral aims could also be made by regarding the first group as short-range political goals, and the second as long-range ideological (or religious) goals. Of course in all wars the two are intermixed with the result that the justness of the aims may be rather difficult to ascertain.

Means Although in the traditional theory of just war it was often sufficient to determine the justness of a war by ascertaining the justness of its aims, it was a matter of only slightly less importance to inquire whether the war was waged in accordance with some standard of conduct; indeed, in the traditional Christian concept of just war, a breach of such rules would sometimes be enough to disqualify a war from being just. [6]

The war must first of all be a legal war, that is, it must satisfy the criteria established in international law. This meant that the belligerents must be sovereign powers with the lawful authority to commit the members of their society to kill and to risk being killed. Moreover, it was necessary for a state of war to exist that the causes and aims of the war be clearly declared.

The main core of the problem of just means dealt however with the way in which the actual hostilities were carried out. The following principles were usually offered as a standard of conduct.

(a) Military means may be applied only when all other non-military means have proven ineffective.

(b) There must be a reasonable prospect that the military means will actually win the war and thereby redress the grievances.

(c) Only those military means and methods which are necessary for achieving war aims may be used. This principle, sometimes called the 'principle of military necessity' or the 'principle of proportionality', has been expressed in several other ways, some broader – the value preserved through war ought to be proportionate to the value sacrificed in war – and some narrower – the weapons employed must be proportionate to the goal undertaken.

(d)   Non-combatants must be immune from military operations.

(e)   The methods chosen for waging war must be kept under strict control. This principle was added to ensure that the first four could be observed.

(f)   Since it was always difficult to justify the horrors of war that occurred even when the above principles were observed, the 'principle of the double effect' was added. It stated that an evil effect that might result from a struggle to achieve some intended good may only be tolerated if three conditions are met: the good intention must be definite, the evil should be a by-effect, and the good achieved must be much greater than the incidental evil.

These principles have generally been considered to contain broad human values acceptable to all ideologies. They have been reflected in several international conventions designed to limit the destructiveness of war, to reduce the amount of personal suffering, and to establish accepted rules for the treatment of prisoners of war.

The ethical nature of wars in the nuclear era

Traditionally, as we have seen, the primary consideration for assessing the moral value of some particular war was that its aims were just, which in principle implied, as it was held, the just character of the applied means. This approach has been challenged on the grounds that the invention of nuclear rocket weapons has completely upset the relation between aims, methods, and consequences. In meeting this criticism advocates of traditional interpretations have been obliged to restate their case. Below the arguments of the critics will first be expounded, and three traditional views most prominent today will then be reviewed.

The critique The main thrust of criticism has been directed at the question of means, and the principle of proportionality or military necessity has been subjected to a thorough re-examination. Some contend that while a balance between ends and means can perhaps be found in strategic plans of known historical wars, in practice what was considered during the heat of battle to be a military necessity was quite another matter. [7] Consequently, the prospect of a nuclear rocket war is hardly made more acceptable by the assurance that no more violence would be used than necessary for even according to present strategic plans, the amount of nuclear weapons assumed to be necessary would cause immense destruction. [8] A defensive operation to prevent being attacked by nuclear missiles or to diminish the results of such an attack would require that a strike be launched against the enemy's nuclear forces, and since these are deployed over the whole of the opponent's territory, the result would be massive destruction. The nature of the new weaponry thus leads logically to a situation in which even the least conceivable use of force would bring about such horrible consequences that, if the principle of proportionality were to hold, the only political goals for which it would be justified to use such means must be unlimited. The critics therefore conclude that the traditional problem of finding means that would correspond to given goals has been made obsolete; in the nuclear era, the means are given, the consequences unavoidable, and if the political aims are to match the means, they must be unlimited, because they imply the toal destruction of the enemy. [9]

Doubts have also been raised about whether it would be possible to observe other principles of the justness of means in a nuclear rocket war, particularly the principle of the immunity of the non-combatants. Of the two main strategies upon which the deployment of American nuclear missiles is based, the counter-city strategy, directly violates this principle. One of the merits of the second, the counter-force strategy, is that it is designed to minimise the number of civilian casualties, but it might be questioned how far in practice a nuclear strike could discriminate between military forces and civilian population. A new feature of a nuclear rocket war would therefore be that the traditional distinction between front and rear would be eliminated, and since the front would be everywhere, the distinction between combatants and non-combatants would lose much of its significance.

Finally, the critics maintain that a nuclear war would have such far-reaching indirect consequences that the whole status of consequences in the moral assessment of war must be seen in a new light. In the first place, it has been established that the genetic damage caused by a thermonuclear war would be immense and irreparable, but what amount of suffering these genetic changes would cause succeeding generations cannot be predicted or appreciated. [10] Secondly, the psychological consequences of war assume much greater importance. It can be argued that the process of preparing and executing nuclear war would destroy moral values much greater than those in whose defence the war is waged. [11]

The traditional theory of just war reaffirmed. Scholars holding that the traditional theory of just war can with some modification be applied today argue that even if a war is very destructive it may be just as long as its aims are of sufficient importance and value. Total war may be very difficult to sanction on moral grounds, but the opposite extreme, total submissions to an aggressor, would hardly be less reprehensible. [12] All possible steps should be taken to limit the evils of war, but if some vital purpose cannot be attained without using nuclear weapons, it would not be immoral to do so. Moreover, one should bear in mind that the total amount of devastation and suffering may actually be reduced if what would otherwise be a prolonged war is brought to a quick termination by an act of great violence. There may therefore be circumstances in which the use of strategic nuclear weapons would be justifiable.

Furthermore, the advocates of the traditional theory of just war maintain that the development of strategic nuclear weapons has had little impact on war as a whole. The new weapons should be reserved for extreme situations, but it is conceivable that states will use military force short of general nuclear war to pursue their rightful interests. They continue to support methods to establish codes of proper conduct of war, [13] but they do not consider them a necessary element in an appreciation of the moral nature of a war. [14]

Neither just nor unjust The position taken by the political realists is somewhat varied. Their basic assumptions have remained unaltered: that war as a political act cannot be subjected to moral assessment; that statemen act according to their interpretation of state interests, which are regarded as the highest value, and not according to some ethical principles; and that governments must strive for power and, if necessary, wage war in order to maintain the security of the nation and protect its way of life. [15] They agree that it may be useful for a government to present a war in which it is involved as a just one because the support of its population and world public opinion is of great value. One should not be deceived into believing that a war is just because its aims are presented as just, however, for a stateman skilled in rhetoric can relate any war to highly esteemed values such as freedom, defence, and the vital interests of the nation. As to means, the political realists regard the debate about the moral conduct of war to be factitious, for those methods which have been outlawed in the past were always secondary ones. No serious proposal has ever been made to ban methods essential to the outcome of war.

The realists draw somewhat different conclusions about how to regard nuclear weapons, however. In the opinion of many scholars, the dread of nuclear weapons is so widespread and so deeply felt that national leaders would have great difficulty in gaining the support of their people for the use of nuclear weapons. Rather than risk dividing the nation and thus weakening its political power, these political realists conclude despite their general amoral position that nuclear rocket war should be condemned. This does not mean that the status of war as a normal instrument of policy has become altered, but total wars must be replaced by limited ones. The same conclusion can be reached if the new weapons are compared with the old on the basis of effectiveness. To succeed in achieving some political goal by war presupposes that the conduct of battle can be kept under rational control, and only limited wars can conceivably meet this requirement. Therefore it is in the interests of states to adopt a strategy of deterrence to prevent nuclear war both because it provides a shield under the protection of which limited war may more safely be waged, and because it should confirm the belief that governments only fight wars that serve the national interest.

A second group of scholars in the tradition of political realism maintain consistently with the principle of a morality that no distinction between nuclear and conventional war, nor between total and limited war has any meaning. No moral or immoral weapons exist, [16] nor are there any more inhibitions against destroying or killing in some particular fashion. Rationality is the only criterion for assessing military action: whatever is most effective for achieving the aims of

142

war, even if this should be the indiscriminate killing of innocent people, is the best way. [17]

In reply to those who assert that nuclear war must in all events be considered immoral, this second group of realists points out how dangerous it would be for a state to abstain unilaterally from procuring weapons of some particular kind on moral grounds. If its national interests were to conflict with the interests of a state that, holding other views on the morality of the weapon in question, had not hesitated to acquire it, the first state would be defenceless. It would not only risk being defeated in war but also be subjected to the tactics of blackmail in more purely political contexts. Like other political realists, these scholars therefore support the strategy of deterrence, but they emphasise that the deterrent effect can only be credible if nuclear war is taken as a realistic alternative and not one which the state can only choose after first overcoming extremely great moral inhibitions.

These scholars add, that to reflect the arguments of the moralists against using some kinds of weapons is not the same as to reject the idea that the amount of violence applied should be limited. It cannot be rational, they point out, to use more violence than is necessary to achieve the aims of war.

The immorality of war To the pacifists, the development of nuclear rocket technology, which gives man the possibility of completely destroying the world, is taken as further confirmation of their belief in the utter absurdity of the whole idea of war. They fight for their views on two fronts by affirming the neccesity to look at war from a moral point of view, and by criticising the misuse of moral arguments. The great ethical problem of our time they say, is the acceptance of the belief that all war is unjust. At the same time as it has become imperative for human survival to rid the world of war, our moral sensitivity to the inhumanity of war has been dulled. We have experienced horrors worse than anything a satanic fantasy could contrive. Millions of men offered their lives to kill each other in the trenches of World War I, and even more millions died in the slaughter of World War II. Within our civilisation we have witnessed how some men deliberately planned and acted to exterminate entire peoples. We know what terrible consequences the atomic bomb had for the citizens of Hiroshima and Nagasaki, yet we continue to prepare for a war in which similar and worse weapons will be used. Recent technological developments have even made the battlefield obsolete; it is now possible for men to strike with inconceivable violence at great distances from each other. War has thus been completely dehumanised, the enemy need not be seen as a group of human beings, only as a distant target, and the attack is purely a mechanical act based on rational calculation devoid of any human feelings of compassion. What else can all this mean, the pacifists ask, except that we are becoming callous to the sanctity of life, that we are becoming demoralised by the destructive forces we are able to unleash. [18]

The ethical problem of political strategy in our time is to reconcile two opposing wills without subduing one to the other through suppression. Since the use of armed force in order to achieve political or other goals is unjust, war as such is unjust. Any considerations about just ends, means, or consequences are therefore invalid, and more than that they are dangerous. It is always possible for treacherous statesmen to rationalise their acts and deliberately misuse moral arguments in their quest for power pure and simple. But even when one sincerely believes oneself to be acting justly according to one's own interests and assessment of moral values, one's behaviour may very well be unjust for the opponent: thus a belief in the justness of one's cause cannot be valid grounds for justifying war. It is therefore dangerous when a state arrogates to itself the right to judge the rectitude of its own case and to use armed violence to prosecute it. [19] Especially dangerous are such ideas as 'holy war', 'the most just war', and 'the war to end all wars', which are usually exploited to justify the use of unlimited violence in the name of high moral values. [20] A morality that condones the extinction of mankind for whatever reasons must be a false morality.

The special case of internal war

In the traditional theory of just war the question of the ethical evaluation of internal war was circumvented in one of two ways: either the concept of war was so defined that internal war was excluded, or else internal war was classed as illegal behaviour and therefore regarded as unjust by definition. On the other hand, in radical political ideologies, internal wars are assumed to be just because they are the vehicle by which social progress moves forward. [21]

In modern times, the problem of assessing the moral worth of internal wars cannot so easily be ignored since internal wars have become a much more important form of political behaviour than they once were. The position of revolutionaries has not changed, of course, as well as the contrary view in which the activity of oppositional groups who resort to armed violence will always be immoral. There is a middle group however, who argue that in some cases, especially if the authorities try to suppress the opposition, it would not in principle be wrong for the opposition to take the law into its own hands. Assuming that the aims of popular uprisings are just, in practice the moral question boils down to who has the greater amount of popular support, those in power or the insurgents. Since it is often very difficult to determine the amount of popular support behind either side during the hostilities, or even afterwards, views are divided concerning propositions about internal war in general. Some argue that to be successful a revolutionary movement must today have popular support; consequently successful revolutions are in general just ones. [22] But others, more sceptical, point out that victory is a better measure of military strength than of popular support.

Less attention is given to the justness of the means used in internal war, probably because there is general agreement that the problem is not acute when hostilities are limited to conventional weapons. However, some scholars draw attention to the extreme hatred engendered in wars between landsmen, and to the inhumane practices related to the tactics of demoralising the enemy, the use of physical torture, of terror, and so on. [23]

Finally, it might be noted that the pacifists are if anything even more severe in their condemnation of internal war for, they argue, within a state there must always be some possibility of reconciling opposing factions.

Notes

[1] Analysed by this author in 1963, 1969.

[2] Aristotle pointed out three kinds of just war: for a stable peace, for the defence of a state against an external aggression and for the achievement or defence of property (*Polityka,* Wroclaw 1953).

[3] Ralph B. Potter, 'The Moral Logic of War, in Beitz, Herman, 1973.

[4] E.g. Balthazar Avala, Hugo Grotius, Baron de Montesquieu, Thomas More.

[5] Writings of Saint Ambrose, Saint Augustine, Gratian, Antonius of Florence, Saint Thomas Aquinas, Francisco Vittoria, Francisco Suarez; see *Fathers of the Church*, New York 1950. Joseph C. McKenna, 'Ethics and War: A Catholic View', *The American Political Science Review*, September 1960.

[6] L. Winowski, *Stosunek pierwszych chrześcijan do wojny*, Lublin 1948; J. M. Bochénski, *Szkice etyczne,* London 1953; Potter, 1973; Kenneth W. Thompson, 'Ethical Aspects of the Nuclear Dilemma', in Bennett, 1962. The traditional approach to the justness of war assigns different senses to the justness of the initiation of war (*ius ad bellum*) and to the justice of its conduct (*ius in bello*) (see Melzer, 1975).

[7] 'That it is normally possible to determine objectively what degree of military necessity justifies what amount of human suffering has frequently been asserted but has never been satisfactorily demonstrated' (Osgood, Tucker, 1967, p.203).

[8] E.g. Osgood, Tucker, 1967; Weizsäcker, 1969; nuclear missile war can destroy the entire international system (John H. Herz, 'International Politics and the Nuclear Dilemma', in Bennett, 1962, pp.41–68).

[9] Therefore nuclear missile war has crossed the threshhold, the point 'at which the objective consequences – the purely quantitative results – of men's actions, rather than the subjective intention of the actor, must prove determinative for moral judgment' (Osgood, Tucker, 1967, p.247); any admission of a possibility that a nuclear war may be moral leads to sanctioning the wholesale killing of entire peoples, 'For this reason I think it is useful to be guided not only by intention but also by a realistic view of the total consequences of an attack. On the basis of intention almost any degree of destruction can be rationalized' (Bennett, in 1967, pp.104–5).

[10] Some scholars propose to increase the set of principles on which 'laws of war' should be based by: (a) the aspect of the survival of mankind which holds, among others, that the survival of mankind prevails over any national interests; (b) the principle that the damage made to the environment should be taken into account, and (c) the principle of the threshold concerning weapons the use of which can trigger an escalation dangerous for the survival of mankind.

The new principles directly lead to the prohibition of the so-called 'dubious weapons', the use of which may cause unnecessary and disproportionate suffering, and indiscriminate effects; nuclear weapons and biological and chemical weapons are here included (Bert V. A. Röling, Olga Suković; *The law of War and Dubious Weapons*, SIPRI, Stockholm 1976, pp.34–75).

[11] For an analysis of the problem of the consequences of war, see Marwick, 1974.

[12] On one horn of the dilemma lies suicide, on the other a moral degradation (Michael Howard, 'Apologia pro Studia Sua', in Howard 1970, p.17); a typical statement in a military journal: 'There is the danger that the West may allow itself to be persuaded that its very struggle for survival is immoral. . . . Westerners are vulnerable to paralysis by their own ethical system. . .' (Anthony Harrigan, 'War and Morality', *Military Review*, June 1964, pp.82–3).

[13] E.g. in a report by a Catholic body the establishment of principles of restrained use of weapons has been postulated; '. . . destruction not central to the total objective of war ought to be excluded. We have found no moral distinction between these instruments of warfare [nuclear weapons], apart from ends they serve and the consequences of their use' (The Christian Conscience and Weapons of Mass Destruction', The Dun Report of a Special Commission appointed by the Federal Council of the Churches of Christ in America, 1950, p.14); cf. Paul Ramsey, 'The Case for making "Just" War Possible', in Bennett, 1962, pp.146–8).

[14] Even 'an atomic war waged within the limits of military necessity may be not only something we are morally permitted to do; it may be something we are morally obliged to do' (John Courtney Murray, after Paul Ramsey, *War and the Christian Conscience,* Duke University Press, Durham 1961, p.234). In military periodicals the idea is expressed openly: 'The weapons choice should be made on the grounds of efficiency and enemy capabilities to retaliate' (Harrigan, ibid., p.83).

[15] Not only the politician who takes the decision, but also the historian and the political scientist, who analyse it *post factum*, cannot discuss war in terms of good or evil. Use of military force in quest for power, which is something morally neutral, is also morally neutral. One can only discuss whether a given war was a rational act of policy, or not, and any inhibitions in the use of violence can be justified only in terms of self-interest (Howard, 1970, p.203 and others). And expressed more simply in a military periodical: 'What is rational is moral. Flexible use of nuclear power, if it would serve a rational policy, would be

moral' (P. R. Schratz, 'Clausewitz, Cuba, and Command', *US Naval Institute Proceedings*, 1964:8, pp.25–33).

[16] There is no possibility to review here the vast discussion on the moral aspects of the use of chemical and biological weapons. Scholars who condemn these weapons stress (a) the difficulty to control their use, to distinguish between combatants and non-combatants, (b) the inevitability of escalation in their use, (c) the immense destruction which they may cause; the advocates of preparation for such a war assert that, on the contrary, (a) they may be used in a diversified way and to assure controlled destruction, (b) they are very valuable in some kinds of war (e.g. counterinsurgency), (c) they may cause less suffering than other kinds of weapons, (d) they add to the deterrence against any war.

[17] Weizsäcker, 1969.

[18] Anthony Buzzard, 'Unity in Defence and Disarmament', *Royal United Service Institution Journal*, 1959, 8, p.392; Bennett, in Bennett, 1962, p.102; B. V. A. Röling, 'National and International Peace Research', *International Sociological Science Journal*, Unesco XVII, 3/1965; Halle, 1973, p.22; Herz, 1962, p.15; Montagu 1968, p.viii.

[19] Buchan, 1968, p.22.

[20] 'The divine right of historical necessity, as it can be glimpsed in some versions of this Marxist view, seems more likely to encourage militant conflict now than to guarantee a long-run peaceful future. Moreover, by involving such grand prospects of future peaceful rewards for present violence, the Marxist imagery has seriously contributed to the rise of totalitarian or quasi-totalitarian ideologies of "holy war" and "last battles" in our century' (Deutsch, 1967, p.96).

[21] Gabriel Bonnot de Mably: '. . . why the internal war couldn't sometimes be in accord with the laws of the most scrupulous morality, as the external one is? Was the external enemy who wanted to conquer a people or who didn't want to redress the wrong, more guilty than the internal enemy of the people who wanted to enslave it or demonstrated that he contempted its rights?' (*Pisma wybrane*, Warsaw 1956, pp.56–7).

[22] The cause of freedom and justice supported by the population is more frequently the goal of internal wars than of interstate ones (Millis, 1961, pp.42–3).

[23] E.g. Hans Speier, 'War and Peace in the Nuclear Age', in: Speier, 1969, pp.9, 13. In his analysis of wars in the nineteenth century, Liddell-Hart contended that, while on the whole the trend toward human limitations in warfare was continued, civil wars 'have tended toward the worst excesses' (1972, p.75). And in discussing the problems of guerila wars in our century, he stated: 'The habit of violence takes much deeper root in irregular warfare than it does in regular warfare. In the latter it is counteracted by the habit of obedience to constituted authority, whereas the former makes a virtue of defying authority and violating rules. It becomes very difficult to rebuild a country, and a stable state, on such an undermined foundation.' (p.82).

'From the standpoint of their perpetrators, revolutions are "total wars". Any and all means are justified to attain desired ends without, as Mao Tse-tung said, regard for "stupid scruples about benevolence, righteousness, and morality"' (Collins, 1973, p.47).

8 War in the nuclear missile era: some conceptual problems.

Many problems relating to the nature of war have been re-examined in the nuclear era. The new structure of the international system, the global struggle between two opposing systems and the revolutionary development of military technology have contributed to the radical changes that have occurred in the way international and internal conflicts are now resolved. On the one hand, some kinds of war, especially total war and global thermonuclear war, have become very unlikely. On the other hand, new forms of military power have increased its versatility as a political instrument; this can be seen in the great importance now accorded to threats of the use of force.

In the wake of these developments, two conceptual problems relating to the nature of peace and war have emerged:

(a) Is war still a feasible instrument for pursuing political goals?;
(b) Is it still meaningful to distinguish between peacetime and wartime on the basis of the means by which political goals are pursued?

The war-like peace

To take the second question first, the view is widespread that the boundaries between war and peace as traditionally defined have begun to become unclear. Terms such as the Cold War or the war-like peace have been coined to capture the essence of the new situation. In the first place, it is argued, the principle political goals of the two main powers are, like those pursued in war, total and uncompromising. [1] Some limited goals more typical of periods of peace can also be noted, but the character of the Cold War is derived from the ambition of both superpowers to expand and to eliminate the political influence of each other. In the second place, the methods by which this policy is conducted in the new situation resemble those used in time of war; in both there is a mixture of diplomatic and military means. This has led some scholars to view the relations between states as a 'spectrum of conflict' ranging from peaceful coexistence, through sublimited and limited war, to thermonuclear war. [2] In this perspective, the simple dichotomy between the concepts peace and war is too coarse be useful. [3] This contention is implicity supported by scholars who, in their investigation of the process by which historical decisions to engage in war were made, conclude that full military engagement is often the end of an upward spiral in the use of violence that begins with threats of war and includes semi-military actions in between. [4]

Although there is general unanimity amongst Western scholars regarding the characteristics of the post-war period (except the sources of the Cold War – see below), there is a great difference between the conclusions they draw. For some the situation in the post-war period confirms them in their belief in the eternity of war. To them war is one of the two phases through which political developments in the international system regularly pass. [5] Some point out, that there is always a war going on somewhere. It should therefore be regarded as a normal state of social life, perhaps even more normal than peace. [6] Since wars are inevitable, the outbreak of the Cold War should give no cause for surprise. The important thing is to realise that it is essentially a war, and to take the political consequences.

Other scholars strongly contest this view. They admit that the situation is very far from an ideal peace, [7] in which no conflicts nor wars occur, or even from a real peace, in which resort to war is excluded as a means of resolving the conflicts that do arise; nevertheless, despite the fighting that occurs almost every day, the situation is some kind of peace, for the world war that seems imminent has not yet broken out. The variety of names that have been proposed to describe the new situation – among them war-like peace, [8] non-war, [9] negative peace, [10] absence of war, [11] absence of total war, [12] brush-fire peace [13] – should perhaps be seen as an attempt to emphasise both that the world is on the brink of war, and that it has not yet overstepped the edge. The prescriptive implications involved in describing the situation as one of war are avoided. [14]

Many political scientists prefer to treat the present state of affairs as a new category, something intermediate between war and peace. One proposal is to regard it as an order over which impends recourse to force. [15] A second is organised peacelessness, [16] which, it is argued, captures the combination of a graduated use of violence to achieve political goals that have nothing to do with political freedom, cooperation between nations, or social justice. [17]

Of these three interpretations of the postwar situation, there can be no doubt that the oldest of them, the theory of the Cold War, has had and still has a great influence on the foreign policy of Western countries. From it has developed the strategy of deterrence to prevent a 'hot war' from breaking out and, more recently, the strategy of crisis management as a means of solving political conflicts without an overt military confrontation.

The theory of the Cold War

The literature on the Cold War has by now, thirty years after the end of the Second World War, swelled to enormous proportions. Here there is only room to outline some of the main ideas related to the nature of war and peace. [18]

The political picture sketched by this theory is quite sharply focussed. The Communist countries, led by the Soviet Union, are bent on world domination. The West is thereby thrown into a defensive position to protect democracy, freedom, and peace from Communist totalitarianism. [19] It is believed that the

Communist offensive does not aim at defeating its appointed enemies through a major war. Rather they have been conducting what has been called a protracted conflict or indirect strategy, consisting of contests for particular political objectives in all fields of social activity, none of which is in itself decisive, but which together add up in the long run to a total victory. The global plan of the revolutionaries is thus to weaken the enemy by a great number of small actions systematically carried out and followed up, and when the situation is ripe, to crush him with a decisive blow or to force him to capitulate without one. [20]

Growing number of scholars however, criticise the official and dominating view of the origin and essence of the Cold War. They attribute to the Western countries an offensive attitude towards the socialist countries that was epitomised by the slogan of 'roll-back'. [21] In some of this assessment it is argued that the West is fighting neither for democracy nor peace, but for the preservation and even enlargement of its spheres of influence. Views have also been expressed that one should see the sources of the Cold War in the policy of both opposing camps. [22]

With regard to the military picture, for most of the Western observers the Cold War represents a new kind of global war, in which the principal antagonists try to avoid a direct military clash. [23] This can be explained by the revolution that has occurred in military technology: the development of strategic thermo-nuclear weapons, which combine incomprehensibly great destructive power with the ability to strike at great distance from the ultimate target, has introduced an era in which the use of at least some available military means has been excluded. There is some disagreement about the highest level at which a military engagement between the superpowers is conceivable; some consider conventional warfare on a small scale to be possible, others believe that the risk of being forced into a position where nuclear weapons might be used inhibits the superpowers from even the slightest military incident. There is general agreement however, that military power looms large in diplomatic contests, and each superpower may become involved in conventional war within its sphere of influence.

There are some observers who do not share this view and who emphasise the military aspect of the Cold War. As in all wars, so in this one, the use of military resources is one of the main means of fighting.[24] That 'hot' war has not yet broken out can be explained by the fact that it has not yet been advisable for either contender to attack. From this fact it should not be concluded that in this protracted war it will always be to their disadvantage to use military means, even thermonuclear weapons. Thus all discussion of the unique military aspects of the Cold War is premature. [25]

Whatever their assessment of the military features of the Cold War, Western analysts share the view that the Soviet challenge obliges the West to develop a broad strategy to coordinate military, political, economic, and ideological fields of endeavour.

According to Western political strategists, the situation at the end of the Second World War was new in two respects: the offensive adopted by the Communists presented the West with a different kind of challenge while at the same time military technology had entered the nuclear era, the full implications of which were not completely clear. A new strategy on the part of the West was required, and what emerged was the strategy of deterrence.[26]

The military meaning of deterrence is 'to discourage the enemy from taking military action by posing for him a prospect of cost and risk outweighing his prospective gain'. In political contexts the concept has been broadened to mean 'to discourage a second party from doing something by the implicit or explicit threat of applying some sanction if the forbidden act is performed'. [27] Both meanings are well expressed in General Beaufre's definition according to which deterrence is discouraging an enemy power from taking a decision to use its arms, or, more generally, from acting or reacting in a given situation. [28] It should be noted that as a political strategy, deterrence may be intended to perform not only such defensive functions as to prevent war, but also such semi-offensive functions as to restrain the opponent from acting, even from responding to provocation. [29]

The idea of deterrence has been one of the keystones in the foreign policy of the United States and its allies in the whole post-war period. [30] According to official declarations, the American armed forces would not be used to start a war, but their strength would deter any potential enemy from making aggressive moves. [31] To begin with, the object was to insure Western Europe against a Soviet invasion, and more generally, to protect the 'free world' by containing Communism. Throughout the 1950s, the United States enjoyed first a monopoly, later a great superiority in nuclear weapons. This period culminated in Dulles's doctrine of massive retaliation, which was supported by a military programme that enabled the United States to deliver a nuclear blow to the USSR from a fleet of airborne long-range bombers. [32] At the same time, bases were built on the territory of European members of NATO and in some other points (Spain, Morocco) from which middle-range bombers and missiles could attack the heartland of the Soviet Union.

The launching of the first Sputnik in 1957 dramatised the fact that the United States had itself become vulnerable to nuclear attack. Deterrence in a situation of nuclear parity needed to be reformulated, for the doctrine of massive retaliation in such circumstances appeared a bluff. Moreover, the historical evidence indicated pretty conclusively that despite the occasional tests of strength, the Soviet Union was not really willing to risk a military confrontation in Europe and that it accepted the *status quo* there for the time being. Its interest had shifted to the Third World where it hoped to exploit the more fluid political situation brought about by the end of the colonial era.

For these reasons, such strategic analysts as William Kaufmann, Robert

Osgood, Henry Kissinger, and many others, were in general agreement regarding the inadequacy of Dulles's policy. They called for a strategy of graduated deterrence, which would allow the United States to meet each kind of challenge with an appropriate response. If the challenge were raised to a higher level, it should be possible to make a corresponding raise in the response. The most detailed theoretical exposition of this basic strategic concept can be found in the writing of Herman Kahn, who describes forty-four different steps on an 'escalation ladder', each of them characterised by a specific conflict situation, defined with regard to the force used, from diplomatic threats, to various crises, to armed conflicts, to a nuclear holocaust. [33]

The official policy planning of the United States was quick to absorb these new ideas. When McNamara became Secretary of Defence in 1961, he set out to broaden the range of military capabilities. The arsenal of nuclear weapons was by no means abandoned: indeed, it was steadily increased and developed to meet the needs not only of a strategic attack but also of tactical employment on the battle-field. More attention was however given to conventional armaments, and particularly to effective measures against guerilla warfare.

It should perhaps be noted that contributions to the concept of deterrence have also been made by strategists from other Western countries. The French have added the notion of 'proportional deterrence': since the basic tenet of deterrence is that an action will not be taken if the losses risked exceed the potential gains, a small nuclear power, which would represent little of potential value, need not have the capability to inflict more than a proportionately small amount of damage on one of the superpowers in order to deter them from attack. [34] A second idea is that of the 'third partner', whereby a small nuclear power allied to one of the superpowers would, by launching an attack on the other superpower, compel its ally to initiate a thermonuclear war; this fact increases the deterrent effect of the small nuclear power. [35]

West German theorists have been critical of the theory of graduated response, for, they argue, reserving nuclear weapons for a response to a nuclear attack increases the likelihood that the Soviet Union might risk a limited war with conventional weapons. To be effective, deterrence must present the presumptive aggressor with an uncalculated risk in the form of a forward strategy from NATO which will foresee a possibility to respond to a minor armed attack on NATO's boundary with a crushing nuclear blow. For much the same reasons the West Germans have also proposed a 'continuous deterrence', in which no distinction is made between nuclear and conventional weapons such that both kinds of armament might be used at all levels of hostility. [36]

The theory of crisis and crisis management

The second innovation made in Western politico-military doctrine since the advent of the Cold War is the theory of crisis management. Given that the chief features of international affairs in the post-war world have been a conflict over fundamental political goals together with a great peril that any attempt to resolve the conflict by military means would exterminate human life on earth, it is perhaps understandable that attention has been devoted to the problem of winning a crisis while at the same time keeping it within tolerable limits of danger and risk to both sides, or of resolving a crisis without resorting to force.

But what is a crisis? [37] Although there exists no unanimity amongst Western analysts who have studied the matter, [38] most definitions indicate that a situation of an international crisis arises when an unexpected turn of events suddenly endangers some high-ranking political values of a state. Since the situation obliges the leaders of the state to take swift action, they have very little time to collect more information about what has actually happened, to plan and assess various alternative lines of action, or to follow the normal channels of decision-making. [39] If they don't take speedy steps to manage the situation, and if they are not resigned to accepting defeat by a *fait accompli*, they may not see any other way out of their dilemma than to resort to military force. [40]

Crisis is thus a situation related to both peace and war. In a historical perspective it might be regarded as a prelude to war, [41] for at a time when war was an acceptable way of settling differences, the switch from peaceful to violent means of pursuing political goals was usually made after a crisis had arisen.

In the modern context, some argue, since war is a far too risky way of pursuing political goals, and the nuclear stalemate has a paralysing effect on diplomacy, the best way of achieving vital political aims may therefore be to generate a crisis. In other words, these analysts argue, crisis has become a substitute for war as the ultimate means of getting one's way. [42] Others, putting significance on the fact that political crises are always accompanied by some use of armed violence, have suggested that they are really a kind of warfare. [43] A fourth viewpoint looks on crisis as a mixture of elements: while war is characterised by coercion and diplomacy by accommodative negotiation, crisis is a mixture of both kinds of behaviour. [44]

From these different shades of meaning it can be inferred that there is some difference of opinion about whether crisis might be used as an instrument to improve a state's political position. There are those who, like the advocates of 'brinkmanship', find it practicable. Others emphasise that surprise, which is the essence of crisis, is a double-edged sword: a state may try to gain an advantage by surprising its opponent with some unexpected move, but the opponent may turn the tables by responding in a way the first state had not foreseen, or had thought very unlikely. Thus, for somewhat different reasons, there is a general consensus in the West about the need for strategies of conflict management. [45] Since it is widely held that the resolution of crises depends to a great extent on relative military capabilities, such strategies have become incorporated into the general strategy of the Cold War. [46]

Among the principles which ought to govern crisis management have been mentioned political authorisation and close control of each military action, a clear and appropriate presentation of crisis objectives, coordination of military moves with political-diplomatic actions, avoidance of military escalation, and slowing down the momentum of events. [47]

At the same time factors impeding a peaceful resolution of crises have been studied and presented, and among them the simultaneous establishment of irreversible commitments, a failure of communication, and an unwillingness to recognise that the opponent is not bluffing.

In order to resolve crises without actually incurring war, special bodies have been set up in some Western countries with special staffs responsible for planning political, economic, and military measures that may be taken to meet possible international emergencies. Such contingency planning is assigned the National Security Council in the United States, and the Bundessicherheitsrat der BRD in West Germany; in France, the function is carried out by the armed forces according to a special instruction containing guidelines for crisis management and providing for a special centre of command in time of crisis. [48]

Research on the conditions under which crises occur, and attempts to create a typology of international crises continue. [49]

War as an instrument of policy: the great debate

The answer to the second question thus seems to indicate that the difference between peace and war has narrowed in such a way that the term war has received wider application. But at the same time as peace has taken on all the attributes of war save a direct military clash, doubts have been raised whether full-scale war will ever again occur. To become involved in a nuclear missile war would seem to be a kind of madness, for the havoc it would make of the world cannot coincide with the political goals of any state, no country could win such a war. To become involved in a conventional war that might lead to a nuclear war would seem only slightly less insane. For those who share this opinion the notion that war may be used as a continuation and instrument of policy has lost its meaning. They have in effect demanded that war be reinterpreted as a social phenomenon. [50] Their views have been presented in three different forms: as an assertion that it is impossible for wars to occur, as a contention that wars are now highly unlikely, and as a prescription that wars must be eliminated.

In defence of the traditional concept of war it has been maintained that limited and local wars are still possible. There is an abundance of arguments to support this view in the great literature on the subject. [51] It is pointed out that since many old conflicts remain and new ones constantly emerge, all of which can only be resolved by the use of armed force, limited wars will inevitably occur. [52] Moreover, they are a practicable means of resolving conflicts since they can be conducted rationally under political control, and they can be won. [53] Advocates of this position often try to clinch the argument by pointing to history: the fact that wars have occurred in the past and occur even more frequently in the present, is incontrovertible evidence that they will also occur in the future. [54]

Most of those holding this view would allow that nuclear missile warfare is so qualitatively different from all other kinds of war that such arguments do not hold with respect to it. Some argue, however, that even such an unthinkable war should be taken into consideration. [55] The balance of nuclear power is fragile, and governments sometimes act irrationally; it is possible that a thermonuclear war will occur. There are also scholars who maintain that even nuclear war may have some beneficial effects, or that it is worthwhile risking nuclear war in pursuit of such high aims as the containment of communism. [56]

The great debate on the nature of war in the era of nuclear missiles thus centres on these two issues: that peace now resembles war, and that full-scale war can no longer serve any human ends. It can be seen as an off-shoot of the long debate on the nature of war that has its roots in the philosophical controversies of the nineteenth century. The central figure in this respect is undoubtedly Clausewitz, for many modern theories of the nature of war have been fashioned to defend, reinterpret, or refute his ideas. [57] Their imprint on the Western political approaches and Soviet Marxism–Leninsim is, despite the many differences of interpretation, quite unmistakeable. Both of the opposing ideological camps use his maxim, that war is the continuation of policy by other means, as a point of departure in the present debate on the nature and future of war.

The Clausewitzian formula and its traditional interpretation

Clausewitz analysed two interrelated problems: what war is in its ideal and real form, and how to win it. In investigating the first problem, which is dispersed in many places in his production but is one of the main topics of *On War*, he concentrated on two aspects of war – as a military activity and as a political act. Clausewitz attempted to comprise the essence of his views within the preliminary definition that 'war is an act of violence intended to compel our opponent to fulfil our will'. [58] A logical development of this idea led him to the complementary formula that 'war is an act of violence pushed to its utmost bounds'. [59] Clausewitz (and his commentators) considered this expression to contain the concept of an 'absolute war', because the logic of reciprocal violent action to destroy the enemy's forces or to disarm him in order to compel him to act in some desired way would lead, he felt, to an extreme form of fighting without any restraints or limits.

Clausewitz commented that 'absolute war' in this sense should be taken as an ideal form of hostilities; it describes what all wars would be if no moderating factors were to influence their course. In reality, no war is ever absolute because all are limited in some way. A state of pure violence cannot be achieved owing to the friction created by some military and psychological factors as well as political aims, all of which set limits to the amount of violence that can be generated. Of these three, Clausewitz emphasised the importance of taking the last into account when analysing real wars; in his most famous interpretation, 'war is nothing but a continuation of political intercourse, with a mixture of other means,' [60] he contended that wars conducted more by military principles than political objectives tend to be nearly absolute.

Together the concepts of absolute and real war constitute what might be called the Clausewitzian philosophy of war. When combined with his views of the international system, the nation-state, and the role of war, they make up his political philosophy. Clausewitz regarded war as a normal phase in the relations between states, and as a normal instrument of state policy. Since states aim at increasing their power at the expense of other states, their interests are always in conflict. War as a rational instrument of national policy he considered to be the proper means of resolving such conflicts and increasing power. [61] To him it was self-evident that war is planned and waged under the direction of the ruler, who represents the whole nation and embodies its spiritual qualities. [62] Clausewitz therefore had no cause to address himself to the relation between war and ethics; war needed no justification beyond its effectiveness for achieving certain desired goals.

These then are the basic elements of Clausewitz's position, if in a highly condensed form. In the traditional debate on the nature of war, three contending views might be defined in relation to the Clausewitzian scheme:

(a)   the militaristic school, which maintains that once wars have broken out, they are uncontrollable by political objectives and must therefore be guided by purely military considerations;

(b)   the school of political realism, which upholds the view that war can be waged under political control; and

(c)   the pacifistic school, which denies the basic tenet shared by both of the other views, that war is an unavoidable reality, and offers an alternative philosophical perspective in which to set political conflicts and their resolution.

The militaristic school When Clausewitz's *On War* was published posthumously in 1831, it was eagerly seized upon by the Prussian military academies, for no other work covered such a wide range of military problems so comprehensively. To the military man of the mid-nineteenth century, the Clausewitzian concept of the ideal war as violence pushed to its utmost bounds summed up well the role he had chosen in life: to conduct wars for his country as efficiently as possible. To him, Clausewitz's writing was all the more valuable in that it contained lengthy sections of practical advice on the best strategy and tactics to adopt the approach the ideal war. It therefore became well-known amongst the military leaders of the great European armies of the nineteenth century, and influenced their official military doctrines. The Prussian generals were the first to incorporate his views, [63] but his idea of war as a massive concentrated attack of all disposable resources to achieve total victory also influenced French and English military thinking and their preparations for war. [64], [65] It is held that it was in fact the basis of their strategy at the beginning of World War I. [66]

Members of this school of interpretation contended that Clausewitz regarded war primarily as a military act, intended to disarm the enemy, thus military victory as the ultimate, decisive, and therefore the only important aim of hostilities; political goals had very little influence on the actual course of fighting. They pointed out that the bulk of Clausewitz's work was in fact devoted to strategic problems and to the methods of winning war. His comment that the whole conduct of military operations should be guided by the political aims of the war they considered to be a secondary remark. Therefore, although wars might arise for political reasons, once they had actually begun political considerations became subordinate. Politicians should therefore be silent the moment mobilisation begins. [67]

Although Clausewitz's work was virtually unknown in the United Sates until after the Second World War, a certain similarity between some of his ideas on absolute war and the American concept of war during the decades around the turn of the century may also be noticed. [68] When the United States went to war, it did so to achieve total victory over the enemy, at least to begin with. The nation was called upon to make sacrifices in many ways in order to ensure the success of the war effort. American military doctrine of the time was based on the assumption that it was necessary to aim for total victory and to subordinate all other goals during time of war to this one objective. Only in some cases the government announced more limited political aims at some later state of the war.

Despite the debacle of the First World War, the militaristic interpretation of Clausewitz survived and was particularly influential in the German army; [69] von Seeckt was perhaps the most prominent of this school during the early twenties.

The publication of Ludendorff's military theories marked a new development in the militaristic tradition, however. Ludendorff was not content to brush aside Clausewitz's assertion of the primacy of politics over war as his militaristic disciples had done; in effect Ludendorff turned Clausewitz on his head by maintaining that war must predominate over politics and that the object of politics must be to prepare for total war.

And, finally, the German fascist war plans, rooted in the whole tradition of the military thought up to Ludendorff, were also influenced by a militaristic interpretation of Clausewitzian ideas of absolute war and destruction of the enemy. [70]

Political realism There is also a long line of thought behind the view which put emphasis on a quite different aspect of Clausewitzian philosophy, namely that war is simply one of the means by which states try to achieve their political objectives, and that it is therefore subordinate to politics. It is now generally recognised that this idea was one of the main themes in Clausewitz's study of war, but his rise to eminence in this school of thought occurred only after World

War II; earlier the embrace in which he was held by his self-appointed disciples in the militaristic school caused him to be generally identified with their standpoint.

But ideas very similar to Clausewitz's political conception of war were often expressed in the second half of the nineteenth century by statesmen, historians, and other men of letters who understood the interplay of state affairs in terms of the balance of power. [71] Their treatment of war as a means of acquiring power, pursuing national interests, and maintaining balance in international relations corresponded to the Clausewitzian idea that war is an instrument, largely determined by power, which states employ in the conduct of their foreign policy because it is efficacious. [72] Many of the notions of the balance of power framework have been inherited and given a modern guise by the school of political realism.

The pacifistic school The third line of political thought about the nature of war, which may also be traced back to the nineteenth century, where it appears in various philosophical and sociological works, might be called pacifistic. With respect to the Clausewitzian system of ideas it is wholly critical, for it is based on a different set of first principles. Underlying the various views that constitute this tradition is the assumption that war must be seen as a pathological and highly undesirable phenomenon that ought to be eliminated from social life. Considered from this vantage, Clausewitz does not seem less militaristic for having claimed that war should be tempered by political goals; the idea that war might be used as an instrument of policy appears militaristic because the approval it confers on war implies that wars are unavoidable, natural, and justifiable a view which is an anathema to the pacifists.

Three ways of developing an anti-war philosophy might be identified. One, the truly pacifistic belief, is grounded on ethical principles. It is highly critical of any suggestion that because war is efficacious it may be used as an instrument of policy for its total indifference to the moral issue of justifying means on their own and not merely in relation to the ends sought. Any theory of war that does not include a moral evaluation of war itself must therefore be rejected as incomplete. Some pacifists have also contended that the instrumental assessment of war in terms of costs and benefits often stimulated the outbreak of war.

Then there is the utopian tradition, which is based on the belief that there exists a basic harmony of interests between all men and all states. The utopians are highly critical of the position taken by the realists that the highest known good is the national interest. This leads them into the fallacy of defining national interest internally on the basis of a selfish ethic, the utopians maintain. This is an error because the nations of the world coexist in a social system; the interest of each nation can have no meaning if defined without regard to the rest of the world of which the nation is an inseparable part. Consequently, the contention of the realists that wars must occur because states inevitably find themselves engaged in insoluble conflicts in the pursuit of their respective national interests

159

is based on a faulty premise. On the contrary, the utopians maintain, if all states were to follow their true national interests, based on a global perspective, they could resolve all possible differences without resorting to violence. [73]

The third group that might be included under the pacifistic heading is the branch of sociology called peace research or conflict resolution, for it too is devoted to eliminating war. Although they consider war to be a political phenomenon, they do not interpret it as Clausewitz does as an act that continues deliberate efforts to fulfil some political purpose; to them it is rather the outcome of an interplay of political forces that creates a situation, the structure of which obliges the decision-maker to conclude that war is the best course of action. [74] These social scientists believe that war can be prevented if steps are taken to create a structural setting in which war is less attractive than other available means of resolving the political conflicts that arise.

War and politics in the nuclear era

Before tracing the views of these three lines of political thought into the nuclear era, I should like to indicate how the boundaries between them have changed in a way that makes it difficult to maintain the typology at all.

The Second World War and the events immediately preceding it demonstrated that the premises of pacifism, especially of its ethical variant, however worthy they may have been as ideals, were too far divorced from reality to be useful for analysing it. These same experiences, together with the advent of nuclear ages and the 'unthinkable' war, also discredited militarism. Only the political realists seemed to have a grip on what was happening. For this reason the main conceptual elements of the political realists became the basis for virtually all Western understanding of international relations in the post-war period. However, such concepts as power and state interests have proven sufficiently elastic to admit interpretations that accord with a wide range of perspectives on the politico-military situation of the world and on trends in its development. It is in this form that pacifism and militarism have survived.

Scholars who search for peaceful ways of resolving conflicts or of creating a viable peace reflect the former pacifists' aim of eliminating war. However, they are much less utopian than the inter-war pacifists, and they share with political realists their general ideas about the essence of politics and war. The initial impetus to peace research as a separate field of study came from scholars with such an outlook. Despite the fact that this type of research emerged in opposition to the hard-boiled brand of political realism that predominated in the 1950s, the profiles of the orientations have become somewhat blurred. What remains of pacifism is a fundamental optimism held by some scholars that peace is possible and would be normal if it were not for the many unnecessary obstacles created by men that prevent it from developing. Moreover, they stress that the interests of states must be interpreted in the light of the needs of world community.

160

Militarism has similarly become more a basic attitude than a school of thought. It is based on the belief that war is inevitable and that one should therefore make the best use of it one can. This perspective, though weaker than in the heyday of militarism, still pervades the thinking of many professional strategists. Inasmuch as these people, like the political realists, conceive of politics in terms of state interests and of war as a means of resolving interstate conflicts, they may be called the right wing of modern political realism.

Finally, even those who feel that they belong to the tradition of political realism have taken different paths in refining its analytical tools. Various schools have sprung up, and they often disagree with each other. What distinguishes the perspective (as opposed to the concepts) of a political realist from that of a militarist on the one hand, or a pacifist on the other is his unwillingness to pass *a priori* judgement on the value of war: states will go to war when they believe it lies in their interest to do so; whether war will break out or not will depend upon the dynamics of the balance of power – so much he is prepared to say, no more.

Consequently, in what follows, I shall not treat the three lines of political thought of the previous section as three separate theories, but rather as three perspectives within the Western study of international relations. The development of thermonuclear weapons has had a profound impact on the debate concerning the relation between war and politics. Both of the main contenders, the militarists and the political realists, have been forced to modify their position somewhat, but both claim that technological developments have demonstrated the basic soundness of their respective views on the nature of war.

The primacy of politics questioned The militarists have argued that the advent of nuclear weapons, and especially ballistic missiles with nuclear warheads, has strengthened the validity of their contention that war is governed by a logic of its own. [75] Such weapons make realisable the Clausewitzian idea of absolute war as violence pushed to its utmost bounds. [76] The technical limitations to pure war have virtually been overcome, and the total destruction of the enemy is consequently possible.

One of the initial conclusions drawn by the militarists as a result of the new situation that emerged after the Second World War was that the West was in a position to wage a preventive war against the Soviet Union. They reasoned that since war with Communist forces would break out eventually, the West should seize the opportunity it had owing to its nuclear superiority to smash the military power of the enemy before he had sufficient forces to venture an attack himself. As the capability of the Soviet Union to launch a nuclear missile attack quickly grew, however, it became clear that the doctrine of preventive war was untenable. The Soviet Union would have time to launch a missile attack on the United States in retaliation before any American missiles reached their targets and would retain a capacity to do it even after the attack. The realisation that a thermonuclear war would risk mutual destruction therefore forced the militarists to retreat from their traditional stand that war is inevitable and desirable to the position that war is probably unavoidable (because of the designs of the enemy) even though it is undesirable.

This change in the militaristic line of thinking has not had very many consequences for the policy recommendations made by its adherents, for they postulate that a sound defence policy must expect the worst of the enemy. [77] They emphasise the danger of being caught at a disadvantage by some technological breakthrough by the enemy and urge investment in programmes to develop more sophisticated weapons. Strategy must be constantly reviewed to keep abreast of new developments. The stockpile of nuclear arms must be built up to prevent the enemy from gaining a strategic advantage.

War as an instrument of policy confirmed The second great camp in the modern debate on the nature of war holds that with the exception of thermonuclear war, the idea that war is a continuation and instrument of policy has withstood the test of time and remains valid. [78] War is still quite a normal phase in interstate relations, which is only natural, since no foreign policy can be carried out unless it is backed up by military capabilities to wage war if the need should arise. [79] The following arguments are advanced in support of this position:

1 The world is still full of political conflicts, some of which can only be resolved by means of armed violence.
2 The main factors that have an influence on the incidence of war remain. The number of independent states has increased, for example; other things being equal, the risk that wars will break out is thereby greater.
3 War is still a practicable perhaps even a necessary way of achieving some political goals, particularly if it is used rationally in proportion to the political aims.
4 The theoretical statement that war is a continuation and instrument of policy must not be confused with the question whether war of some particular kind is useful as such an instrument. If war breaks out, it will be a continuation of policy; if not, even when the conflict is a major one that cannot be resolved by peaceful means, the reason is simply that there is no war of any kind that is a suitable instrument for pursuing the policy. Thus far from invalidating the Clausewitzian formula, the unlikelihood of full-scale nuclear war actually confirms it: such a war is unlikely for the very reason that it could not serve any political purpose. However, in this respect nuclear war is unique.
5 The corresponding argument is that governments continue to resort to limited wars as a means of policy, as is witnessed by the frequency with which such wars still occur. Moreover, there are many indications to show that limited wars are in fact guided by political objectives: strategy and tactics stress flexibility, to meet a variety of situations, and the hostilities are intended not so much to kill as many men as possible as to demoralise the enemy's soldiers. Some even argue that in modern revolu-

tionary wars, the ultimate aim of which is to win the hearts and minds of men, the military operations are not merely limited by, but are directly governed by, political directives.

6 To conclude, mention should be made of attributing to Clausewitz the idea of limited war. This concept is 'explicitly based on the principle, drawn from Clausewitz that the conduct of war should be scrupulously disciplined by over-all political considerations', i.e. that war should be a carefully restricted instrument of policy. [80]

On the other hand the realists have been prepared to admit that if a thermo-nuclear war were to break out it could probably not be controlled by political objectives and that to this extent the concept of war has changed. [81] They contend however, that such a modification cannot challenge the validity of the basic idea of war as a political act. Because such a war would, like the Clause-witzian absolute war, be purely destructive, it should be regarded like absolute war as a hypothetical case. Since nothing of value can be achieved by releasing so much destructive power, to launch a nuclear war cannot be a real alternative. An 'unthinkable' war must be a highly improbable war. For all other kinds of war, the real wars, the primacy of politics has not been altered.

Nevertheless, the political realists have been obliged to concede that absolute war no longer belongs to the realm of ideas alone; it has become a physical possibility. Therefore, although the realists draw different theoretical con-clusions from the militarists, there is great similarity in the military policy they each prescribe. Since thermonuclear war is now possible, it must be prevented, and this can only be accomplished by building up a nuclear arsenal sufficient to deter the enemy from being tempted to use its nuclear weapons. Such a strategic capability would also act as a shield under the protection of which limited wars guided by political aims may continue to be waged.

The anti-war tradition Representatives of the third traditional line on the nature of war have also made their voice heard in the modern debate. They maintain that the shattering experience of total war in the Second World War and the prospect of total annihilation in a thermonuclear war clearly show the grotesque-ness of the militarist position. The view held without any reference to ethical norms by many realists of war as a normal and everlasting tool for realising political goals the pacifists condemn as extremely dangerous; it directs efforts towards maintaining war as an instrument of policy and creating rational forms of war instead of towards searching for ways of completely eliminating war. [82]

The traditional notion of war questioned Finally, it should be noted that within all three main outlooks are individual scholars who doubt the adequacy of the standard conception of war for an understanding of war in the nuclear era. Amongst the observations on which they base their uncertainty are the following: [83]

1 If a thermonuclear war were to break out, it would be quite unlike any traditionally conceived war not only because of its uselessness as a

political instrument, but also because it would be a process of mutual destruction without any combat.

2 Paradoxically, also in a quite different kind of war, the so-called protracted war, the main aim is to undermine the opponent's military, economic, and political power, not to destroy his armed forces. [84]

3 Unlike past wars, which were all the result of at least some amount of political planning and deliberation, it is very likely that the next major war will be unwanted and unintended and break out accidentally.

4 Wars at the national level, which were once the most common type, occur much less frequently now in relation to subnational or internal wars; the standard war has in other words changed character.

5 Put another way, this observation could point to the tendency for modern wars to be unconventional in their means and ends if compared with the conventions of war as it is traditionally conceived. In the old paradigm, war was fought by clearly defined military forces in a limited number of battles in order to settle a dispute over the policy the combatants would allow each other to conduct. Many modern wars have the less restricted aim of gaining control over the socio-political system of the enemy through protracted warfare involving both regular troops and civilians.

6 If one of the main conceivable kinds of war, the nuclear rocket war, cannot be regarded as instrument of policy, a re-examination of the whole concept of the nature of war is required.[85]

7 The obliteration of the boundaries between peace and war, a tendency noted above, confirms the previously mentioned conclusion.

The great debate of the nature of war in the nuclear era is thus very complicated; it is difficult to draw the lines between the various positions with any great degree of precision.

Notes

[1] War and peace differ not in the goals pursued, only in the means used to attain them (Barbera, 1973, p.1); 'Clausewitz's formula – war is the continuation of policy by other means – has been replaced by its opposite: policy is the continuation of war by other means. But these two formulas are, formally, equivalent. They both express the continuity of competition and the use of alternately violent and non-violent means toward ends which do not differ in essence' (Aron, 1966, p.162).

Since World War II, peace, to adopt Clausewitz's celebrated dictum, has been little more than the continuation of war by other means. Peace was a phase of (or a variation on) the power struggle waged by the Kremlin with non-violent means (Kintner, 1973). Lenin saw that the modern war was fourfold: it was political, economic, psychological, and military. From this he inferred that a campaign might be fought and decided without an overt use of armed violence. 'Therefore he inverted Clausewitz's well known dictum that "war is a continuation of State policy by other means", and substituted it for "state policy is a continuation of war by every means".' This meant the establishment of a state of continuous warfare until Lenin's aim of world revolution was accomplished, when the Soviet Imperium would embrace the whole globe (F. C. Fuller, 'Our War Problems', *Marine Corps Gazette*, November 1960, p.10).

[2] On a chart drawn in the Naval War College a 'spectrum of conflict' was presented, from peaceful coexistence to a thermonuclear war (after Eccles, 1965, pp.36–8).

[3] '. . . the nature of war itself has changed. In particular, there is no longer a dividing line between a state of peace and a state of war' (Eccles, 1965, pp.36–8); behind both phenomena, war and peace lies the same dimension of power (Barbera, 1973, pp.1–2).

Many 'political realists' point out that the common basis of policy in both peace and war, namely the search for power, makes them two inseparable parts of the same social activity. Blainey, 1973, contends that the causes of war and peace dovetail into one another (pp.1–5, the main theme of his book). ('War and peace are not separate compartments. Peace depends on threats and force; often peace is the crystallization of past force'.)

'In a system of power politics there is no difference in kind between peace and war.' (G. Schwartzenberger, 'Peace and war in international society', *International Social Science Bulletin,* 1950:2, p.336).

[4] Stephen Withey, Daniel Katz, 'The Social Psychology in Human Conflict', in McNeil, 1965, pp.78–80; Holsti, ibid., pp.155–71; 'It is probably more useful to think of relations among states as lying along a continuum where one finds complete amity and absence of violence at one end, and total violence without restraint on the other', specific interstate relationships placed always somewhere in the continuum and seldom if ever reaching the extremes (Tanter, Ullman, 1972, Ch.16, p.362).

[5] '. . . hot or cold, nuclear or "conventional", war is still war' (Strausz-Hupé, 1958); '. . . war, be it fought with military hardware or with nonviolent political and psychological instruments, is a unity. "Hot" and "cold" are phases of intensity in one and the same war' (Cottrell, Dougherty, 'The Larger Strategic Vision', in Hahn, Neff, 1960, p.120).

[6] Frankel, 1973, pp.141 ff., p.169; Anthony Leeds, at the General Discussion on the theme 'Primitive and Modern War', in Fried et al., 1968, p.101.

[7] Ideal peace would be the highest stage in a graduation, where the lowest stage is the contemporary peace, when some fighting occurs almost every day, the higher stage means that conflicts exist but are not resolved by means of armed violence, finally the highest stage is a world without both conflicts and wars (Hanna Newcombe, Alan Newcombe, 'Peace Research Around the World', Canadian Peace Research Institute, Oakville, Ontario 1969, pp.1–2).

[8] Aron 1966, p.162. Also: 'Peace has hitherto appeared to be a *more or less lasting suspension of violent mode of rivalry between political units'* (p.151).

[9] Bert Röling: 'es ist ein Nicht-Krieg' (1970, p.87).

[10] Kjell Goldmann: 'negative peace is defined as absentia belli' (Bipolarity, Bipolarization, and War: An Outline for a Proposed Research Project', Stockholm 1973, p.1.).

[11] Modelski, 1972, p.289.

[12] Carl-Friedrich von Weizsäcker, 'The Ethical Problem of Modern Strategy', in *Problems of Modern Strategy,* 1970, p.131.

[13] C. L. Sulzberger, 'Limited Peace and War', *International Herald Tribune,* 16 February 1974.

[14] Several classifications of peace have been presented. For instance, Aron lists three kinds of peace: equilibrium, hegemony, empire (1966, pp.151 ff.); Gerhard Wettig proposes a classification including: (a) hegemony (hegemonialer Frieden), with two variants, that of the American–Soviet condominion, and that of the Soviet dominion, (b) the 'positive' peace, (c) 'co-existential' peace (Kriterien der Friedenssicherung in Europa, *Beiträge zur Konfliktforschung,* 1974:3). In addition, many scholars generally state that the years ahead will see neither complete peace nor complete war (Theo Sommer, 'Detente and Security: Options', *NATOs Fifteen Nations,* December 1970–January 1971).

[15] 'the term "order" is more fashionable than peace; peace cannot be true, when there is such an imminent probability of recourse to armed force (Michael Howard, 'Apologia pro Studia Sua', in Howard, 1970, p.14).

[16] Dieter Senghaas: 'organisierte Friedenslosigkeit' (1972 (a), p.5). Peace and war have now no unequivocal contents (Sachverhalte), and the divisional line between them has disappeared. Also: Helmut Plessner, 'Über die gegenwärtige Verhältnis zwischen Krieg und Frieden', in *Zwischen Philosophie und Gesellschaft,* Bern 1953, pp.318–34.

[17] Senghaas writes about 'abgestufter Gebrauch der Gewalt' (1972 (a), pp.61–77). Gantzel sees the contemporary peace as a state of affairs when policy uses violence and coercion of various kinds, size, and intensity for achieving various goals. An open armed violence is not used but a true peace doesn't exist (1972, p.81).

Galtung: 'such "negative" peace is unsatisfactory because it doesn't exclude the use of indirect forms of violence' ('Friedensforschung' in Krippendorff, 1968, pp.531 ff.).

[18] Reviewed by this author in 1963, 1966, 1971.

[19] From semi-official statements: 'The cold war arose in a clash between two antagonistic and, so far at least, incompatible views as how the world has to be shaped. It developed into a struggle in which the survival not only of the United States, but of the free world as well, was at stake' (United States Foreign Policy, prepared under the direction of the Committee on Foreign Relations, United States Senate, vol.1, September 1960, P O Washington 1960, p.686). From the

innumerable statements in military periodicals: 'The most important fact of political life in the 20th century is the presence of militant communism. . . By its very nature a state of war exists between it and any opposing system' (Robert K. Cunningham, 'The Nature of War', *Military Review,* November 1969, p.49). from scientific conferences: the authors of the preface to Abshire, Allen, 1963, present the 'unanimous opinion' of all participants of the conference on the goals of Communist countries as follows: '. . . of significance is the consensus that the Communist goal or expectation of world domination is unchanged, and that the Comunists are, as one participant put it, arguing as rival morticians over the best ways to bury us' (p.xviii).

[20] The main theme of Hahn, Neff, 1960, esp. Council on Foreign Relations, 'The Basic Aims of United States Foreign Policy', and Robert Strausz-Hupé, 'The Protracted Conflict'; also writings of Strausz-Hupé, Kintner, Possony, Cottrell et al. (see bibliography). International Communism attacks many countries on two fronts, either physically on their outer or military front, or psychologically on their inner or national front. These two fronts are complementing and of equal importance. Nuclear weapons cannot deter two forms of war: revolutionary and conventional wars. 'The aim of the first is to rot an enemy internally and undermine his will and economy, and of the second to defeat him in battle in the traditional way.' (J. F. C. Fuller, 'Our War Problems', *Marine Corps Gazette,* November 1960, p.15). 'Never occurred so many guerilla-wars, subversive actions or other military conflicts, fought by great powers through their representatives ("by proxy"); the revolutionary *indirect* strategy has become the main kind of combat applied by Moscow' (Miksche, 1976).

[21] W. A. Williams, *The Tragedy of American Diplomacy,* World Publishers, Cleveland, Ohio 1959; Fleming, 1961 (2 vols.); D. Horowitz, *The Free World Colossus; A Critique of a Foreign Policy in the Cold War,* Hill and Wang, New York 1965; G. Alperowitz, *Atomic Diplomacy: Hiroshima and Potsdam,* Vintage Books, New York 1967.

[22] The Cold War was the product of a conflict of irreconcilable interests between great powers, raised to high levels of hostility by misperceptions, and predispositions of both opponents (Booth, 1973, pp.13 ff.). He adds, however, that the determination of the Soviet Union 'to hold ground' and to secure its positions in the Eastern Europe was the initial catalyst of the hostile mutual relations.

[23] Ihno Krumpelt, 'Der Kalte Krieg', I–III, *Wehrkunde* 5 July 1963; Andrew M. Scott observes that cold warfare, as a special kind of conflict, is well known in history, the informal attack being the main form of struggle ('Internal Violence as an Instrument of Cold Warfare', in Rosenau, 1964, pp.154–7). 'We live in an age of "wardom" initiated by the Russian and nuclear revolutions, in which the competitive preparations for war is the real war, permanent and unceasing' (J. F. C. Fuller, 'Our War Problems', *Marine Corps Gazette,* November 1960, p.15).

[24] 'Within the overall category of "peace" there has been a series of propaganda campaigns, political warfare operations and economic wars which impinged and fed upon each other, and which themselves were meshed with the chain of military and violent conflicts' (Strausz-Hupé et al., 1959, p.3).

[25] 'The growth of Soviet nuclear power, together with the maintenance of huge conventional forces in the Communist bloc, has compelled the United States and other free nations to be prepared for a wide variety of military moves the Communist powers might make, from the fomenting of civil conflict to the launching of all-out war' ('The Basic Aims of U.S. Foreign Policy', in Hahn, Neff, 1960, p.9). Up to now, innumerable books and articles on the Cold War as 'World War III' have appeared. Of the recent ones, publications under the heading 'Conflict Studies' (edited by the Institute for the Study of Conflict, London) can be pointed out, e.g. Brian Crozier, Security and the Myth of 'Peace'. Surviving the Third World War (no. 76, October 1976); the author states: 'The point of departure is that the Third World War has been in progress for a long time. . .'. (p.1). It is fought mainly by non-military techniques (pp.2 ff.).

[26] Snyder, 1961, p.3. 'Deterrence is preventing someone from doing something he otherwise would do, for fear of consequences' (Barber Jr, 1975, p.8); cf. Kaufmann, 1956; Osgood, 1957; Brodie, 1959: Schelling, 1963; Kissinger, 1957; Hopkins, Mansback contend that in contradistinction to defence which is a strategy for coping with an adversary's attack after it has occurred, i.e. it is a passive strategy, deterrence is an attempt to dissuade an opponent from an action before it is undertaken (1973, pp.380–1).

[27] Snyder, 1961, p.9; Schelling wrote that to deter a potential enemy is to persuade him to abandon a certain path of activity by making it appear to him to be in his own interest to do so (1960, pp.6 ff). In another, even more general, wording: 'Deterrence. A measure or a set of measures designed to narrow an opponent's freedom of choice among possible policies by raising the cost of some of them to levels thought to be unacceptable' (Luttwak, 1972, p.82).

[28] Beaufre, 1964.

[29] 'Si la dissuasion se limite à empêcher un adversaire de déclencher sur soi-même une action que l'on redoute, son effet est *défensif,* tandis que, si elle empêche l'adversaire de s'opposer à une action que l'on veut faire, la dissuasion est alors *offensive.'* (Beaufre, ibid.). Ken Booth writes about two types of the threat of force; first, deterrent threats concerning a certain and unacceptable damage should an enemy initiate a specified action; and second, more positive coercive threats, concerning a certain and unacceptable damage should an enemy fail to take a specified action. The notion of deterrence is here used, we see, only in relation to the first type ('The Military Instrument in Soviet Foreign Policy 1917–1972', Royal United Service Institute for Defence Studies, London–Aberystwyth 1973, pp.5–6).

[30] This author has analysed the theory of deterrence in: 'Teoria

168

"Odstraszania". Implikacje spoleczno-polityczne', *Studia socjologiczno-polityczne,* 1957:24, and in other writings, cf. bibliography.

[31] Many scholars contend that deterrence has become the main function of armed forces. 'The peacetime deployment and political use of military force became the central doctrinal issues for most advanced states, and warfighting doctrines generally took a back seat.' (Horelick, 1973, p.195); among the tasks which face the military establishments 'war prevention has, by and large, superseded victory during hostilities as the main objective of the nuclear powers' (Phil Williams, 'Deterrence', in Baylis et al., 1975, p.67).

[32] Nash, 1975, Smith, 1955, Reinhardt, 1955, Peeters, 1959, Snyder, 1961; the strategy of massive retaliation was widely understood as being 'designed to deter an attack from the Sino-Soviet powers by drawing a line around their periphery and creating the pointed implication that instant devastation would rain upon Moscow or Peking if the line were violated. . .' (Charles G. Andres, *Force and Strategy in a New Environment,* Univ. Microfilms, Ann Arbor, Mich., after Battreall Jr., 1975, p.67).

[33] The main theme of *On Escalation* and Kahn's other writings.

[34] Writings of Pierre Gallois.

[35] Writings of André Beaufre.

[36] Kai-Uwe von Hassel postulated a strategy based on an 'incalculable risk' and 'continuous deterrence' in 'Der Bundesminister der Verteidigung: Abschreckung muss unteilbar sein', *Bulletin des Presse- und Informationsamtes der Bundesregierung,* no. 105, 20 April 1963.

[37] Some definitions and descriptions may be quoted. Holsti, 1972: Crisis is 'a situation of unanticipated threat to important values and restricted decision time' (p.9); Corall Bell: Crisis occurs, when 'the conflict in relationship rises to a level which threatens to transform the nature of the relationship' (1971); Obermann: 'Crisis means in this terminology the period of political conflict among two or more states, after one side has challenged the other in a definite question and the decision of the latter as the answer for the challenge has to be found' (1971, pp.158-9); Young: an international crisis is 'a set of rapidly unfolding events which raises the impact of destabilizing forces in the general international system or any of its subsystems substantially above "normal" (i.e. average) levels and increases the likelihood of violence occurring in the system' (1967) and it is 'an acute transition in the state of a certain system' (1968, pp.6-7).

In a more general interpretation: '. . . the essence of the crisis in any given relationship is that the conflicts within it rise to a level which threatens to transform the nature of the relationship', e.g. from peace to war, or from alliance to rupture (Bell, 1971, p.9).

[38] Definitions and descriptions of crisis may be divided into three groupings: (a) 'procedural', which point out the generic characteristics of crises without regard to their concrete politico-military and social subject or substance (a severe threat of high-priority goals, surprise and lack of plans, short decision time, decreased control over events, danger of war, etc.); (b) 'substantive' which

enumerate various types of crisis according to their politico-military and social content (Kahn's 44 steps on the ladder of escalation, or 7 kinds of crisis situations in Soviet foreign policy, described by Triska, Finley 1968); (c) 'systemic' which treat crisis as the disruption of a system. 'A crisis is a situation which disrupts the system or some parts of the system and more specifically that creates an abrupt or sudden change in one or more of the basic systemic variables' (Charles F. Hermann, 'International Crisis as a Situational Variable', in Rosenau, 1969, p.411). Cf. Hermann, 1972. Another distinction is that between international and internal crises (Horst von Zitzewitz, 'Gesamtverteidigung', III, *Wehrkunde,* 1973:3, pp.131–2; Janssen, 1974, pp.144–5). Several general descriptions of crisis present it as a clash of national interests and objectives, leading to an increase in international tensions and to a risk of war (Osgood, Tucker, 1967, p.150).

[39] 'It is the essence of a crisis that the participants are not fully in control of events; they take steps and make decisions . . . in a realm of risk and uncertainty' (Schelling, 1966, p.97).

[40] P. Williams, 'Crisis Management', *International Relations,* The Journal of the David Davies Memorial, Institute of International Studies, May 1973, p.262; Anthony J. Wiener, Herman Kahn, *Crisis and Arms Control,* Hudson Institute, New York 1962; Ole R. Holsti, 'Time, Alternatives, and Communications', in Hermann, 1972.

[41] Crises lie at the crucial juncture between peace and war; before the missile age they often led to war, because there was no perception of common interest in avoiding war; nowadays the aim of avoiding war is interacting with the aim of achieving some national interests (Phil Williams, 'Crisis Management', in Baylis et al., 1975, pp.155–6).

[42] In future acute conflicts, even if a *casus belli* were to occur between nuclear power and military power were to be used, there would be a period of threats and limited acts of violence instead of war in the classical sense (Howard, 1970, p.206).

[43] Luard, 1968, p.88; Kahn, 1965; Armin Zimmermann, quoting Prof.Grewe in 'Die Rolle der Streitkräfte im Rahmen des Doppelkonzepts Verteidigung und Entspannung des Nordatlantisches Bündnisses', *Wehrkunde,* 1973:2, p.60; Bell, 1971, writes that crises are characterised as conflicts in which armed forces are used either for demonstrations or for limited actions; Michael Howard observes, however, that during crises the opponents will has to be changed without large scale hostilities, and if possible without any hostilities ('Realists and Romantics', *Encounter,* April 1972, p.36); in military periodicals the term "crisis" has been sometimes used as encompassing all military actions short of a general war (Leilyn M. Young, ' "Win". Its Meaning in Crisis Resolution', *Military Review*, January 1966, pp.30–9).

A military scholar observes that, since World War II, the United States has encountered some 28 international crises, and that some degree of force (including demonstrations) was used in all but five of the cases. He concludes that (a) this gives every reason to expect one crisis each year, and (b) strategy of managing crises and meeting military threats connected with them should be worked out (Edward B. Atkeson, 'International crises and the evolution of strategy and forces', I–II, *Military Review,* 1975:10–11).

Phil Williams makes a parallel between crisis management and limited war; both aim at securing national objectives and both aim at ensuring that the situation doesn't escalate to a nuclear war ('Crisis Management', in Baylis et al., 1975, p.155).

[44] Glenn H. Snyder, 'Crisis Bargaining', in Hermann, 1972. 'Techniques of coercive diplomacy and the "manipulation of the shared risk of war" lie at the heart of the tough bargaining process that is an integral feature of crisis interaction' (Phil Williams, 'Crisis Management', in Baylis et al., 1975, p.152). Here the coercive diplomacy is one aspect of process of negotiation.

[45] 'There is no strategy any longer, there is only crisis management' (Robert S. McNamara, after the Cuban crisis 1962, quoted by Bell, 1970, and Howard, 1972); crisis management can be defined as 'reaching a solution acceptable to both sides without resorting to force' ('Crisis Management or Crisis Prevention?', *NATO-Letter,* Aug.–Sept. 1966, p.14); crisis management may be defined as 'winning a crisis while at the same time keeping it within tolerable limits of danger and risk to both sides' (W. R. Kintner, D. C. Schwarz: *A Study of Crisis Management,* Foreign Policy Research Institute, Univ. of Pennsylvania, app. B, p.21, after Williams, 1975, p.263, see note 37); crisis management has become an art of solving conflicts below the threshold of an atomic war (Prof. Grewe, after *Wehrkunde* 1973:2, p.60). For an analysis of crisis management, see Buchan 1966.

Cf. André Beaufre, 'Krisenbeherrschung in Europe', *Wehrkunde* 1966:7; Gerd Schmückle, 'Die Krisenbeherrschung', *Wehrkunde,* 1966:5, 'Krisenbeherrschung durch eine Allianz', *Wehrkunde,* 1967:8; Walter Schütze, 'Praxis und Grenze der Krisenbeherrschung', *Wehrkunde,* 1966:6; the main theme of Jansses, 1974; Fernand Th. Schneider, 'Zur Suche nach einer neuen Strategie des Abendlandes', *Wehrwissenschaftliche Rundschau,* 1967:10.

[46] Crisis management is treated as a new function of armed forces; cf. Zimmermann, 1973.

[47] E.g. Harland Cleveland, 'Fünf Leitsätze der Krisenbeherrschung', *Wehrkunde* 1966:8–9; Alexander L. George, 'The Development of Doctrine and Strategy', in George, Hall, Simons, 1971, pp.8 ff.; Buchan 1966.

Coral Bell adds one general principle: '. . . in situation of crisis, political ends should maintain ascendancy over military means' (1971, p.2), which reminds us of the similar principle of conducting wars.

[48] For a review of crisis management measures Zitzewitz, 'Gesamtverteidigung I–III', *Wehrkunde,* 1973:1–3; Janssen, 1974.

[49] E.g. Coral Bell presents two main sorts of crises, adversary (between the opposing dominant powers) and intramural (within the walls of an alliance), and consequently four categories of crises: (a) Adversary crises of the central

balance (like Cuba); (b) Intramural crises of the power spheres or alliance systems of the dominant powers (Cyprus, Czechoslovakia); (c) Adversary crises of local balances (Kashmir 1965, crisis of the Israeli–Arab balance 1967); (d) Intramural crises of regional alliances or organisations (Biafra and OAU) (1971, pp.7 ff.).

[50] Such assertions were presented as well by scholars, well known as active fighters for disarmament, as by politicians and military arguing for limited wars; the former expressed them, however, in a more general and categorical form. 'War can no longer serve its greatest social function – that of ultima ratio in human affairs – for it can no longer decide.' (Walter Millis, 1961); for the first time in history war doesn't profit anyone (Margaret Mead, 'Die Psychologie des Menschen in einer Zeit ohne Krieg', in Krippendorff, 1968, p.147).

[51] Writings of Robert E. Osgood, Henry A. Kissinger, William W. Kaufmann and many others.

[52] Osgood and Tucker write, that several conflicting issues have still remained, e.g. power, prestige, position, economic interests, strategic values, ideological influence etc. (1967, p.21); J. F. C. Fuller contends that wars with limited aims will always be fought (1964, p.11); H. Halperin states, that 'although major world war has been avoided, the amount of violence in the world has not been appreciably reduced. Nuclear weapons, then, have not abolished wars; war very much remains an instrument of policy' (1971); for many other opinions, see Pietzcker, 1972.

[53] '... the principle justification of limited war lies in the fact that it maximizes the opportunities for the effective use of military force as a rational instrument of national policy' (Osgood, 1957, p.26); 'The prerequisite for a policy of limited war is to re-introduce the political element into our concept of warfare and to discard the notion that policy ends when war begins or that war has goals distinct from those of national policy' (Kissinger, 1957, p.141); 'Today, threatened or applied, force is a rational instrument of policy only if it is used with restraint' (Charles A. Lofgren, 'How New Is Limited War?' *Military Review,* July 1967, p.18).

[54] E.g. Luard, 1972.

[55] Kahn's writings; Klaus Knorr, Thornton Read, eds, *Limited Strategic War,* Praeger, New York 1962.

[56] 'If we should elect to abstain from nuclear warfare under all circumstances, the most rational course would appear to be not to fight at all, but surrender on the most advantageous terms.' Security of the North America, maintenance of Free World Positions, and control of the seaways, the airways and outer space are the minimum objectives. 'We must seek to obtain them, even at the risk of general nuclear war. Such a war is "thinkable".' (Strausz-Hupé, Kintner, Possony, 1961, pp.109–10).

[57] E.g. Raymond Aron begun his exposition of the theory of international relations with a comprehensive recapitulation of Clausewitzian ideas (1967, ch.1, pp.11 ff.); he wrote a two-volume analysis of Clausewitz's impact on the

philosophy of war: *Penser La Guerre: Clausewitz, t.I, l'Age européen, t.II l'Age planétaire,* Gaillimard, Paris 1976; Anatol Rapoport presented his ideas on the nature of war, and on contemporary theories of the subject in the framework of an analysis of Clausewitzian ideas (Editor's Introduction to Clausewitz, *On War,* Pelican Books, London 1968); Dieter Senghaas begun his analysis of problems of war, peace and deterrence with 'A Look at Clausewitz' (*Rückblick auf Clausewitz*) in 1972, and Bernard Brodie wrote: 'The central idea of this look I have borrowed from Clausewitz . . . that the question of *why* we fight must dominate any considerations of means' (1974, vii). Cf. Turner, Challener, 1960, ch.I.

[58] Carl von Clausewitz, *Vom Kriege,* 17 ed., Ferd. Dümmlers Verlag, Bonn, 1966, pp.89–90, Translation: *On War,* Pelican Books, London 1971, p.101. All subsequent quotations are taken from the latter edition.

[59] Ibid., p.103.

[60] Ibid., p.402.

[61] In a dictionary of modern war, the view of war as a permanent manifestation of social life in the context of a world of nation-states is regarded as 'essentially the classical view of war as defined by von Clausewitz' (Luttwak, 1972, pp.212–3). In order to protect their interests and extend their spheres of influence, states apply a whole battery of instruments: formal diplomacy, propaganda, political warfare, economic warfare, threats of war, and finally war itself.

[62] 'That policy unites in itself, and reconciles all the interests of internal administration, even those of humanity, and whatever else are rational subjects of consideration is presupposed, for it is nothing in itself, except a mere representative and exponent of all these interests toward other States' (ibid. p.404).

[63] F. N. Maude: Clausewitz's book learned how to use force, unrestrained by any law save that of expediency, for attainment of political goals; therefore it gave key to the interpretation of German policy, past, present, and future (in introduction to an English edition, 1908, reprinted in Pelican's edition 1971, p.83); H. Rothfels: 'Clausewitz may be regarded as a foremost exponent of "Prussianism" and the "battle-mania" of the nineteenth century, and his book as a textbook for Sadowa and Sedan' ('Clausewitz', in Earle, 1943, p.93); O. J. Matthijs Jolles: undoubtedly the conception which regarded war as an exercise of unrestrained force most strongly influenced the minds of the German nation and its military leaders: 'Famous military leaders and writers of Germany in those days, like Moltke, von der Goltz, von Blume, Meckel and many others, declared themselves to be pupils of Clausewitz and said that Germany owes to him her success on the battlefield.' (in the introduction to the first American edition of *On War*, after *Military Review*, November 1957, p.61); the main thesis of Alfred Vagts' 1957, concerning German militarism in the nineteenth century and beginning of the twentieth century, was that Clausewitz was its promoter, and that Germany prepared for an 'unconditional and absolute war' in the light of Clausewitzian teaching (p.185).

'In subsequent generations generals and statesmen have made this concept [the absolute concept of war] a virtual reality by blindly following the unlimited principle suggested in Clausewitz's striking premise – to introduce into the philosophy of war a principle of moderation would be an absurdity ... War is an act of violence pursued to its utmost bounds".' (Leonard, 1967, p.7).

[64] Dallas D. Irvine, 'The French Discovery of Clausewitz and Napoleon', *Military Affairs*, 1940:IV; Jay Luuvas, 'European Military Thought and Doctrine', in Howard, 1965.

[65] Luuvas, op.cit.

[66] '... World War I, probably the most bloody clash of all history ... was a struggle between belligerents who all followed the same philosopher of war' (Smith, 1955, p.50); Clausewitz's teaching highly influenced both the causation and the character of World War I (Liddell-Hart, 1967, p.357).

[67] Helmut von Moltke, Field-Marshall of the German army, after Parkinson, 1971, p.337.

[68] Smith, 1955; Weigley, 1973.

[69] One of the main themes of Liddell-Hart, 1967, esp. Part IV.

[70] The theme of: Clausewitz, Jomini, Schlieffen, The United States Military Academy, West Point 1951 ('Clausewitz's influence is not dead. The philosophy of *On War* is the philosophy of Bismarck's *Blood and Iron* and the philosophy of *Mein Kampf*).

One of the themes of: Wilhelm Ritter von Schramm, 'Die Studien des Generalobersten Beck und ihre wehrpolitische Bedeutung – Ein Beitrag zum 20. Juli 1944', *Wehrkunde,* 1974:7, p.343. Adoption of Ludendorff's ideas by Hitler were one of the causes of Germany's defeat, since total war couldn't serve any political goals.

[71] The similarity is sometimes striking. E.g. Clausewitz regarded European states as a 'community of nations' with 'collective interests', in which 'the whole relations of all States to each other serve rather to preserve the stability of the whole than to produce changes'. This 'tendency to stability' or tendency to maintain the existing state he conceived to be 'the true notion of a balance of power, and in this sense it will always of itself come into existence, wherever there are extensive connections between civilized states'. If single states want to effect important changes for their own benefit, the whole usually prevents it, if it doesn't do it, this means that the action of the tendency in favour of stability was not powerful enough at the moment (*On War*, Routledge and Kegan, London and Boston, Eighth Impression 1968, vol. II, book VI, pp.160, 162).

Clausewitz contends that the disappearance of Poland as an independent state doesn't contradict the idea of 'political balance', because Poland was in fact not independent, it was not a homogeneous member of the community of nations in Europe; with its 'turbulent political condition', and 'unbounded levity' it had ceased to play any independent part in European politics for a hundred years, it was a 'Tartar State', which inevitably must have become a

Russian province. 'Poland was at this time politically little better than an unin-habitated steppe.' (pp.163–5).

[72] Peter Paret, 'Clausewitz and the Nineteenth Century', in: Howard, 1965; cf. Paret, 1976.

[73] Angell, 1910, the main theme ('. . . it is a logical fallacy and an optical illusion in Europe to regard a nation as increasing its wealth when it increases its territory', p.36).

[74] To characterise this approach, many findings of students in conflict theory 'product an image of heavy environmental constraint on nations' behaviour, of the serious degree to which nations' leaders are prisoners of the situations in which they find themselves, and especially of their past policies. This emphasis on the structural properties of national systems, the global international system, and of conflict sub-systems should sober us. The world seems to be deter-ministic. . .' (Hanson, Russett, 1972. Introduction, p.17).

Some scholars describe the approach as negating a rational decision on war: 'Wars are seen as the outcome of political forces and structural facts which "generate" conflict, rather than the result of conscious rational decisions. While the classical Clausewitzian view sees war as a chosen instrument of politics willingly used ... here war "happens" because a given situation is inherently unstable' (Luttwak, 1972, p.214).

[75] 'In these conditions war can no longer be a continuation of policy but rather its negation' (S. O. Tiomain, 'Clausewitz: A Reappraisal', *Military Review,* May 1963, p.79).

[76] Absolute war is now not a hypothesis but a real possibility. 'The fact remains, that "unlimited war" is accepted as a rational possibility in modern military thought' (Forrest K. Kleinman, 'The Pied Piper of Modern Military Thought', *Military Review,* November 1957, p.60).

The abstract formula now may become reality (P. R. Schratz, 'Clausewitz, Cuba, and Command', *US Naval Institute Proceedings,* August 1964, p.26).

The idea of 'Blitz-Krieg', to which Clausewitz was an early contributor, may now be accomplished, the enemy can be destroyed in a single battle waged by using maximum violence (Bruno J. Rolak, 'Fathers of the Blitzkrieg', *Military Review,* 5/1969, p.73); roots of the massive retaliation doctrine go back to Clausewitz, who emphasised the massive attack, instantly, at the critical point of enemy strength (Smith, 1955, p.46 a.o.); both main enemies can now realise the Clausewitzian idea of crushing the opponent's forces: the Soviet Union can do it in regard to Western Europe, combining the use of nuclear and conventional weapons, and the USA can use its strategic nuclear weapons against the Soviet Union, combining them with land operations of its allies, and the united action can annihilate the enemy's forces, result in an occupation of its territory and crushing its will to resist. Thereby nuclear weapons enabled to realise the Clause-witzian concept of war (Ihno Krumpelt, 'Die strategischen Atomwaffen und der Kriegserfolg', *Revue Militaire Générale,* November 1972, p.9). Collins emphasises the Clausewitzian comment that 'if bloody slaughter is a horrible

175

spectacle, then it should only be a reason for treating war with more respect, but not for making the sword we bear blunter and blunter by degrees from feelings of humanity' and states that serious students of grand strategy find it difficult to apply Clausewitzian teachings to US concepts for limited war (1973, pp.xxiv–xxv).

[77] 'When an enemy is dedicated to destroy us by any means, it seems perfectly right, if not morally imperative, to utilize any conceivable weapon against him in self-defense' (Smith, 1955, p. 189).

Von Schramm states that since war in the twentieth century may be very similar to the Clausewitzian concept of an absolute war, in which the political aim disappears, one ought to take into consideration what is politically purposeful and militarily possible (*Die "Studien" des Generalobersten Beck und ihre wehrpolitische Bedeutung,* pp.340–1).

[78] The main theme of 'political realism' writings.

The West-German admirers of Clausewitz (Werner Hahlweg, Wilhelm Ritter von Schramm, Friedrich Ruge, W. Gembruch and many others, see esp. *Wehrkunde* and *Wehrwissenschaftliche Rundschau* in the 1950s, 1960s and 1970s) stress, that he was the great political classicist of war, and his teaching on war as an instrument of policy is always vivid. 'We owe to Carl von Clausewitz the inclusion of war into a philosophico-political system, with the primacy of policy' (Egon Overbeck, 'Militärische Planung und Unternehmungplanung', in: *Clausewitz in unserer Zeit*, Verlag Wehr und Wirtschaft, Darmstadt, 1971); cf. Schramm, 1965, 1973. A broader analysis see Lider, 1971, pp.40–72).

[79] The main theme of writings of Kissinger, Osgood, Tucher, Aron and others. Kissinger: 'as Clausewitz argued, war grows out of the existing relations of states, their level of civilization, the nature of their alliances, and the objectives in dispute' (1958, p.65); Osgood: 'war is a normal instrument of pursuing national interests and cannot be outlawed' (1957); Slessor: 'nations resort to war as a means of policy because they want to realize their aims, if they cannot convince the opponent to submit to their will by diplomacy' (1954).

To date, the proponents of 'graduated deterrence' and 'limited war strategy' continue to present their concepts. Although the arguments vary, the main idea remains in principle the same: by preparing for a war corresponding to the challenge, one achieves two aims: the global deterrence increases, and the destruction – if war breaks out – diminishes. See: 'Problems of Modern Strategy I–II', *Adelphi Papers*, nos. 54 and 55, 1969; cf. John Baylis, 'Limited War', in Baylis et al., eds, 1975.

[80] Clausewitz is often quoted as a proponent of limited war, esp. in military periodicals. George W. Smith: 'Clausewitz's compelling words seem of greater impact today than ever before', and nations have to limit political objectives, and accordingly, also dimensions of force applied ('Clausewitz in the 1970s', *Military Review,* July 1972, p.92); William D. Franklin: '. . . according to Clausewitz, there can be wars of all degrees from one of extermination down to a mere state of armed observation.' ('Clausewitz on Limited War', *Military*

Review, June 1967, p.25). Also Byron Dexter: Limited war was one of two main Clausewitzian concepts of war ('Clausewitz and Soviet Strategy', *Foreign Affairs,* October 1950, p.42). 'He actually drew an almost perfect blueprint of modern-day irregular operations and his theories have exerted tremendous influence on almost all major strategic thought – both in the West and of the Communists. . . . His work was an early blueprint for insurgency and counter-insurgency of the type being conducted today in Vietnam' (Franklin, 1967, pp.24, 29).

Cf. Robert E. Osgood, 'Limited War' in: *The International Encyclopedia of the Social Sciences,* vol. IX 1968, pp.302, 305. Earlier: Gordon B. Turner, 'Classic and Modern Strategic Concepts', in Turner, Challener, 1960, pp.10–11. Turner contends that the idea of extreme violence means that soldiers should fight their battles with the utmost effort, but what is right with regard to battles cannot be related to the war as a whole. Pellicia writes that Clausewitz did not deny the possibility of using even extreme violence in a rational way, i.e. subordinated to policy (1975–6, p.62). Cf. Walters, 1974, pp.77–80 (Clausewitz's view of war, as guided by political aims, is a concept of limited war).

[81] John Strachey: 'It is a deeply disturbing reflection that none of Clausewitz's three "modifying factors" apply to nuclear war' (1962, pp.74–6); E. J. Kingston-McCloughry: 'nothing would seem further from the truth in the event of a nuclear war, than the pronouncement, that war is the continuation of policy by other means. Such a war if unleashed, would be the end of all policies and an utter mutual annihilation.' (1957, p.248); nuclear weapons have made the total war impossible as a rational political act, 'the existence and possibility of the use of these weapons have today made senseless the Clausewitz's thesis on war as a continuation of policy by other means.' ('Krieg', in *Sovjetunion und demokratische Gesellschaft. Eine vergleichende Enzyklopädie,* C. D. Hernig, ed, Harder, Freiburg, Basel, Wien 1969, Band III, p.1042); Theodor Ebert: 'Some theorists don't see in war [between nuclear powers] still the continuation of policy with admixture of other means, but they see the outbreak of war as a failure of policy' ('Wehrpolitik ohne Waffen. Das Konzept der sozialen Verteidigung,' *Beiträge zur Konfliktforschung,* 1972:2; for similar opinions, see George Kennan, *Russia, Atom, and the West,* London 1958; Finletter 1960; Fuller, 1961; André Beaufre states that nuclear weapons make the Clause-witzian law of the growing violence in war a nonsensical one ('Konflikte der Zukunft', *Österreichische Militärische Zeitschrift,* 1973:1, p.4) because such an escalation of violence reduces its utility to the absurd ('Konflikte der Zukunft', *Österreichische Militärische Zeitschrift,* 1973:1, p.4).

[82] For a comprehensive critique of these views, see Rapoport, 1964, 1968/1971.

[83] Strachey: an exchange of nuclear warheads would be no combat in the classical meaning of the term (1962, pp.73–4); an open use of armed force ceases to be a normal instrument of policy, prepared in advance, wars may break out only accidentally (Knorr, 1966, 1973; Schelling, 1966; Hoffmann, 1973); new

military technology has transformed the nature of war, because it enabled to achieve directly and immediately strategic objectives (Horelick, 1974, p.194); national states cease to be the main actors in wars, and being 'national' is one of three key concepts of the Clausewitzian paradigm of war, as well as 'rational' and 'instrumental' (Rapoport, 1968/1971, p.13); main kinds of contemporary wars are non-Clausewitzian, Sebastian Haffner states this on guerilla war (Einleitendes Essay, in Mao Tse-tung, *Theorie des Guerilla-Krieges,* Rowohlt, Reinbek bei Hamburg 1966, pp.5 ff.), and his main contention is: 'War in its conventional form has become . . . unusable (unbrauchbar) as a means of policy at least now. Simultaneously, however, a new kind of war has arisen, so different from the conventional one, that a new term is needed for it'; similar views on the guerilla variant of revolutionary war as a new kind of war, see Raymond Aron ('Clausewitz et la guerre populaire', *Défense Nationale,* January 1973, pp.3–10); Samuel P. Huntington states the same on 'insurrectionary war' which is 'asymmetrical' and to which the Clausewitzian formula is not applicable (1962, p.20), and Hans Speier on 'people war' which Clausewitz did not consider a kind of war (1969, pp.59–60).

[84] Thus Atkinson, 1974, contends that Clausewitz's emphasis on the destruction of enemy's armed forces as the 'centre of gravity in war' is now a faulty strategic assumption. The new war is a 'social war' and the main aim of military operations is to destroy enemy's social base.

In other words, the aim is to destroy the enemy's economic (in the first line – industrial) capabilities, and not armed forces. Thus some contend that Clausewitzian formulae are now irrelevant (e.g. the main theme of Miksche, 1976, Introduction, pp.10–11).

[85] 'The Clausewitzian dictum that war is an extension of politic by other means has been fundamentally altered by the strategic nuclear equation' (Richard B. Foster, 'The Emerging U.S. Global Strategy: Its Implications for the U.S.–European Partnership' in Foster, Beaufre, Jushua, 1974, p.20).

Miksche contends, that both Clausewitzian formulae, that of war as violence pushed to its utmost bounds, and that of war as a continuation of policy by other means, has nowadays become obsolete (zeitfremd), invalidated (aufgehebt) by the 'over-dimensional' impact (überdimensionale Wirkung) of the nuclear weapons; nuclear war cannot bring victory, and this invalidates Clausewitzian formulae (1976, p.10).

Part II

THE SOVIET ANALYSIS OF WAR

9 War as a continuation of policy

Some general ideas

The basic principle of Marxist ideology, its outlook on the world and society, its method of analysing physical and social phenomena, has not changed. But principles of analysis can only be applied to a concrete epoch, and since the world has changed, so also has the Marxian interpretation. The modern world with its two opposing political systems and nuclear technology does not resemble the world of the nineteenth century. Politics, war, revolution in our time are quite different from those that occurred at the time Marx and Engels wrote. As a result of new laws and regularities of the new world, new aims of the Communist movement and the socialist states, which did not exist at the time the classics were written, changes in Marxism and the political-military doctrine of the Soviet Union have inevitably followed.

Lenin developed and changed Marx's ideas during a period in which the important changes were the rapid growth of industrial monopolistic countries, the social revolution in Russia, and the beginning of socialist construction in the Soviet Union. Stalin in turn changed the ideas and adapted them to the needs of the only socialist state as it became a world power. After his death further changes in the theory were made to bring it in accordance with the diversified character of the postwar world,. To describe Marxism at each of these four stages goes beyond the aim and scope of this study, but some of the important changes that have occurred in the development of the Marxist view of the nature of war will be mentioned in the following presentation. Here a view that is termed Marxist–Leninist has been officially adopted in present-day Soviet doctrine.

The Marxist–Leninist approach to the nature of war has three important characteristics. In the first place, war is distinguished as something quite different from conflicts in the physical or organic world. War is a social phenomenon. Secondly, war is also quite different from conflicts in the human psyche or from conflicts between or among individuals. Such conflicts are considered to have a biological or psychological nature, while war is a conflict between or among social groups. Thirdly, war is always seen in the form of a political conflict between classes. When the contradictions and conflicts between classes, or between class-states, or between nations led by classes, increase, expand, and sharpen, war may be a method of resolving them. Thus war is regarded as both the most acute form of social conflict and as the method by which this is resolved. The point of departure of the Marxist–Leninist definition of war is accordingly a view in which war is seen as a continuation and instrument of policy, whereby policy is meant the conduct of relations between classes and states.

According to Lenin, 'war is simply a continuation of policy by other means (namely based on violence). Such was always the view of Marx and Engles, who considered all war to be a continuation of the policy of the given states involved and the various classes inside these states. . .'. [1] In the modern classical work on the theory of war the concept of war is defined as 'armed violence: organized armed conflict between different social classes, states, groups of states, or nations, in order to achieve definite political goals'. [2] And in the first East German military dictionary, war is defined as 'an armed conflict between states (coalitions) or classes in order to achieve their political and economic goals. It is the continuation of policy by violent means'. [3]

The similarity between these and many other definitions is striking in spite of the fact that a half century divides the newest from the oldest. [4] The common ideas seem to be the following:

(a)   The parties involved in a war are classes of states; sometimes nations are added. [5]

(b)   The specificity of war as armed struggle which distinguishes it from other social phenomena, and in particular from the peace time political struggle, is always underlined.

(c)   It is always pointed out that civil wars are a kind of war. [6]

(d)   In addition to such a narrow definition of war, a second broader interpretation usually complements it to explain the class roots of war and its transitory character, and to indicate that imperialism is the main source of war in the modern age.

Policy, class, and war

Politics

One of the crucial elements in the Marxist-Leninist theory of the nature of war as a continuation of policy is the notion of politics, [7] just as the difference between Marxist–Leninist and many non-Marxist interpretations of the term is characteristic of the difference in their respective interpretations of war.

The meaning of the term politics According to the general definition, politics is that activity of organised social groups that is directed towards other social groups; it is generated by, and expresses the mutual relations between groups. One of the basic tenets of Marxism–Leninism is that all presocialist societies, apart from the so-called primitive society, are class societies, that is, they consist of classes with contradictory interests. The relations between classes are expressed as politics which reflect first and foremost the relative economics of the classes. These may fight to protect their interests by various means, one of which may be to seize and preserve political power. The main object of the political struggle is to obtain control over the authoritative organs of government, for only state power enables a given class to establish and maintain

182

a socio-economic and political system that favours its interests. For the ruling class the political struggle involves resolving all contradictory economic and other interests in its favour by using the political power at its disposal; for the governed class, it involves tearing down the existing superstructure and creating conditions for the transformation of the whole socioeconomic structure. Thus in an official version, politics is the activity by which a particular social class aims to seize, maintain, and make use of the authoritative institutions of the state. [8]

Four kinds of politics Politics in this sense relates to the internal life of a society. Since social groups are organised into nation-states, the mutual relations between states constitute a second kind of politics. These are the two kinds of policy traditionally recognised in Marxist doctrine. After World War II, two further kinds of politics were indicated and analysed: relations between different socio-political systems, and relations between peoples or nations living within one state or in several states. [9]

All four kinds of politics reflect ultimately the economic interests of the classes struggling against each other. All are a 'manifestation of economics', as Lenin put it. But the central link in the chain is the policy of the state, for all four kinds of politics are conducted by states or groups of states. [10] The next step in this presentation is therefore to see how Marxism–Leninism has interpreted the nature of the state as related to the class struggle and to politics.

The class-state and its politics

The class-state According to Marxism–Leninism, the state appeared at a point in man's social development at which antagonisms between classes could no longer be reconciled. The purpose of the state was therefore to establish some sort of order by legalising and perpetuating the domination of one class and by precluding any violent opposition from the governed classes. The governing class, which has at its disposal the state apparatus, is thus able to defend its interests both inside and outside the country. In doing so it conducts the policy of the state.

The state apparatus is thereby an instrument for resolving class conflicts according to the interests of the governing class, and for preventing conflicts from exploding into violent revolution which might transform the existing social order of production and the whole socio-economic and political system. The state apparatus is thus an aid by which the governing class can suppress the governed classes and defend its interests. Of course, being a special political organism in some sense separated from society, and the struggling classes, and 'endowed with a movement of its own' (Engels), the state apparatus may appear to play an independent role, at least to some extent. Since the governing class can always exert decisive influence over the state apparatus by economic or other means, however, in reality the state apparatus is an instrument of the ruling class. [11]

Functions As an instrument of class domination, the state pursues its internal policy in three areas: it defends the existing political system, it strengthens the economic domination of those in power, and it imposes their ideology on the entire society. The foreign policy of the state also realises the interests and aims of the governing class, either by expanding the size of territories, markets, or capital resources, or by protecting the monopolist position of the rulers in the country against foreign competition. [12]

As a corollory it is often strongly emphasised that internal and external policy complement each other. Nevertheless, since one of the principal objectives of the internal policy of the ruling class must be to maintain political power, it will subordinate its external policy to this goal if need be. Therefore of the two the internal policy of the state is the more basic; it determines the character of the external policy. On some occasions, and especially in time of war, the foreign policy of the state may appear to be primary, but even then it is rooted in the internal needs of the ruling class. [13]

The socialist state The socialist state is included within the general definition of the state, but its essence and functions are said to be quite different from those of the presocialist states. Since it is created as an instrument of the working class, who comprise the majority of the population, the policy of a socialist state represents the nation as a whole and defines national interests. Gradually, as the power and influence of the overthrown classes diminish and the class struggle fades away, the repressive functions of the state are reduced; those that remain are directed against the external enemy. [14] During the period of transition to socialism, when the reactionary classes attempted, through an external intervention, to regain power, internal war was always possible. Once Socialism was in principle built, however, this danger disappeared, and in 1961 the Soviet state was declared to be no longer a dictatorship of the proletariat but to have become a non-class state of the whole people. As such it wages the class struggle only in the international arena and not within the country. Therefore it only prepares for possible international war. [15]

The foreign policy conducted by the socialist state is consequently presented by Soviet theory both as a class policy and a national one. Its declaratory aim is to defend the interests of the nation and the socialist coalition and to support the class struggle both internationally and within particular capitalist countries.

Class character of the foreign policy The class character of the foreign policy of both capitalist and socialist states may not be apparent at the present time, because the international struggle is not carried on directly between the antagonistic classes but is waged on four partly intersecting fronts. [16] In some instances it takes the traditional form of a conflict between different ruling classes conducted through the state apparatus. Secondly, it is waged between coalitions, which although made up of many states, act as a unit and represent antagonistic socio-political systems. [17] Thirdly, it is carried on between these two blocs indirectly for influence in the countries of the Third World, which must decide now upon their socio-political and economic orientation for development, a choice that is partly one between socialism and capitalism. Finally, since relations between nations and classes are maintained not only through governments and state organs, but also through other political bodies, parties, political and public organisations, international political movements, etc., all of these relations constitute an additional front of international struggle. Moreover the capitalist and socialist countries share some common interests, such as the search for peace, a fact which may also obscure the outlines of the international class struggle. Nevertheless, the struggle is continuing, and the position of each state is decided by its class character.

The class character of war

The conclusion of these two assertions, that war is a continuation of policy, and that policy is always related to class interests, is thus that all wars are class wars, and that war is an instrument of class policy. It is important to distinguish, however, between the notion that war has a class nature and that war is a form of the class struggle. The idea that all wars are class wars means that the interests of classes represented in the policy of parties and states, determine the aims of all wars. In some, the class interests either appear directly, e.g., in civil revolutionary wars, or are obvious, as in wars between socialist and capitalist countries. But in other wars they are less apparent. When wars are fought between countries governed by exploiting classes, the class aims can be discovered, but when antagonistic classes unite temporarily to wage wars in the name of common interests, for example wars of national liberation, the class aims may be very difficult to discern, although such unity is unstable and contradictory, and in fact each class fights for a mixture of national and selfish interests. Therefore, although all wars are rooted in the basic class contradictions of society, not all wars are a direct form of the class struggle. [18]

War as a continuation of policy

In the Marxist–Leninist interpretation, the formula that war is a continuation of policy is based on two complementary assumptions: firstly that the aims of war are a continuation of the aims of peace, and secondly that war implies a change of method.

Continuation of aims

According to Marxism–Leninism, each class, state, or nation attempts during peacetime to attain its aims by the most diverse means. The ruling class, which in a class-society comprises the owners of the means of production, does everything in its power to maintain and perpetuate its rule, and to achieve the best economic and political conditions for exploiting the working classes and establishing outlets for their products. The working classes on the other hand fight for their liberation and for the overthrow of the existing system; in their everyday struggle they strive to improve their working and living conditions. In capitalist society, the administrative apparatus, the economic system, and such ideological instruments as political parties, the Church, and the educational system, are used to maintain and strengthen the power of the ruling class. To counter this policy, the working classes, led by their own members, use strikes and demonstrations and propagate revolutionary ideology. If the ruling class cannot maintain its power by relatively peaceful means, including non-armed violence, it resorts to armed violence, that is, to what is called in Marxism–Leninism counter-revolutionary military action, in order to disarm and destroy the revolutionary forces. [19] And, on the other hand, when the working classes consider their conditions of existence to be unbearable, they may also resort to armed struggle. [20] Usually, however, they resort to arms only in answer to armed violence applied by their exploiters. In either case, the resort to violent means would not reflect any change in the political objectives of the antagonistic classes, for these remain the same. If the revolutionary uprising succeeds, it becomes necessary to defend, by military means if necessary, the newly won position of power against counter-revolutionary uprisings.

According to Marxism–Leninism, war at the international level is analogous to internal wars. The ruling classes use the state apparatus at their disposal to further their interests beyond the borders of the state. [21] The relations between states are thus essentially class relations, for rarely do the interests of the ruling class converge with the interests of the nation. In the previous historical period, the struggle between capitalist states for colonies, and the inter-imperialist struggle for the redivision of empires, were the main fronts on which the ruling classes competed. With the advent of the socialist system of states, however, the international class struggle has mainly taken the form of a conflict between the socialist and capitalist systems. There is a great variety of means short of war by which the international struggle can be conducted; some examples being the building of alliances, a demonstration of military strength, diplomatic persuasion, and economic sanctions. Despite this formidable arsenal of instruments, states must always take into account the possibility that the conflict cannot be kept within peaceful bounds, that war may break out. They even consider war to be a legitimate means for achieving their goals should peaceful means fail. All states must therefore prepare for war, either with the object of starting it, or to force back an aggression, or to dissuade a potential enemy from attack. [22]

Change of methods

In the Marxist–Leninist analysis, the goals for which states strive remain unaltered with the outbreak of war. What does change is the means by which the struggle is conducted. The most important change is that armed combat is introduced. This has far-reaching consequences for the use of military operations has its own special requirements (troops, armaments, supplies) and its own special objectives (the destruction of the enemy's military potential). [23]

Armed fighting must have an organised character before it is regarded as war. This means that regular armed forces must be engaged on at least one side, and that a certain amount of continuity in the armed clashes is ensured by planning an organised activity on both sides. [24]

However, the laws of war cannot be reduced to the laws of armed combat. The struggle between states continues on diplomatic, economic, and ideological fronts, and the course and outcome of war is dependent on the complete potential and the total activity of the fighting countries. Victory, which is the end for which wars are fought, is more than success on the battlefield, and must be related to the political objectives of the fighting parties. This notion is formulated in the aphorism that 'politics is the whole, war is its part', in which war is understood as warfare.[25] This phrase has also been interpreted to mean that war is only politics of a special form in which other forms may also be used, or that war is only a stage in a permanent political intercourse. To support the contention that war only implies new priorities, Marxist–Leninist analysis cites examples in which all methods including military ones are used during peacetime to carry out some policy.[26]

To be war, military action must be taken by both sides. A one-sided employment of military force is not enough to constitute war, for it may occur when the attacked object is unable or unwilling to resist. In international relations, such actions are called unresisted invasions; in the domestic sphere they are called punitive campaigns, counter-revolutionary policy actions, which may be supported if need be by armed force, armed *coups d'état*, or popular uprisings unresisted by the government. [27] Since armed violence is often used by one class or another during peace-time to further its interests, this qualification may be an important one for distinguishing between war and peace.

Some modifications in the idea of continuation

In the nuclear era some modifications have been made to the Marxist–Leninist interpretation of the mechanism of continuation: one relates to the policy that is continued in war, the second to the description of the causes and conditions of the outbreak of war.

The Marxist–Leninist interpretation of the notion that war is the continuation of policy by other means, was formulated at a time when the Communist Party desired to mobilise all forces against the participation of Russia in the First World War and to transform the War into a proletarian revolution. The interpretation therefore emphasised two main points: that the World War was a continuation of the prewar policy of all its participants, i.e. ruling classes of the fighting states, and that it was also a continuation of the class struggle going on within each of the participant nations as well as internationally (the

187

international working class versus the international bourgeoisie). According to this interpretation, the policy of the imperialist powers had three components, which together led to the unleashing of war: one was their mutual relations, the second was the class struggle within each country, and the third was the international class struggle.

In the post-Leninist analyses of war, actual or potential, these three aspects of policy, which reflect the three fronts of the socio-political struggle, were variously emphasised, depending on the needs of the current policy. Nevertheless, Lenin's basic principle, that to discover the basic socio-political character of a given war, one must examine the prewar policy of its participants, has always been stressed.

In the nuclear era, the main aspect of politics is said to be the international class struggle between the two principal forces of the contemporary world, the world working class and the world bourgeoisie. Of the many fronts on which this confrontation is fought, the antithetical relationship between the two global antagonistic systems, is said to be the primary one, 'the main contradiction of the epoch'; all other international and internal contradictions, as the national-liberation struggle of the oppressed peoples against the colonialists, the democratic struggle of the popular masses against monopoly associations, the revolutionary struggle of the proletariat of particular countries against the bourgeoisie, as well as the struggle between capitalist countries for strengthening the positions of the monopoly capital, are affected by it. Accordingly, the main current of policy in which wars in our epoch flow, springs from the struggle between imperialism and the world socialist system. All actual wars, even those which seem to be far removed from the struggle between the two systems, in some way fit into the designs of imperialism. Each war can be regarded as a link in the chain of the world-wide struggle, and to a greater or lesser extent, as a continuation of the imperialist policy for world domination. [28] If a particular war is not a means for achieving some immediate concrete goal in such a policy, it may nevertheless be used for preparing conditions for a new world war. [29] All wars may therefore be seen as a continuation of the policy of two different kinds of actors, the direct participants and world antagonistic camps. The idea that imperialism may generate wars has various meanings, depending upon the particular circumstances. Imperialism may initiate a war but war may break out without any apparent links with imperialism. For example, the intertribal quarrels in Nigeria, and the separatistic tendencies of some leaders in the Eastern provinces, may be regarded as the direct cause of the civil war there. But a more fundamental, if indirect, cause was Nigeria's long history of colonial occupation, for it was imperialism that endowed the country with its contradictions and quarrels, and it was imperialism that afterwards aggravated these conflicts in accordance with the principle of divide and conquer, by financing and supporting the separatist tendencies in Biafra. In general then, modern Marxist–Leninist analysis emphasises that the analysis of war as a continuation of policy cannot be confined to the particular members of the international system, but must also be related to the system as a whole.

There are also indications that a second shift of emphasis in the Marxist–Leninist analysis of particular wars has taken place, although this is more uncertain. When analysing the causes of modern international conflicts, some Soviet scholars have tended to weaken the emphasis previously placed on the internal class struggle as the primary and deepest source. [30] Others do point to the importance of analysing the internal situation of the countries participating in conflict, but they do not mention such traditional sources of war as the worsening of internal conditions or the tendency to 'redirect' internal conflicts. Perhaps this shift of emphasis from the analysis of internal to international direct causes of particular wars is connected with the main focus of the present foreign policy of the Soviet Union, namely peaceful coexistence, which is directed mainly at the prevention of international wars the source of which, it is charged, lies in the global policy of imperialism. [31]

Moreover, at a deeper level Marxist–Leninism allows for the possibility that class policy, which is common to all the three mentioned activities leading to war, may change as the capitalist class becomes stratified. The modern capitalist state acts on behalf of the whole bourgeoisie only on some issues, such as the defence of the capitalistic economic system and the political power of the capitalists; but there are circumstances in which the interests of monopolies do not coincide with those of the remaining part of the bourgeoisie. [32] Here the state is used to implement violence not only against the working class but also against the non-monopolistic strata of the bourgeoisie, which may then participate in the anti-monopolistic struggle. It is therefore possible that part of the bourgeoisie may oppose wars fomented by monopolists in a number of countries in the hope of making vast profits if the wars bring immense losses and destruction for the whole population. [33]

In sum, what is meant in Marxist–Leninist analysis by the continuation of policy in war seems to have been modified in two respects. From the former focus on the prewar policy of the states that participate directly in the war, Marxist–Leninist analysts have shifted to an emphasis on the global policy of the two competing systems; the importance of domestic policy as the determinant of all policy in the state has in the analysis of particular wars been diminished.

The change that has taken place in the nuclear era according to the Marxist–Leninist interpretation of war as the continuation of policy, can also be discussed in relation to the causes and conditions at the outbreak of war. On this point Marxist–Leninist analysis has become considerably more sophisticated, the idea of a cause being treated as a complex notion with many sub-categories.

For example, in a modern study of the causes of international conflicts and wars, five main kinds are analysed in detail: [34] the main causes, which underly all wars of some epoch; the essential causes, which are related to particular types of war; the final causes, which are decisive for the outbreak of a given instance of war; the direct causes, which include the circumstances leading directly up to the outbreak of war; and finally, the reason or occasion for war. These causes are all interrelated. The essential cause of a revolutionary war is, for example, the aspiration of the suppressed class for liberty, but various final and direct causes may influence the ultimate decision to go to war at the same time as the main cause is the existence of the system of exploitation. Similarly, the main cause of wars of national-liberation is the existence of the colonial system, while their essential cause is the desire of a suppressed people to obtain freedom.

A second study offers much the same categories, [35] although an original feature is a category called simply additional causes. For the most part these refer to the political requisites, such as an increase of contradictions, a sudden expansion of the class or national-liberation struggle, etc., that are necessary before the political leaders of the class or nation can overcome the resistance of those opposed to war and take the final decision. Included in this study, however, is a discussion of accidental causes, as the appearance of a pretext for war, the personality of those who take part in the decision-making process leading up to a declaration of war, mistakes made by statesmen so that they misinterpret the situation, actions taken by one state that force a second state to begin war against its own will, and so on. Amongst accidental causes are sometimes included such spontaneous events and processes as financial crises. It is suggested that the atmosphere of the Cold War, with its diplomacy based on crisis, and the growth of nuclear stockpiles, are two factors that have increased the importance of accidental causes in the nuclear era.

To conclude this section, it may not be out of place to speculate a little on some reasons for these changes in the Soviet analysis. They may reflect a more general tendency to free the description of socio-political processes from automatism; they facilitate at least some theoretical changes, which would seem to be necessary should the doctrine that certain kinds of war are inevitable be renounced. A second possibility is that these changes have been made to provide an alternative to the theories of decision-making and correlates of war that have been developed in the West. The explanation that seems most plausible to the present author is that Soviet analysts have striven to provide a framework for the control of conflict and the prevention of war. This they have done by arranging the causes of war hierarchically. The transition from a conflict to a crisis, and from a crisis to war does not take place automatically, but is the result of decisions that are in turn influenced by lower order causes. Thus if the effect of these causes can be counteracted, the outbreak of war may be prevented.

The relationship between politics and warfare

Marxist–Leninist analysis asserts the primacy of politics over war. It is the political goals for which a war is fought that determine its character, whether it is progressive or reactionary, whether it is just or unjust. The strategy with which a war is fought is dependent on political guidelines. Moreover it is the policy followed by a government that determines whether it is able to muster material necessary for the conduct of war, or whether it can stir the population to make the necessary sacrifice. [36]

Nevertheless, the Soviet scholars do recognise the fact that the course of war may also limit political alternatives, and that the policy followed by a government may in some part be dependent upon war.

The influence of policy on war

In Soviet military doctrine, 'the objective of military strategy is the creation by military means of those conditions under which politics is in a position to achieve the aims it sets for itself'. [37] The connection between strategic (or as they are sometimes called politico-military) goals in war and political goals is clearly very close. [38] Two aspects of this relationship can be discerned. First, the contents of the political goals defined by the political leaders, the socio-political changes they pursue in the interest of the class they represent, has direct bearing on concrete strategic objectives, for example, the occupation of enemy territory, the defence of one's own territory, access to some important economic or strategic area. In the second place, great importance is attached to the character of political goals. If these are just, the military goals will be fought for, with great determination and with the endorsement of the whole people, which will improve the likelihood of their being accomplished. Political goals of fundamental significance for the class struggle are especially important in shaping military strategy. In the Soviet view, should the conflict between the two opposing social systems develop into war, the acute class character of the collision would preclude any sort of compromise; since the ultimate political goals of both classes would be at stake, all military means would be used, and unprecedented violence would ensue. [39]

In the nuclear era, the task of defining correct and purposeful strategic goals on the basis of the political aims of the state has become much more difficult. Because of the intensification of the class struggle the goals of modern warfare would not be limited to the destruction of the armed forces of the enemy, as in previous times, but would aim at the total destruction of his economic potential and the state apparatus. The development of new military techniques, and the mobilisation of the immense numbers of people in military operations, which have made it possible to achieve these goals, have introduced many new laws and regularities that greatly complicate a proper definition of strategic goals. The politicisation of strategic goals has also made more difficult the tasks of the political and military leaders, for no decision can be taken in a purely strategic context, but must always be related to the political circumstances at each moment. In the imperialist countries, the leaders are confronted with additional difficulties: their goals are inconsistent with the general direction in which the world is evolving, which they are unable to perceive because they lack the correct

191

scientific methodology. Moreover, the inner contradictions of capitalism become manifest in the domestic struggle between various political parties and influential groups within the administration and military establishment and give rise to difficulties in coordinating divergent views between members of a coalition or alliance (e.g. NATO). [40]

The influence of war on politics

Although Marxism–Leninism asserts the primacy of politics over military strategy, it does recognise that under certain circumstances politics may be influenced by military considerations. [41] Since the basis of both war and politics is conflict, many of their basic principles are similar: to gather forces (either revolutionary or military) at the right moment at the right place, to strike at the weakest part of the enemy's defence, to adapt the means used in accordance with the situation, to manoeuvre skilfully with reserves, to follow the general line of the plan (of battle or of the party), and so on. [42] Stalin was fond of making a point about political strategy by drawing an analogy with military strategy and by using military examples.

Soviet analysts recognise that during the course of a war the demands of military strategy may influence the policy followed by the government. It must take into account the real military alternatives, and their leaders must plan their activities in order to maintain control of the military operations. Moreover, as the war brings victory or defeat and thus increases or diminishes the alternatives, the politicians must continually make adaptations in the goals they attempt to reach.

War also has very far-reaching consequences on the domestic conditions within a fighting country: even here, there is a need for the politicians to adapt to the requisites of war. One of the most general effects relates to the economic policy of the country, for changes must be made in the structure of production in order to provide war materials, and more generally, to meet all needs of the conduct of war.

But in capitalist countries, war also has a profound impact on the political and ideological struggle. The governing class attempts to limit the democratic rights of the population both by taking direct legislative and administrative measures, and by exhorting all citizens to social solidarity against the enemy. The response of the working classes will depend on the nature of the war: if it is an aggressive war, which is the usual case, they intensify the class struggle; if it is defensive, they increase their productive contribution to the war effort. In multi-national countries, the demands for cohesion usually lead to greater oppression of national minorities. Should such a country become occupied by an aggressor during the hostilities, an opportunity is provided to commence the national-liberation struggle. In socialist countries such problems do not exist. There, according to Soviet scholars, war increases the cohesion of society.

The nuclear era has introduced two special aspects to the problem. In the first place, the need to adapt political objectives to military realities has increased; the latter influence policy already in peacetime. The immensely destructive capability of the new military technology compels all states to prepare for the possibility of a new world war. For many states, the need to seek the protection of a stronger ally or for obtaining modern armaments may make certain political concessions necessary. Moreover, the political strategy of war prevention and the military strategy of limited war that have followed from the development of military technology have greatly changed the nature of politics in peacetime. Thus even during time of peace, the political goals and the political strategy of a state must be commensurate with its military potential and that of its allies.

Secondly, the impact of military strategy on policy would increase during a thermonuclear war, for the results of military operations come so quickly that little time is available to consider more than the immediate political strategic aims. One solution to this problem is to increase the control of the political leaders over military operations by requiring their authorisation for the use of weapons of mass destruction.

Civil war as a continuation of policy

The Soviet general ideas on the nature of war have been reflected in the study of civil and international wars, and so-called wars of national-liberation.

In the traditional Soviet theory, civil war is defined as 'armed struggle between antagonistic classes within a country for political power'. [43] Although this definition has not been changed, its interpretation has been modified in recent times to encompass not only war between the working class and the bourgeoisie, which has been named revolutionary civil war, but also that between the progressive masses and the extremely reactionary forces of monopolistic capitalism, called democratic civil war.

Revolutionary civil war

In revolutionary civil wars, a combination of political and military means are used in the hope of achieving a social transformation, supported politically by the masses of the population. Essential though they are, the political elements do not necessarily overshadow the military measures taken. Indeed, the military struggle may be very widespread and intense. [44]

There are three situations in which revolutionary (or counter-revolutionary) wars may occur. The first is the pre-revolutionary situation. The ruling classes may then regard the growth of the revolutionary movement as a threat to the existing system and attempt to forestall the revolution by taking the offensive in a counter-revolutionary war. Another possibility in the pre-revolutionary situation may arise if in a democratic country the exploiting classes, fearing that the government will introduce measures to effect a social transformation of society, decide to unleash war to obtain political control over the state (Spain 1936).

The second situation is that of the revolution itself, for the overthrow of the existing order is usually performed with the use of armed force.

Thirdly, once the revolution has been accomplished, the new order must be defended against counter-revolutionary action that may be taken by the overthrown exploiting classes in order to regain power (Russia 1918–20).

In all three situations, revolutionary civil war is a continuation of the preceding political struggle between classes. It is considered to be at hand when the class struggle has become greatly sharpened. [45] The governing exploiting class becomes unable to maintain its power according to the existing order, and therefore attempts to defend its position by using armed violence. The oppressed masses on the other hand, no longer able to tolerate their ever-deteriorating living conditions, increase their ideological and political activity significantly. In such circumstances, revolutionary civil war becomes a logical continuation of the previous political situation. Under the leadership of the revolutionary party, a prolonged ideological, economic, and political struggle of the working classes inevitably leads to a revolutionary uprising. Naturally, the defence of the new order is a continuation of its establishing.

Democratic civil war

Not all civil wars are revolutionary wars, however. It is possible that the inherent contradictions of capitalism will lead to break the power of the monopolistic cliques without destroying the system itself. Since the ruling strata will defend its absolute rule however, an armed conflict can break out between it and the majority of people. This is called democratic civil war. [46] The main difference between such war and revolutionary war lies in the aims: in democratic civil war the object is to introduce democratic, not socialistic reforms, or to defend the democratic system. The progressive forces thus include part of the national bourgeoisie as well as the workers and peasants, and although the working class takes an active part, it does not always lead the struggle. [47]

Civil war and internal war

In the Soviet theory of war, the term internal war applied only occasionally, denotes a general concept that encompasses both types of civil war as well as two other kinds of war: wars of national-liberation, and wars conducted by groups of people with tribal or religious identification for independence or autonomy. Only civil wars, therefore, have a purely social base: the other two kinds of internal war may be related to the class struggle, but their main aims are either national or religious.

194

The study of international war

The treatment of international conflict and war as separate subjects is in many ways a new departure. [48] The main focus is on the various kinds of causes of such a war mentioned above and on the conditions under which these causes may actually lead to war.

Three types

Three types of international conflict and war are usually presented. [49] First are those which are most directly related to the main contradiction of the epoch, the international class struggle. From the contradiction between the two global antagonistic systems are generated conflicts between the two superpowers, between the coalitions of states they are members of, or between individual members of these alliances. Wars may break out when a member of the capitalistic bloc or the bloc as a whole attempts to intervene in the internal affairs of a socialistic state or commits an act of aggression against such a state.

The so-called wars in defence of socialist countries have been paid the most attention; in fact, the focus in the whole Soviet study of war is on these wars, and the preparation for them is the core of Soviet military doctrine and policy. The above notion is interpreted very broadly: it includes not only the possible wars directly waged by the socialist countries against their principal antagonists but also, in the Soviet terminology, assistance to other states in rebuffing aggression and assistance to the working class, to the peoples of colonies and dependent countries, and to the young national states in their liberation struggle. Thus war waged by the Soviet Union for such aims can also be classified as one of this type. [50]

The second contradiction, between forces of imperialism and national-liberation movements, gives rise to conflicts and wars between countries in the Third World and imperialist powers. In addition, imperialism foments war amongst Third World countries. Thirdly, owing to the contradictions inherent in capitalism, wars can occur between the imperialistic powers as well as between medium-sized and small capitalistic states.

Stages in the change of method

One of the main problems dealt with in the new concern for international wars is the mechanism by which political differences develop from peaceful conflict to violent conflict. Seven stages are usually distinguished: [51]

1 The basic contradictions – economic, ideological, political and other – arise.
2 Based on the contradictions, a concrete difference of interest arises in the relations between two (or more) parties.
3 The parties determine their aims and decide on the strategy to achieve

195

them; this may include taking hostile action, which will usually provoke similar reaction.

4 Other states, in particular the superpowers, become involved through the alliances or agreements they have with the parties involved in conflict.

5 The struggle escalates to its most acute political form – an international political crisis.

6 Armed forces are introduced either as a show of force or to take some limited military action.

7 International war breaks out. Although the military operations may be limited to begin with, the situation is pregnant with global nuclear war.

The course of development taken by any given conflict and the exact nature of its particular phases will depend on contemporary changes in the international situation and on the scale and intensity of the means used by both sides. It is not suggested that the boundaries between the different stages are sharply defined nor that they necessarily follow the given sequence; some new stage may arise or one of the suggested ones may not appear at all. Moreover, the sequence is not intended to suggest a process that unfolds with inevitable force; the escalation may stop at some point and de-escalation may take place.

Crisis as the crucial stage

According to this view, crisis is the crucial stage, lying as it does at the watershed between peace and war. [52] Crises do not occur by accident. On the contrary, it is an important part of the grand strategy of imperialism to generate crises in order to achieve some partial political aim or to prepare for a new world war by depriving the enemy of some important assets. In other words, crisis has to some extent become a substitute for war as the main means by which imperialism seeks to achieve its objectives. As such it is now the most acute of the actual confrontation between the superpowers and the alliances led by them.

One important theme stressed by Soviet scholars in their study of crisis is the risk of world war that is involved. With modern communications systems and military–technical facilities, weapons of mass destruction can be concentrated on the crisis area in an unprecedented short time; there is therefore a danger that escalation will take place rapidly. Local crises may appear to be less dangerous since the great powers are not directly involved. There is reason to believe, however, that just because of this fact local wars are more likely to break out as a result of some crisis, since aggressive forces may calculate that through swift action they can achieve their goals before peace-loving forces have time to react. Moreover, one cannot discount the danger that an imperialistic country attracted by the power vacuum might intervene in the hope of improving its power base. Such a step would of course give a local war a more far-reaching international significance. [53]

Nevertheless, the main emphasis in the study of crisis seems to be placed on the possibility of avoiding an automatic escalation into war. Although Soviet scholars characterise the international situation in terms of the three contradictions mentioned above, which are a permanent source of potential conflict and war, they distinguish clearly between such a situation of conflict [54] and actual conflicts, crises, and wars. [55]

Wars of national-liberation

The Soviet study of war encompasses, besides civil war and the three types of international war, wars of national-liberation. The principal political actors in such wars are the oppressed people of a colony or dependent country on the one hand and the imperialistic power, often allied with domestic reactionary forces on the other. The policies the opposing sides pursue in war are respectively to achieve national-liberation and to preserve colonial rule.[56]

Wars of national liberation are treated as a category of war intermediate between international and internal war or as a mixed category. They resemble international wars in that the colonial power is an external state; but the political regime it establishes, and against which the oppressed people struggles, often represents the interests of domestic reactionary forces as well.

The number of wars of national liberation since the Second World War testifies to their importance in the modern era. This accounts for the status of nation (people), together with class and state, as an actor in the Soviet theory of war. The nature of the national-liberation struggle has changed as more than seventy new states have come into being in the last three decades. Now it takes two additional forms.

In the first place, many states that have obtained independence must struggle to defend this much as well as to achieve a truly independent political status. This consists primarily in struggling to build up an independent economy free from the exploitation of foreign capital, and to withstand pressure to become involved in military blocs and to allow foreign powers to set up military bases. This struggle can take the form of armed violence, not least because of the possibility that imperialists intervene militarily. Wars of newly independent states against imperialistic aggressors are in Soviet terminology often combined with the traditional kind of national-liberation wars and given the common name of wars between colonialists and peoples fighting for their independence.

The second new form of national-liberation war, internal in essence, may occur in the course of the struggle for social progress, especially when social transformations are to be made. The social struggle is usually conducted without the use of arms, but if the domestic reactionaries resist violently or especially if the imperialists interfere, the political struggle may escalate into war. Intervention from the imperialist would mean that an essentially internal struggle had also become an international war. Such a step could in turn intensify the internal conflict. [57]

The relationship between civil and international war

Although civil wars, or for that matter any internal wars, are very unlike international wars in many respects – the conditions pertaining at the outbreak of violence, the nature of the conduct of war, and features of the outcome – according to Soviet theory they both have the same primary roots: directly or indirectly, class contradictions and the class struggle underly all kinds of war. This thesis follows from one of the most basic tenets of Marxism that all politics spring from the dialectical relations between classes within particular countries and in the international arena.

The effects of civil and international war on each other

Owing to this fundamental unity, one kind of war may influence the outbreak or course of the other. There are, however, many ways in which this influence may in practice be exerted.

To take the impact of civil war first, when conditions within a country reach the stage where civil war is on the verge of starting (or perhaps has already begun), it is possible that an international war may result. The ruling class may attempt to redirect the tension and dissatisfaction that is mounting either by unleashing an external war or by stepping up preparations for a war it had previously been planning. If one or both of the combatants in a civil war are supported by a foreign country, it is also possible that relations with these countries will deteriorate and lead to armed conflict. Should a foreign country decide to intervene in a civil war, an international war would very likely result.

In other situations, civil war may have an opposite impact on international war. It may make it impossible for the ruling class to start an international war or, alternatively, force it to terminate one which is under way. An international war may also be transformed into a civil war and thereby interrupted.

Similarly, international war may have a stimulating or restraining effect on civil war. By inciting hatred against the external enemy and playing on the national feelings of the population, a government may exploit an external war to weaken the internal struggle and prevent a civil war. But participation in an international war may aggravate the internal situation and weaken the government; in such a situation an internal war may result (e.g. the October Revolution).

In some cases, no direct relation between a civil war and an international one may be apparent. A civil war may break out relatively isolated from the current of international politics, and an international war may have no direct impact on the internal political struggle of the participating countries.

Apart from such direct relationships as these, it is also conceivable that some indirect relationships between international war and internal war may exist. Without actually causing civil strife, international war may nevertheless provide opportunities in which an internal armed struggle may more easily be carried on. The ruling class of a country engaged in an international war may be more vulnerable when its regular forces are sent abroad. Or again, in a country occupied by the enemy, the internal struggle may gain impetus initially as part of the struggle for liberation. Furthermore, in the long run, all changes in the internal situations of particular countries or in the international situation will have indirect effects on future conflicts. Soviet scholars hold, for example, that the success of socialist revolutions reduces the likelihood of international wars by eliminating some of their potential causes.

No uniform relation

It would appear that in the Marxist–Leninist view, the relation between international and civil war cannot be reduced to one general formula. It is accepted as a fact that both types of war spring from similar class roots, but the relation between the two types must be decided from case to case. Generally, the impact of one kind of war upon the other will in the modern era be dependent upon the importance attributed to it by each of the opposing camps and upon the balance of power in the internal and international realm. When waging a civil war, a working class will first take into account its prospects within its own country, but also the impact of the war on the international revolutionary movement; whether or not a country will intervene in an internal conflict in a second country will depend upon its estimation of the consequences of various outcomes of the war, on its real possibilities of suppressing its opponent and so on. Thus, the relation between the two kinds of war in each individual case will depend upon a great many factors.

The mutual impact on the course and outcome of war

If there are no regularities between one kind of war and the outbreak of the other according to Soviet theory, the course and outcome of a civil war will influence and be influenced by the permanent international conflict between two global antagonistic camps. Because of the competition between the two camps, each supports one of the parties in a civil war, that which closest represents the same class interests. The imperialists, who export counter-revolution, support reactionary forces in every way they can, including armed intervention that frequently is undertaken under a collective banner. The socialistic countries have a duty to support peoples struggling for liberation when asked to do so, but they avoid intervening directly with armed forces. The change of the balance of forces in the world in favour of socialism has greatly decreased the usefulness of open armed intervention for the imperialists; the possibility of such moves cannot be excluded, however.

On the other hand, the outcome of civil wars has a direct effect on the international correlation of forces, and thereby constitutes part of the course of the world-wide conflict.

The Chinese variant

The same basic assumptions about war as a continuation and instrument of policy also underly the Chinese concept of the nature of war. War is a form of the class struggle and imperialism is its deepest source in the present epoch. However, the Chinese differ from the Soviets in their assessment of the main contradictions of the epoch and of which policies war is now a continuation. Consequently, they see the role of the Soviet Union in quite a different light. These points of disagreement rest on a third issue, namely the necessary stages countries must go through to achieve the socialist goal. In orthodox Marxism, it was foreseen that with the beginning of industrialisation would come a democratic revolution led by the bourgeoisie, to be followed at a later stage of development by socialist revolution under the proletarian banner. On the basis of their own experiences the Chinese assert that this process can be telescoped by countries with a pre-industrial economy. The first stage, the 'new democratic revolution' is led not by the bourgeoisie but by the revolutionary classes, which thereby create a new kind of state that is an instrument of these classes together. The state develops into a republic under the dictatorship of the proletariat and is then used to help carry out the socialistic revolution. National and social revolution thus become two stages of one revolution. The success of the Chinese Revolution in linking the two stages together is offered as evidence of the validity of the thesis.

The Chinese go so far as to maintain that presently the most important aspect of the world struggle against imperialism is the struggle for national liberation. The wave of revolutions for national liberation is described as the greatest historical development of our epoch since the formation of the world socialistic system. More than the contradiction between the proletariat and the bourgeoisie, or internationally between the socialistic and imperialistic systems, the struggle for national liberation reflects the main contradiction of our epoch. The ultimate victory of socialism is dependent upon its successful outcome. It is, however, an autonomous part of the global revolutionary struggle against imperialism; that its course and outcome are determined by the role of the socialistic system and especially of the Soviet Union in the world correlation of forces is strongly denied by the Chinese.

In a further elaboration of these ideas the Chinese have suggested that the struggle for national liberation is a manifestation of the conflict between the 'world villages' and the 'world cities'. In the original version of this theory the 'villages' consisted of the underdeveloped countries of Asia, Africa, and Latin America, which were thus in a struggle for national liberation and social revolution against the capitalistic countries of North America and Western Europe. The victory of the 'villages' would constitute the first stage of the world revolution, the second being social revolutions in the developed countries.

In a later version, it is both the 'world villages' and the lesser capitalistic countries that are the objects of competition between the two superpowers. The

Chinese maintain that the Soviet Union has ceased to act as a socialistic state and become 'social-imperialistic'; the contradictions between the superpowers are no longer based on the ideological struggle between socialism and capitalism but on the imperialistic aim of both to dominate the rest of the world. Where their aggression succeeds, they create oppressed states. This aggression breeds wars by which the oppressed peoples try to win national independence, or by which the remaining states try to preserve theirs.

To sum up, the Chinese assert that wars in the modern epoch have three dimensions: firstly, they are a continuation of the policy of the direct participants; secondly, they are part of the political confrontation between the two superpowers; and thirdly, they are part of the struggle of the peoples of the world for national and social liberation. These three dimensions may coincide as in the great struggle of the countries of the Third World.

Notes

[1] *Polnoe sobranie sochinenii,* Moscow, vol. 26, p.224.

[2] *Voennaya Strategiya,* V. D. Sokolovii (ed.), 3rd. ed., 1968, pp.209–10. All following quotations have been taken from this edition.

[3] *Deutsches Militärlexikon,* Deutscher Militärverlag, Berlin 1961, p.222. 'This German military dictionary is the first attempt to define the most important military and military-political terms on the basis of the Marxist–Leninist military science', from Preface, p.6.

[4] 'War is an armed violence, an organized armed struggle between various social classes or states for the achievement of definite political goals' (D. Palevich, 'Kharakter i osobennosti sovremennoi voiny', *Kommunist Vooruzhennykh Sil,* 1962:21, p.77); '. . . the essence of war is politics continued by means of violence' (Kozlov, 1971).

'. . . Every war is a link in one political chain, an extension and continuation of the policy of a particular class and state at a given time' (*Problems of War and Peace,* p.64).

For an analysis of war as a continuation of politics, see Rybkin, 1973, 1974; Tyushkevich, 1975; *Marxism–Leninism on War and Army,* 1972, Milovidov, Kozlov, 1972 (cited in other notes as *Filosofskoe nasledie V.I. Lenina. . .*).

[5] Each war as a continuation of policy is class in its essence (Kozlov, 1968).

[6] 'War: armed fighting (stolknovenie) between states (coalitions of states) or between antagonistic classes in the state (civil war) for achievement of their political goals. In its substance each war is a continuation of policy of the interested states and their ruling classes by violent means.' (*Tolkovyi slovar' voennykh terminov,* P I Skubeda, Voenizdat, Moscow 1966, pp.99–100).

[7] Azovtsev, 1971, p.23; *Marksizm–Leninizm o voine i armii,* 5th ed., 1968, p.13; *Marxism–Leninism on War and Army,* 1972, pp.8–9; the main theme of Rybkin, 1959.

[8] Bolshaya Sovetskaya Entsiklopedia, 2nd ed., vol. 33. It is a type of relations between classes and peoples which clearly expresses the inter-class relations with respect to the state-authority. This aspect of relations is the most fundamental, it corresponds to the basic interests of these classes (A. Vishnyakow, 'Politika KPSS: nekotorye teoreticheskie voprosy', *Kommunist,* 1973:10, p.52).

[9] *Voennaya strategiva,* pp.207–9; P. I. Trifonenkov, 'Voennoistoricheskoe nasledie V.I. Lenina i sovremennost'', *Voenno-Istoricheskii Zhurnal,* 1967:11; G. E. Glezerman, 'Klass i natsiya', in *Sotsiologicheskie problemy mezhdunarodnykh otnoshenii,* Izd 'Nauka', Moscow 1970, pp.175–87; '. . . politics is the relations between classes, states, and nations' (Y. Sulimov, 'Nauchnyi kharakter wneshnei politiki KPSS', *Krasnaya Zvezda,* 20 December 1973); 'Politics . . . covers relations between the classes, nations, states, between the different social systems' (T. Kondratkov, 'War and Politics', *Soviet Military Review,* 1971:10, p.4).

[10] 'Politics expresses relations between different classes, between classes and their organizations and establishments that represent their interests, of which the state with all its agencies is the main one.' (A. Sergeyev, 'Bourgeois Pseudo-Science about the Future' *International Affairs,* Moscow, 1972:2, p.83); cf. F. Konstantinov, 'Sotsiologiya i politika', in *Marksistskaya i burzhuaznaya sotsiologiya segodniyya,* Moscow 1964, p.15).

[11] Engels' functional definition of the state is well known: 'an organization of the possessing class for its protection against the non-possessing class' ('The Origin of Family, Private Property and State', in Marx, Engels, *Selected Works,* II, p.291); 'As the Marxist–Leninist science indisputably proved, the dictatorship of the economically governing class is the essence of each state; it (the class-JL) uses the political power for the protection of the existing economic system and for the suppression of the resistance of its class antagonists' (*Istoricheskii materializm i sotsialnaya filosofiya sovremennoi burzhuazii,* p.378); the capitalist state is an instrument of the capitalist class, and in the modern capitalism it is an instrument of the monopolist groups (Gus Hall, *Der amerikanische Imperialismus in der Welt von heute,* Dietz Verlag, Berlin 1973, p.103).

[12] Foreign politics is the extension of the internal politics of the state beyond its territorial boundaries (Rybkin, 1959).

[13] Kondratkov, 1971, p.4.

[14] J. Stalin, *Zagadnienia leninizmu* (Problems of Leninism), KiW, Warsaw 1949, p.605.

[15] E. Nikitin, V. Konovalov, 'Dostoyanie mirovogo revolutsionnogo kvizheniya', *Kommunist Vooruzhenykh Sil,* 1973:12, pp.13–14; Kh. Shaknazarov, *Sotsialisticheskaya demokratiya,* Politizdat, Moscow 1972; some of these ideas are repeated in the critique of Chinese theories; the latter are said to hold that class struggle is the motivating force also in a socialist state and that such a state retains its repressive functions (V. Sidikhmenov, 'The Anti-Leninist

Essence of the "Theory of the Continuation of the Revolution under the Dictatorship of the Proletariat"', *Problemy Dalnego Vostoka*, 1973:3.

For an analysis of the development and transformation of the functions of the socialist state, see the study material in the Soviet armed forces: 'Sovetskoe sotsialisticheskoe gosudarstvo – glavnoe oruzhie postroeniya kommunizma i zashchity ot imperialisticheskoi aggressii', *KVS*, 1976:24; it is held that the internal function of armed forces has completely disappeared.

[16] V. Gavrilov, 'The Soviet Union and the International Relations System', *Mirovaya Ekonomika i Mezhdunarodnye Otnosheniya*, 1972:12; D. Tomashevskii, 'The Class Nature of the Soviet Foreign Policy', *Krasnaya Zvezda*, 1 September 1972.

[17] The term 'coalitionary policy' is frequently used. See V. Zemskov, 'Peredovoi kharakter sotsialisticheskoi voennoi doktriny', *Voennaya Mysl*, 1971:1; 'The countries of the socialist community have a common coordinated policy practically on every principal question of world politics' (L. Brezhnev, to the 15th Congress of Soviet Trade Unions', *Daily Review*, 21 March 1972).

[18] 'Each war has a class nature, but not each one is a type or a form of the class struggle. Thus, the numerous Russo–Turkish wars, the "reconquista" in Spain ... and also the First World War did not represent a conflict between classes.' (Rybkin, 1973).

[19] 'When antagonisms were sharply aggravated, they (classes, states, nations – JL) resorted to forms of armed conflict – to war' (*Voennaya Strategiya*, p.209).

[20] '... civil war is the most acute form of class conflict when a series of economic and political clashes which repeatedly occur, gather momentum, spread, become more intense, and are transformed into an acute and armed conflict' (V. I. Lenin, *Sochineniya*, vol. 26, p.11).

[21] Nazi–Germany politics is quoted as an example. When the strategy of capturing other countries without war failed (Poland), Germany resorted to war.

[22] '... wars were the most organized and purposeful undertakings the spontaneously developing societies ever carried out.' (*Marxism–Leninism on War and Army*, p.14).

[23] 'During war, the armed forces and armed combat are the basic and decisive instruments of policy. All other instruments are directed primarily toward supporting the armed forces and the other military formation, created by extensive enlistment of masses of the population, in order to attain political goals by armed violence.' (*Voennaya strategiya*, p.212).

[24] Kende, 1972, p.11.

[25] Azovtsev, 1971, p.26; Marxism–Leninism on War and Army, pp.11–12; V. Zemskov, 'Voiny sovremennoi epokhi', *Voennaya Mysl'*, 1969:5; '... in spite of the variety of wars waged under the most different historical circumstances and independently of the character of means and measures applied in armed combat, all of them [was] were a part of a whole – of politics, and its continua-

tion'. (V. Kozlov, 'Kharakter i osobennosti sovremennoi voiny', *Kommunist Vooruzhenykh Sil,* 1969:19, p.73); 'war is not synonymous with politics in general, but comprises only a part of it ... politics, in addition to war, commands a large arsenal of various non-violent means, which it can enlist to attain its goals, without resort to war' (*Voennaya strategiya*, p.210).

[26] After the Nazis' victory in Germany the Soviet Union started a many-sided struggle against them – ideological, diplomatic, economic, and military. Thereby all means, including military, were applied (V. Baskakov, 'O sootnoshenii voiny kak obshche stvennogo yavleniya i vooruzhennoi bor'by', *Kommunist Vooruzhennykh Sil*, 1971:1, p.40).

[27] 'War always includes the armed struggle of two sides and thereby essentially differs from all other political acts in which force is employed.' (Rybkin, 1973).

[28] S. Tyushkevich, 'Politicheskie tseli i kharakter voiny', *Kommunist Vooruzhennykh Sil,* 1969:7; Marxism–Leninism on War and Army, pp.14–15. B. Trushin, R. Lipkin, 'Imperialism – istochnik voennoi opasnosti', *Voennyi Vestnik,* 1975:11 (recommended as study material for commanders); 'Under the modern conditions imperialism is the only source of war' ('Voina' [war], in *Sovetskaya Voennaya Entsiklopediya*, vol.1, 1976, p.283).

[29] One of the main themes of *Mezhdunarodnye konflikty,* 1972.

[30] Dimitrij Jermolenko, 'Zur Frage der Methodologie der Erforschung internationaler Konflikte', in Bredow, 1973, p.106.

[31] *Mezhdunarodnye konflikty*, pp.32–6. 'Whatever the particular or local character of a given international conflict, war between the two systems, which has become the core of modern international relations, is reflected in it.' (p.32).

[32] Sovremennaya epokha i mirovoi revolutsionnyi protsess, p.15.

[33] But the working class has not been stratified, and the following two principles of class struggle have remained: (a) The leading role of the working class, which steadily increases, (b) The determining role of its class struggle which remains the main front of social struggle, and also the role of national-liberation movement (V. Yevgenev, 'Class Analysis vs. Anti-scientific schemes', *Rabochii Klass i Sovremennyi Mir,* 26 January 1973).

[34] E.g. Yermolenko, ibid. The German terms are: Hauptursachen, wesentliche Ursachen, nächsten und die Endursachen, unmittelbaren Ursachen, der Anlass.

Tyushkevich, 1975, presents a similar classification of causes, but connected with the trinitarian concept of the nature of war (the unity of the general, particular and individual (see ch.xii.3) Each war is generated by three kinds of causes: general, rooted in the economic and political conflicts characteristic of all class antagonistic formations; particular, specific to the given socio-economic formation, and individual, generated by the complex of concrete objective and subjective factors in the given circumstances. These three kinds of causes are reflected in the activity of the socio-political forces which are interested in war (ch.II, esp. p.5).

[35] Rybkin, 1973.

[36] 'Politics prepares and generates war, formulates its goals, determines, where and when to start, wage, and to terminate it. It affects the choice of methods, of ways and forms of armed combat, its size and intensity, it constitutes the basis of the military art, of strategy and tactics. Politics is materialized in the class character, distinctive features and historical designation of armed forces, it defines their organizational structure, the principles of education and training of the troops.' (Shelyag, Kondratkov, p.10); 'Political influence on [war] is manifested in the determination of general and particular strategic aims, in the general nature of state strategy and in the selection of the methods and forms of waging war' (*Voennaya strategiya*, p.26).

[37] S. Kozlov, 'Nekotorye voprosy teorii strategii', *Voennaya Mysl'* 1954:11, p.23.

[38] The more decisive goals at which armed forces aim, the more intense and sharp is the struggle. Policy also influences such concrete strategic and operational decisions, as the deployment of armed forces, their moves, the direction of the strategic strikes, the use of armament, etc. (N. N. Azovtsev, I. Lenin – osnovopolozhnik sovetskoi voennoi nauki', *Kommunist Vooruzhennykh Sil*, 1968:18, p.18); for similar statements, see Pukhovskii, 1965, ch.6 'Politika i voennaya strategiya'; 'Politics not only determines the character of a given war, but also affects the methods of conducting military operations and of war as a whole. The policy of a given state in war determines the character of strategy, of tactics, of organization and preparation of armed forces and formulates the concrete goals. Political strategy, on which the course and outcome of war depends, is the basis of military strategy and tactics.' ('Voina' in *Bolshaya Sovetskaya Entsiklopediya*, 2nd ed., vol. 8, Moscow 1951).

[39] All Soviet scholars repeat this assertion. 'War unleashed by imperialists against socialist countries will be a decisive clash of two opposed socio-political systems and the basic contradition of our epoch, that between socialism and imperialism, will be resolved by it. War will assume a fierce, acute class character, which will predetermine the extreme decisiveness of goals and plans, methods and forms of its conduct. It will take on an unprecedently widespread scope, and intercontinental character . . . it will be characterized by unprecedent destruction and by annihilating actions.' (V.I. Lenin i *Sovetskie Vooruzhennye Sily*, Voenizdat, Moscow 1969).

[40] '. . . unified plan of strategy among capitalist states in a coalition war can be achieved only by compromises, mutual concessions, or dictation by the strongest countries. American dictation creates the unified strategy of present-day imperialist coalitions, whose primary purpose is to satisfy American military and political aims. Quite understandably the irreconcilable contradictions inherent in capitalist society make it impossible to achieve complete strategic unity in imperialist blocs and coalitions.' (*Voennaya Strategiya*, p.35).

[41] *Voennaya strategiya*, pp.24–37; S. Ivanov, 'V. I. Lenin i sovetskaya

voennaya strategiya', *Kommunist Vooruzhennykh Sil,* 1970:8, pp.53–4.

[42] Lenin made reference to the political strategy when he spoke about the military strategy and vice versa (Ivanov, ibid.). Shelyag and Kondratkov write, that from Lenin's definition of war as a continuation of politcs one can conclude that there is a very close connection betwee.i the political and military strategies (ibid., p.12).

[43] N. Khmara, Nekotorye osobennosti grazhdanskikh voin v sovremennuyu epokhu, *Kommunist Vooruzhennykh Sil,* 1971:16, p.17; cf. *Marxism–Leninism on War and Army,* pp.78 ff; 'Civil war: armed hostilities between the progressive and the reactionary class forces within a country for the (state) power (*Kleines Politisches Wörterbuch,* Dietz Verlag, Berlin 1967, p.110); 'Civil war is the armed struggle between the antagonistic classes of a country, a struggle for the state power by means of violence' (*Marxism–Leninism on War and Army,* p.79).

[44] The extremely offensive character of all military actions together with the permanency of combat and quick change of methods are considered to be characteristic of such a war; it was repeatedly stated that defence means the death of the uprising.

[45] Lenin is quoted as saying that it is impossible to seize political power until the struggle has reached a certain stage which will be different in different countries and in different circumstances.

[46] Other definitions or descriptions: civil war 'between the people and the regime of extreme reactionaries, of facist type, monarchic type, etc.' (Rybkin, 1970, p.11); 'civil wars between the peoples of particular countries and reactionary forces within them' (S. Tyushkevich, 'Politicheskie tseli i kharakter voiny', *Kommunist Vooruzhennykh Sil,* 1969:7, p.32); wars between peoples and regimes of extreme reactionaries, based on monopolies (K. Stepanov, Y. Rybkin, 'O kharaktere i tipakh voin sovremennoi epokhi', *Voennaya Mysl',* 1968:2, p.68).

[47] In some studies democratic civil wars are included into the classification as a quite new type ('Marksizm–Leninizm o voine i armii', 1968; Tyushkevich, 1969, p.32) in others, however, they are treated as one variant of civil wars (Rybkin, 1974). The authors of *Marxism–Leninism on War and Army* write about 5 types of main political contradictions, but about 4 types of war.

[48] The authors of the first study devoted entirely to the theory of international conflicts write of themselves as 'the only researchers who apply the methodology of international conflict analysis' and they consider their work as the beginning of a long-ranged study of the subject (*Mezhdunarodnye konflikty,* p.10). International conflicts were the main topic of the VIth International Sociological Congress in Evains; Soviet studies prepared for the congress were published in: Sotsiologicheskie problemy mezhdunarodnykh otnoshenii, Izd. 'Nauka', Moscow 1970 (esp. D. V. Yermolenko, 'Mezhdunarodnyi konflikt kak obiekt sotsiologicheskikh issledovanii', pp.103–26); international conflicts were also the topic of an international symposium in Vienna (15–17 December 1972), and the Soviet papers were published, among others in von Bredow, 1973. Cf.

Voennaya sila i mezhdunarodnye otnosheniya, 1972; V. Zhurkin, 'Détente and International Conflicts', *International Affairs,* (Moscow), 1974:7.

[49] *Mezhdunarodnye konflikty,* pp.34–6. In another study five groups of international contradictions as sources of international conflicts and crises are listed, contradictions between imperialists and countries of the Third World, as well as between these countries themselves are treated separately from the contradiction between imperialism and national-liberation movement (Zhurkin, 'Internationale Konflikte und Krisen, Wege zu ihrer Verhinderung und Regelung', in von Bredow, 1973, p.59). In both the above studies it is held that a non-typical conflict may originate from the activity of the nationalistic and chauvinistic forces which have risen to power in some socialist countries (China). (See also Zhurkin, 1974, p.90.)

[50] *Marxism–Leninism on War and Army*, Chapter 3.

[51] *Mezhdunarodnye konflikty,* pp.50–4.

[52] Also in other models of the 'conflict ladder', containing three degrees: conflict, preceded by the emergence of contradiction of interests – crisis (and/or limited armed conflict) – world war.

[53] International political crises were in the past preludes to great wars, but the latter might have been unleashed in various other ways; now, however, crises have become the main means of unleashing a global nuclear missile war, since neither a conscious long-planned surprise attack, nor an accidental or catalytic war is probable (Zhurkin, 1973, p.63–4).

[54] The Russian term 'konfliktnost' (*Mezhdunarodnye konflikty* p.30).

[55] This is inherent in the definitions of conflict. 'Generally speaking, conflict is a case of a sharp aggravation of contradictions. Social conflict, carried on to the extreme, results in armed struggle (armed combat, revolutionary upheaval, civil war etc.). Conflict in international relations means worsening of relations between two states or groups of states, or worsening of international relations resulting from a radical change within one or some states.' (Yermolenko, 1970, p.105). As we see, the author emphasises the internal roots of international conflicts.

[56] *Marxism–Leninism on War and Army,* ch.2, 4.

[57] For the literature on the subject, cf. note 12 in ch.7.

10   The historical character of war: war as a transitory phenomenon

Marxism–Leninism defines war as an instrument in the political class struggle, and consequently contends that war is a social phenomenon coexistent with class society. In other words, wars first occurred once class societies had emerged, and they will disappear when class society is eliminated. [1]

The origin of war as a social phenomenon

Marxism–Leninism thus asumes that there was no war in primitive societies. [2] There were undoubtedly quarrels amongst primitive peoples regardless of whether their economic structure was based on herding animals or hunting them, and armed clashes must sometimes have broken out. But such conflicts do not meet the two conditions necessary to classify them as war. In the first place, the armed clashes themselves took place spontaneously; they were not part of a general political strategy, for according to Marxism–Leninism politics is essentially the pursuit of class interests and at this time there were not yet any classes. In the second place, such conflicts engaged entire tribes. There was no separate organisation distinct from the tribe as a whole charged with the job of fighting.

The first wars were therefore not fought until the first class state emerged in history. The socio-economic foundation of this state was slavery, and war quickly became a means by which economic interests of the exploiting class were pursued. Wars were one of the main ways by which the slaves, which constituted the human means of production at that time, were recruited.

In all subsequent class societies, war continued to be a normal and indispensable instrument by which the exploiting classes preserved and strengthened their position through expansion.

A few single studies have proposed some modifications of this classical exposition of the origin of war. One point concerns the way in which war first appeared as a social phenomenon. One proposes to regard the development between social stages as a long process in which classes, states, and armies are gradually formed. [3] It is therefore impossible to give the exact point in time when the first war was waged. What can be said is that when class society definitely emerged, war had developed as a normal instrument of political behaviour.

A second proposal suggests that armed clashes in primitive society must be considered to be a predecessor of war in their technical-military aspects. [4] Although they were not fought by a special military organisation separate from the rest of society, the actual fighting resembled the conduct of war. This is the reason Marx and Engels sometimes refer to war in primitive societies. [5] To be exact, it is only the birth of war as a socio-political phenomenon that is contingent upon the emergence of a class society. This same idea is expressed in a somewhat different form when it is suggested that fighting in primitive societies is an initial form of war; [6] or again, in an expression used in a very authoritative study, that with the division of society into classes, war became a 'permanent phenomenon in the life of nations' – the use of permanent instead of new may well indicate the author's intention to suggest that some wars did actually take place in primitive societies. [7]

The traditional view is still the prevailing one, however. [8] Most new studies emphasise that fighting in primitive societies cannot be considered a war *sensu stricto*, i.e. a continuation of policy or an indispensable instrument for achieving political and economic goals, [9] nor as a regular and normal function of a state performed by special institutionalised armed forces.

The future of war

The classical exposition

The classical exposition of the second assumption is just as simple and categorical: war will disappear when class society disappears, that is, when all states have become communist ones. No reason for war can exist once classes, in whose interests all wars are fought, cease to exist. [10]

When the classical doctrine of the class nature of war was modified to take into account the national question, [11] a supplementary proposition to explain the disappearance of national sources of war was added: at the same time as states mature into communism, all nations will become equally sovereign and will thereby acquire equal opportunities to develop in all respects, including economically and culturally. Differences between nations with regard to their level of development will therefore disappear. [12]

The orthodox Marxist view about the future of war is thus an optimistic one, at least in regard to man's ultimate destiny. But what about the future of war before the ultimate victory of socialism is declared?

During the first half century after the triumph of the first socialist revolution, the orthodox belief that communist society would be free of war was considered to have the corollary that before the ultimate victory or socialism in the whole, or virtually the whole world, wars are inevitable. With the complex of new political, technological, and psychological conditions ushered in by the nuclear missile era, and the subsequent new policy of the Soviet Union according to which peaceful coexistence is the main basis of relations between countries with different social systems, has come a need for a new approach to the role that war plays in a mixed world consisting of two opposing social systems plus the Third

World countries. Although the general principle that there can be no war in a communist society could not be changed, the problem of the inevitability of war before the ultimate victory of socialism required reexamination.

The process leading up to the abandonment of the theory of the inevitability of war is the subject of chapter 12. Here the modification of the orthodox position will be set in relation to the modern interpretation given to the concept of the transitional epoch.

War in the Mixed World: a gradual process of disappearance

It is the basic tenet of Marxism that man's historical progress passes through five socio-economic stages beginning with primitive society and culminating in Communism, and that the transition from one stage to another is accomplished by revolution. These basic ideas are still accepted, but the notion of transition is now interpreted less as a single revolutionary act than as a process, or what is more, as a complex of processes differing greatly in various parts of the world and together constituting a transitional epoch. [13]

It is asserted that in such a transitional epoch, elements of the new socio-economic and political structure will attain a different level of development at different times in different countries; it is conceivable that some countries have fully matured into the new system while others are still at an earlier stage and yet others have not even begun. The pattern of military activities in such an epoch will accordingly be very diversified.

When these general ideas are applied to our own time (called by the Soviets the transitional epoch from capitalism to socialism) a great variety of stages in the revolution can be discerned; in some countries socialism is already victorious, in others it has been temporarily defeated, and in still others the capitalist system itself has only reached an early stage of development. Despite this great array of shapes in which the two basic socio-economic systems coexist, the tendency towards socialism is clear.

It follows that if the transition from capitalism, in which war is indispensable, to socialism, in which war is eliminated, is a process, then war should gradually disappear in this transitional epoch. The stronger the socialist system becomes, the greater the number of new countries it includes, and the greater the number of socialistic elements in embryonic form within the capitalist countries, the fewer will be both the number and variety of wars. It is thus possible to argue that war between the two main systems is now unlikely, that national-liberation wars have gradually become unnecessary, that imperialist colonial intervention has become difficult to perform, and even that revolutionary war may in some circumstances also be unnecessary.

In addition, Soviet politicians and scholars note that the transition is being marked by the growth of a new moral and psychological climate in which war is becoming unacceptable as a means of resolving international conflicts. A new awareness of the folly of trying to prevent war by means of the arms race is becoming more widespread within political circles and amongst the broad mass of people. This change in attitude can in part be explained by the fact that nuclear war cannot lead to any desired political results.

These developments have important consequences for the struggle for peace. It is still essential to be alert to all instances of potential war and to act to prevent each concrete war. But in the transitional era, the very source of war, imperialism, has begun to decay, so the risk of the outbreak of war has also lessened. Whereas the struggle for peace previously could only succeed in delaying the outbreak of some war or occasionally in preventing one, now it is possible to eliminate all wars even before imperialism is completely destroyed. Although the occurrence of war remains a theoretical possibility, in practice it is possible to eliminate war from social life. This conviction has also been reflected in the proposals for peaceful coexistence that the Soviet Union has put forward. Initially it was suggested that a ban be put on one type of war, namely world war; now the Soviets believe it is possible to go a step further and ban all interstate war.

War within the socialist system

According to the orthodox Soviet view, all domestic sources of war in some country will disappear once the socialist revolution has successfully been accomplished there. It is possible that the counter-revolutionaries may instigate war, and one socialist state may send brotherly assistance to a second socialist state to help it quash subversive elements. But war between socialist states is unthinkable. When Soviet scholars discuss the possibility of war between China and the Soviet Union, they note the situation would be an atypical one since in China the anti-socialist circles have temporarily usurped power. [14]

To digress, some non-Soviet Marxists have pointed out, however, that in the transitional period there will be many contradictions and conflicts both within and between socialist societies, and that there is always a risk that these will lead to war.

Within socialist states the material inequalities between citizens will take a long time to eliminate. Therefore the old contradictions between individuals, between individuals and the state, and between specific interest groups will remain. They may even be aggravated by the increased activity of the state in economic affairs and other social spheres. For the same reason, new conflicts will be generated. The process of building socialism and afterwards communism, will require that the old system of values and relations be radically changed which takes time and generates conflicts. Furthermore, since the first step towards building the new society will probably be taken after the collapse of the old system following a civil war or an economic breakdown, the first period of socialism may be plagued by material shortages, which may simply add to existing difficulties.

Similarly with regard to relations between socialist states, the process of evening out national inequalities will be a long one. In the meantime, expressions of national egoism and tendencies for one state to dominate over others or even to exploit them will continue to occur. Paradoxically, as the capitalist system becomes weaker and weaker and the need for a united policy from the socialist countries becomes less urgent, the contradictions between socialist countries may will become more open. The possibility of armed conflict between socialist states has been discussed by Yugoslavian scholars. Their interest in the question was aroused at a time when relations between their country and the Soviet Union had deteriorated owing, they argued, to the hegemonial policy of the USSR. [15]

Some theoretical implications of peaceful coexistence

Implied in the concept of a transitional epoch in which war will gradually disappear, is a somewhat modified view of the essential nature of the capitalist state. Orthodox Marxism would have it that the capitalist state, like all class states, uses violence as an instrument to maintain both internal rule and continuous external expansion; warring is as natural, regular, and vital for such a state as breathing is for human mortals. To suggest that capitalist states might abide by and even support actively a policy of peaceful coexistence, that is to say that they cut themselves off from one of their life-giving functions, is to admit that they have become changed in some essential respect.

Soviet scholars who describe peaceful coexistence usually deny that any fundamental change has occurred in the capitalist states and argue that it is the change in circumstances (technology, public opinion, the superiority of the socialist states etc.) that has made war inadvisable for the capitalist states. But it is difficult to reconcile this argument with the historical materialism on which Marx constructed his whole social analysis: since all social phenomena and institutions are rooted in the contemporary material conditions and circumstances, essential changes in the latter cannot fail to be reflected in essential changes in the former.

The problem also has wider ramifications. In orthodox Marxist theory, the historical character of war is closely bound up with the historical character of classes, states, and politics. The concept of class is the base in relation to which the other three terms are defined: the state is an instrument of a class, politics is the activity of a class state, and war is a means by which such policy is carried out. But class is itself a social phenomenon with definite temporal limits, for man's social progress is moving inexorably towards the classless society. When there is no class, there can be no state, no politics, nor any war. However, if the emergence or disappearance of classes is treated as a process, then the same must be done with states, politics, and war. That being the case, the concrete manifestations of state, politics, and war in the present transitional epoch will differ to a greater or lesser degree from the state, politics, and war of the pure capitalist system.

To sum up, the Soviet position is optimistic about eliminating war not only when man reaches the end-state of his social progress but also now. The process of eliminating war is not an automatic one, however. The so-called reactionary elements in capitalist countries are said to create difficulties on the road to *détente,* disarmament, and elimination of all interstate wars, but as capitalism declines these circles may lose their political influence and more progressive elements may take over.

Notes

[1] 'Voina' (war) in *Bolshaia Sovetskaya Entsiklopedia,* 2nd ed. vol. 8, 1951: 'Voina', in *Sovetskaya Istoricheskaya Entsiklopedia,* vol. 3, Izd. *Sovetskaya Entsiklopedia,* Moscow 1963, pp.619 ff. 'Voina' in *Sovetskaya Voennya Entsiklopediya,* vol. 1., Moscow 1976, pp.282–5. F. Khrustov, 'Marksizm–Leninizm o voine', in *Ob istoricheskom materializmie,* F. Konstantinov, ed, Akademiya Nauk SSSR, Moscow 1950; E. Usenko, *Prichiny imperialisticheskikh voin,* Voenizdat, Moscow 1953; *Marxism–Leninism on War and Army,* 1972, pp.5ff., Prokop' ev, 1965; Pukhovskii, 1965; *Tolkovyi slovar' voennykh terminov,* Voenizdat Moscow 1966, p.100.

[2] 'The theory of Marxism–Leninism has proved, that in the society of the primitive community, where no states and classes existed, no wars occurred, as an organized armed struggle for political goals. When state appeared wars also appeared. War became a function of states, which represented interests of the governing strata; it became a means of strengthening and extending their domination and oppression of other peoples' ('Voina', 1951). 'Marxism–Leninism considers war as a socio-historical phenomenon proper only to the class socio-economic formations. In primitive society war in the proper meaning of the term, as an organized armed struggle of a political character, didn't exist.' ('Voina', 1963, p.619).

[3] 'War emerged as a socio-political phenomenon at a definite stage of social development, namely with the disintegration of the primeval system and the emergence of the slave-owning mode of production, when private ownership of the means of production appeared, when society was divided into antagonistic classes, and the state emerged.' (*Marxism–Leninism on War and Army,* p.6).

[4] *Marxism–Leninism on War and Army,* p.6.

[5] Dela, vol. 21, p.99. Engels writes about wars conducted by Indian, German and other tribes before the emergence of the class-society (The origin of the family, private ownership, and state).

[6] 'There was a war in its initial form as a social phenomenon in the tribe

societies, but it developed only when state and division of the society into classes have appeared. When means of production developed and as well as the contradictions which resulted from them, the ends of war became bigger and so also means of its conduct' (*The Soviet Encyclopedia of 1969*, vol. 2, p.245).

[7] *Voennaya strategiya*, p.12. In the first edition the expression inevitable was used (p.7); afterwards it was replaced by permanent. In another publication war was said to become necessary, when the first class-society emerged (which was not identical with the adjective new) (*Marksizm–Leninizm o voine i armii*, 1968, p.10), but in the next edition the subtlety disappeared. (*Marxism–Leninism on War and Army*, pp.5–6).

However, in *Sovetskaya Voennaya Entsiklopediya*, 1976, the statement that Marxism–Leninism regards war as a socio-political phenomenon specific only to the class socio-economic formation is followed by an assertion that in primitive society there was no war 'in the modern meaning of the term' ('Voina' vol. 1, p.282).

A non-Soviet scholar has put forward a far-reaching proposal that one should distinguish between two senses of the idea of war, a narrower one confined to armed violence between states, nations, or classes, and a broader one encompassing fighting between all forms of 'global societies', i.e. classes, tribes, and states (Karel Kara, 'On the Marxist Theory of War and Peace', *Journal of Peace Research*, 1968:1).

[8] 'War is a social, concrete, historical phenomenon. Wars emerged with the appearance of the private ownership of the means of production, when society was divided into antagonistic classes. In the primitive society wars didn't exist, nor special organisation of armed people. Armed clashes between particular tribes [rody i plemiena] which sometimes occurred in that society, couldn't be called wars, because they didn't stem from the character of social relations. These armed clashes differed by their character principally from wars, characteristic of the antagonistic class formations.' (Strokov, 1965, p.648):

'The history of mankind testifies that wars originated only when the means and instruments of production became privately owned, when society was divided into antagonistic classes, giving rise to states. The exploiting classes – slave owners, feudal lords, and capitalists – have always regarded the apparatus of coercion, primarily the army, as an indispensable attribute of power, and war as a 'legitimate' method for plundering other peoples and consolidating their own positions inside the country and in the world arena' (K. Bochkariev, 'Wars: Their Sources and Causes', *Soviet Military Review*, 1972:5, p.8). 'Marxism–Leninism regards war as a socio-historical phenomenon. Wars appeared with the advent of private ownership and the division of the society into hostile, antagonistic classes: slave owners and slaves, feudal lords and serfs, capitalists and hired labourers. Those who owned wealth made every attempt to multiply it. Those who were exploited and plundered rebelled against their exploiters. It was the exploiting classes that legalized organized armed struggle for the sake of profit, enslaving the peoples and buttressing their rule.' (T.

Kondratkov, 'War as a continuation of policy', *Soviet Military Review,* 1974:2, p.7).

[9] Hostilities broke out only when tribes searched for food and pastures. In a sense it was a supplement to the productive work performed by the whole collective body, a way for achieving means of existence, and not an instrument of increasing wealth and of domination (Rybkin, p.17). Marx's utterance, that war was one of the first kinds of labour, was frequently quoted (*Formy predshestvovavshie kapitalisticheskomu proizvodstvu,* Moscow 1940, p.24).

[10] Wars occur only in class societies; they will disappear together with classes (Strokov, 1965, p.648); 'for its class nature war is a phenomenon inherent only in class society' (Rybkin, 1973, p.1); cf. *Filosofskoe nasledie V.I. Lenina . . .* p.37 'It is only with the abolition of private ownership of the means of production, the division of society into classes, and the exploitation of man by man and of one nation by another that social antagonisms will be removed and all wars will be made impossible.' (Kazakov, 1975, p.II).

[11] I.e. national oppression was included as one of the main causes of wars. This reflected a more general proposition: presocialistic socio-economic formations were considered as characterised by the enslavement not only of one class by another one but also of one nation by another nation (Barsegov, Khairov, 1973, p.74).

[12] Nikolskii, 1964, pp.149 ff.

[13] Istoricheskaya nauka i nekotorye problemy sovremennosti. Stat'i i obsuzhdeniya, Izd. 'Nauka', Moscow 1969; A. M. Kovalev, ed, *Sovremennaya epokha i mirovoi revolutsionnyi protess,* Izd. Moskovskogo Universiteta, Moscow 1970; M. Khalipov, 'Sovremennaya epokha i ee osnovnoe protivorechie', *Kommunist Vooruzhennykh Sil,* 1975:9; Yu. Sumbatyan, 'mirovoi revolutsionnyi protsess, ego cherty i dvizhushchie sily', *Kommunist Vooruzhennykh Sil,* 1976:21.

[14] *Mezhdunarodnye konflikty.*

[15] Kardelj, 1961 ch. 'War and Socialism'; Kara, 1968; Brucan, 1971; 'History reveals, that violent social revolutions which alter the class structure of society do not result, even after decades, in changes in traditional national behaviour in international relations.' (Brucan, p.84).

11 The assessment of war: unity of three characteristics

The three kinds of assessment

In the Soviet study of war, the first problem to be settled when forming a judgement on some particular war is to assess its political character – which social forces are waging the war and for what political aims. A more complete picture might be obtained by relating the particular war to the main types of war to which the characteristic contradictions of the epoch give rise. The political assessment would thus include a description of the two warring parties defined in socio-political terms (usually classes or states, sometimes nations), and the war aims of each, for example, an interimperialist war for the redivision of colonies, or a war between the proletariat and bourgeoisie respectively to seize and to maintain state power. [1]

The second step is to determine the socio-historical characteristics of the given war, that is to say its role in the history of man's social progress. Here too the terms used are ostensibly descriptive, but actually no answer can be given this question before a criterion of social progress has been chosen. Marxism–Leninism posits that man strives to attain a socio-economic formation in which his human qualities are allowed free reign for development (that is, communist society). To achieve this purpose he must pass through various stages of socio-economic development, each stage being a prerequisite of the next. Transition from one stage to another is thus not only inevitable: it is also desirable since it means that man moves closer to the best of all possible worlds.

Each pre-socialist socio-economic formation is defined according to the governing socio-economic and political system and is characterised by some central contradiction (in accordance with the dialectical view of history). For each formation there is one class that is regarded as the carrier of progress because it is most interested in struggling for change. In the life-span of a formation, divided into epochs, [2] the main progressive class may be aided for a time by other social groups. In our epoch the working class is the carrier of progress together with nations struggling for their independence. [3] Therefore, the progressiveness of modern wars is determined by their influence on the struggle for socialism or for national liberation. [4]

The final and perhaps most difficult part is to appraise a given war in relation to ethical standards. It is important to note that the terms just and unjust do not have a purely moral sense in Marxism–Leninism, for the criterion of justness is essentially the same as the criterion of progress, namely whether the object serves the interests of the working class. The main moral value is, according to

216

the Marxist–Leninist ideology, man's happiness, consisting in the availability of all material and cultural goods that he needs, and in the all-round development of his personality. Liberation from poverty and economic exploitation, from political and national slavery is the main condition of man's happiness. Since the socialist revolution is to bring about such liberation, all which is good for revolution is good for man, and is thereby moral.

Thus the ethical character of war is related to a set of moral values that are not purely moral because they are connected with a socio-political cause and its carrier; social progress is the criterion of morality. Wars are thus simultaneously just in relation to the behaviour of the progressive forces and unjust in relation to that of the reactionary ones [5] (there are some exceptions, such as inter-imperialist war in which both belligerents are reactionary). [6]

In orthodox Marxism these three different types of evaluation have always been closely interrelated. The agent of progress identified through scientific study would have progressive aims, and since the historical dialectic brings the ultimate victory of the progressive forces, any war in which they are involved will as a rule take mankind a step closer the final goal. And from the viewpoint of the progressive forces the war would be a just one by definition.

Typologies

On the basis of these criteria two different typologies of war have been proposed, one political, the other ostensibly ethical.

Typologies based on a political assessment

In any given epoch there are according to Soviet teaching some fundamental socio-political contradictions that generate war. The types of war characteristic of the epoch are therefore determined by the interplay between the main socio-political forces and the aims pursued by the particular parties involved. Since the basic socio-economic formations change and thereby usher in a new epoch, the set of wars typical of an epoch also undergo change. It is important to note that in the assessment of wars means and the consequences of war are not treated separately but are rather considered to flow logically from the aims of the belligerents.

The first classification of wars according to the assessment of the political situation was presented by Lenin. It reflected his interpretation of the contradictions characteristic of the final stage of the imperialistic epoch and of the beginning of the transitional epoch during which capitalism is transformed into socialism, that is to say the first decades of the present century. He proposed first a major division into three classes: imperialist, national, and revolutionary. By national he meant what are now called wars of national liberation; revolutionary war refers of course to proletarian revolution. The class of imperialist

217

wars he sub-divided into inter-imperialist wars, counter-revolutionary wars against nations fighting for independence, and counter-revolutionary wars against the proletariat or a socialist country.

In the Stalinist era war between capitalist countries and the Soviet Union was defined as a separate type of war. When after the Second World War several socialist states appeared, this type was replaced by so-called war between the two antagonistic systems. Thus in orthodox Soviet theory there was a standard set of four types of war: (a) wars between capitalist states, (b) civil wars between the proletarian and bourgeois classes, (c) wars of national liberation, and (d) wars between the capitalistic and the socialistic systems. Khruschev attempted to modify this typology by dividing all wars into world wars, local wars, wars of liberation, and popular uprisings, and thereby mixing the political features with other ones. [7] But for the most part the standard set is the one that has been most frequently used in all sorts of analyses. [8]

Some more recent developments can be noted although it is as yet uncertain whether these will become generally accepted. Two additional types of war are sometimes mentioned: democratic civil wars, like the Spanish Civil War, in which a broadly based alliance of popular masses fights against reactionary forces of monopoly-capital for general democratic aims; and wars between countries of the Third World for new territory or to reunify peoples, separated by colonial rule. [9] In addition, it is occasionally remarked that the principal types of war rarely occur in a pure form. Each instance of war may contain features of several types, and the political characteristics of a war may shift as the war unfolds; for example, a civil war may be combined with a war of national liberation, or against a foreign intervention and an interstate war may generate civil war. [10]

Typologies based on an ethical assessment

In a second kind of typology wars are first classified as just or unjust, [11] which is equivalent to being progressive or reactionary respectively. Within these two categories, wars are then defined according to their political characteristics. Amongst the just wars are usually included civil wars between the proletariat and the bourgeoisie, wars of national-liberation, and wars in defence of a socialist country against imperialist aggression; unjust are wars against socialist countries, imperialist wars to redivide the world, colonial wars, and counter-revolutionary wars. [12] Thus wars placed in different categories in fact are the same wars.

Some difficulties related to combining the characteristics

In the political typologies

This brief description of the typologies proposed by Soviet scholars should serve

218

to show that the key to passing judgement on a particular war is the identification of the social forces the belligerents represent together with their aims. From this follows everything else – the progressiveness and justness of the war, and the evaluation of its means and consequences. There are, however, several difficulties involved in trying to uphold the unity of these disparate concepts.

Take the correspondence between the political and socio-historical assessments for instance. A war is progressive according to Marxism–Leninism if one of the belligerents is the agent of progress; in the capitalistic epoch, for example, a war between proletarian and bourgeois forces respectively to seize and maintain political power would be progressive. Here war is called progressive because the aims of one of the belligerents are progressive. This does not seem to be an adequate interpretation, for war is a social phenomenon that encompasses not only the aims of the opposing sides but also the resolution of their conflicting positions. For a war to be progressive, it may be necessary that the aims of one of the belligerents are progressive, but it is not sufficient. The crucial point is whether these progressive aims are in some measure realised through war. If the war ends in the victory of the reactionary forces, it would seem strange to call the war as a whole progressive. For example, there is no doubt that the ambitions of one side in the Spanish Civil War were progressive in Marxist–Leninist terms; there is as little question that the war as such was a serious setback for progressive forces.

Thus the unity of the political and socio-historical characteristics rests on the questionable premise that progressive aims always lead to progressive outcomes. This should not be confused with a faith in the ultimate victory of progressive forces, which is quite another matter. In the short-run, progressive aims can lead to progressive consequences only if the agents of progress have the necessary means.

These remarks also have implications for the correspondence between the socio-historical and ethical assessments. As was noted above, the two criteria are virtually indistinguishable in the orthodox position, where they both relate to the aims of the opposing sides. [13] If when describing a war as a whole, progressive more aptly refers to the consequences of war, as is argued here, then both unjust progressive wars and just regressive wars are quite conceivable. The Spanish Civil War was just in terms of proletarian morality even if it failed to further man's social progress. The wars waged by the Tsars for conquest in Asia were clearly unjust in their conception, but the incorporation of these underdeveloped areas into the Russian empire unquestionably hastened their economic growth. [14]

The final combination, between the political and ethical assessments, is not problem-free either. The political typologies include both sides in the struggle, while the ethical appreciation evaluates the behaviour of each side separately from the standpoint of the working class. For example, some particular war might belong to the political type 'war between the bourgeoisie and proletariat'; in ethical terms, as said before, it is simultaneously a just revolutionary war waged by the proletariat and an unjust counter-revolutionary war waged by the bourgeoisie.

There are further complications, however. Soviet scholars indicate that the political character of a war may change in the midst of hostilities (as did the Second World War upon the entrance of the Soviet Union), that there may be progressive and reactionary tendencies intermingled within one of the opposing camps, and so on.[15]

Insurmountable though all these difficulties seem to be, the orthodox unity of social characteristics is upheld not only in general political declarations, propaganda, and educational publications,[16] but also in the work of reputable theorists of war. One of these, T. Kondratkov, combines what he calls the 'complex socio-political and moral-legal characteristics' of war, and even includes an appreciation of means and consequences. In his presentation,[17] aims determine the political assessment, methods the ethical assessment, and consequences the socio-historical assessment. The interrelation between the three features is presumed to be very close. Progressive classes wage just and progressive wars with moral means and methods; revolutionary aims, moral means, and progressive consequences are inseparably tied together. Reactionary classes wage wars that hinder progress and use means that are cruel and barbarous. In his view it is therefore sufficient to identify one link in the threefold evaluation, for the other two follow automatically.

In the ethical typologies

There are also great difficulties involved with the second kind of typology, in which the main division distinguishes between just and unjust wars. One of the fundamental principles of scientific measurement requires that the categories of a classification be mutually exclusive. However, in these ethical typologies wars can be both just and unjust at the same time, since they can be given two different descriptions depending on which antagonist is being considered.[18] For example, a war waged by a capitalist state against a socialist state is unjust, while a war waged in defence of a socialist state is just; yet a war in which the first description would apply must also be an instance of the second type, for there would be no war if the socialist state did not defend itself.

Some scholars favour diminishing the importance of the moral characteristics when discussing theoretical problems of classification of wars. Of the four criteria on which it has been suggested in one proposal that a new typology of wars be based, none is ethical. The only mention the author makes of moral issues is a sentence stating that the problem of who is right and who is wrong cannot be ignored.[19]

Nevertheless, the appeal of the ethical classification which has become one of the fundamental concepts in Soviet theory of war, remains too strong for most Soviet scholars to resist. They commonly solve the logical difficulties involved by presenting the two classifications parallel to each other. For example, in an article entitled 'Just and Unjust Wars', Prof. Y. Rybkin begins with a discussion of moral questions, but the various types of war he uses as examples are taken from his political typology, and there is no independent criterion by which these are linked to his ethical assessment. It appears therefore that although the division of wars into just and unjust remains, and its importance is still stressed, it is in fact the classification according to political characteristics that provides the basis for scientific analysis. [20]

Modifications in the orthodox position

The orthodox answer to the question of how to appraise particular wars has thus proven resilient, perhaps surprisingly so in view of the problems it has brought in its wake. Some changes have been introduced into the authoritative standpoint, however, and there are indications that further modifications are being called for, particularly with regard to the place of means, methods, and consequences in the total assessment of war, and to the interpretation given the notion of just war.

Methods and consequences

In the orthodox Soviet view, the justness of a war is determined by its political characteristics, the actors and their aims. Those guided by just aims would use just methods, and from their progressive aims would come just consequences. The problem of evaluating means and results was therefore of negligible importance.

The advent of the nuclear era has obliged Soviet scholars to pay greater attention to the assessment of means. Initially, when the United States had a monopoly of atomic weapons, their use was condemned on the grounds that they were instruments of mass destruction unusable for achieving any politico-military goal. Once the Soviet Union had acquired a nuclear capability of its own, however, the orthodox position that means and methods have no relevance to the justness of war was reiterated, often categorically: 'it would be wrong to confuse the evaluation of war as just or unjust, with its aspects, such as ways, means, and the scale of military operations, which are different questions'. [21] The use of nuclear weapons by the socialist armies for the purpose of self-defence would therefore be justified. [22] (A similar argument has been used with regard to biological and chemical weapons, which it is believed the enemy would be prepared to use.) [23] The opinion expressed by some individual authors that a war in defence of a socialist country would cease to be just the moment nuclear weapons were employed has been denounced as dangerous and deeply erroneous. [24]

Nevertheless, it is more widely accepted now that the character of the means used in a war affects the character of the war as a whole, and the problem is included for special consideration in most analyses of the characteristics of war. [25] Such is the case, for example, in Kondratkov's presentation, discussed briefly above. For him, the ethical question has special application to the means employed in war, while the political assessment pertains to aims, and the socio-historical assessment to consequences.

However, by linking revolutionary aims with moral means and progressive consequences, Kondratkov reaffirms the orthodox unity of the three character-istics and the pre-eminence of the political features. In practice, the main new addition, in his proposal is that unjust wars (i.e. wars waged by capitalist states) can be even more unjust if certain kinds of weapons are used. And this implies that the struggle fought by the socialist states against war is correspondingly more worthy.

The development of weapons technology has not left the problem of conse-quences untouched either. [26] In general, there is no change in the basic position that if the agents of progress find it necessary to wage a war, the outcome will usually be a step forward for mankind. However, there is a marked tendency for Soviet scholars to be cautious about advocating armed violence as an instrument for furthering man's social progress. Beneficial though the results of a war may be, the direct costs required in terms of human lives and material damage have increased enormously. It is even questioned whether civil war is necessary for progress within countries and that being so, whether the peaceful alternative is not preferable in view of its lower cost.

The problem can also be set in a broader perspective. Most wars up to and including World War II are considered to have had progressive consequences since class contradictions have been aggravated, and some socialist revolutions have succeeded. Nuclear technology has subsequently made available such destructive force, however, that a possible nuclear world war may very well delay the final victory of socialism by checking the advances made in countries where the people have already begun to build socialism and by disrupting the growth of the world revolutionary movement. Moreover, a nuclear war would bring with it new problems of biological and genetic damage and disease, whose ramifications can now only be dimly perceived. A potential world war may therefore retard the whole of man's social progress by generations.

New shades in the meaning of the term just

A second modification made in the orthodox view of war concerns the inter-pretation given the term just, which is of course crucial to any appreciation of war.

In the first place, its use in relation to the aims of the progressive forces has been broadened. In the original sense, just aims were those which clearly furthered the interests of the working class in its struggle for socialism. Soon was added the cause of nations fighting for their liberation, which was considered to be indirectly in the interest of the working class. Perhaps because of the difficulties involved in trying to assess World War II, in which the Soviet Union waged a just war side by side with capitalist countries, or recent wars in the Third World, where the class criterion is often difficult to apply, it is now more common to interpret just aims in relation to the long-range interests of socialism. Thus, the aims of those who work for the 'cause of peace and independence of nations' are now included in the category just. A good measure of flexibility is thereby obtained; should a comparatively small and weak capitalist country be attacked by an imperialist power, or a relatively democratic country by an extremely reactionary and expansionist state, its struggle might be viewed as just; and if the Soviet Union would undertake some military action to support the just cause, it would of course be presented as a just war.

In the second place, the term just is now frequently complemented by the additional value terms legitimate and moral. No distinction is made between these different ideas, so the reason for their introduction must remain a matter of speculation. The inclusion of the word moral and, occasionally, defining the term as man's happiness, possibility of meeting all man's cultural and material needs, and his freedom and dignity may perhaps be interpreted as a confirmation of the tendency to stress that through furthering human progress in a historical materialistic sense, war may also advance mankind toward high moral values. [27]

As we have seen, the traditional criterion for the division of wars into just and unjust was not directly ethical, since it was identified with the place of war in social development, in the struggle for socialist revolution. The assessment of war as a means of achieving moral values was usually implied in the determination of political characteristics of a war. By application of moral criteria in the proper sense of the word, and by putting more emphasis on the moral characteristics of means and methods, the just-and-unjust typology is now directly becoming an ethical one, although it is, as before, closely related to the political assessment.

The character of peace

In Soviet analysis peace is assessed in precisely the same way as war. The key is to correctly identify the political character, the classes and states whose interests are served by peace and the policy they follow to maintain it. Once this step is accomplished, the socio-historical and moral features of the peace follow automatically.

According to the writings of Soviet scholars, there are two main kinds of peace. In one, both unjust and reactionary, the dominant political forces aim at maintaining the *status quo*. [28] Such peace is unjust because it preserves social and national inequalities and permits imperialism to hold sway over a part of the globe. The socio-political structures it upholds do not meet the requirements of our times, of rapid development in science and technology. Instead of directing this new knowledge towards creating better material conditions for mankind in general, the imperialistic peace wastes resources on war preparations. It is a fragile thing, an absence of open hostilities, and a pause between wars.

The dominant political aims in the second kind of peace, which is just and progressive, strive towards social development, the building of socialism, the liquidation of all inequalities and injustice, and the establishment of eternal peace after the final victory of socialism. [29]

Two non-Soviet positions

The Soviet solution to assessing the character of war has been challenged, at times heatedly so, by analysts in other socialist countries. Of these, the Chinese and Yugoslavian interpretations are perhaps the most interesting.

The Chinese view

In their official proclamations the Chinese have unswervingly followed an 'absolutist' line with regard to the justness of war. Three contentions are specially emphasised: that all wars fought by the popular masses are just; that all just wars are progressive; and that even a world war between capitalism and socialism would be just, regardless of the immense destruction incurred, since it would result in the ultimate victory of socialism.

In practice the actual political position of the Chinese on these questions is much more moderate. Although just, a world war is considered to be undesirable. Moreover, the Chinese say that revolutionary wars should be avoided in countries not yet ripe for revolution even if such a war would in principle be just. [30]

The Yugoslavian variant

If in defending the purity of the orthodox position the Chinese criticise the Soviet view of just war for compromising the cause of socialist revolution, some Yugoslavian scholars go to the other extreme of throwing out the whole Soviet theory as unscientific and non-Marxian. [31] The brunt of their polemical attack has been directed at the contention that any war waged by a socialist state must be just. Some contrary examples have been put forward:

(a) Wars of conquest to widen the sphere of the socialist system. For the attacked country, such a war would be a war of national independence. The working class there may therefore fight in the name of independence in defence of a capitalist system.

(b) Wars to impose a more backward political form of socialism on a much more socio-economically developed country.

(c) Armed intervention against a socialist country whose socialist character is denied (a type of war reflecting the Yugoslavian fear of Soviet attack), or in

224

which it is claimed that counter-revolutionary forces have returned to power (Czeckoslovakia).

The Yugoslavian scholars have also denied that there is any conceptual congruence between the justness of a war and its role in history: not every just war is a progressive one, nor does progress always imply justice.

Since the conclusion of the Soviet theory can be shown false, the theory itself must be invalid. The error of Soviet scholars has been to take some statements made by Lenin as grounds for regarding the justness of war as a criterion for scientific or political analysis and classification. According to the Yugoslavian scholars, it was rather Lenin's intent to assess particular wars and to provide the party and the working class with a guide to political action in the concrete situation.

The attitude adopted by the working class or a socialist country to some certain war should instead be based on the role the war plays in the whole complex international development and on its true progressive or reactionary character, not on an abstract and arbitrary criterion of justice.

Notes

[1] A broader term: 'the political content', was sometimes applied to indicate that the assessment included 'a sum total of the subjective plans and objective tendencies which express the trend of development of the social systems taking part in war and their possible change in the course of the armed struggle' (Rybkin, 1973, p.5).

[2] *Filosofskoe nasledie V.I. Lenina* . . . pp.16–19; 'Every historical epoch is marked by specific contradictions differing as regards social content. . . Not only individual antagonistic formations, but also different periods of the same formation (for example, capitalism) are marked by specific contradictions' (*Marxism–Leninism on War and Army,* pp.69–70).

[3] 'In characterizing the present epoch, Marxist–Leninists emphasize the new fact, that the present is not the epoch of imperialism and war, but the epoch of the decay of imperialism as a world system, the epoch of revolutions and the triumph of socialism and communism on a world scale. This fundamental characteristic of the epoch is decisive for explaining the basic problems of war and peace' (*Voennaya strategiya,* p.218); cf. S. Tyushkevich, 'Politicheskie tseli i kharakter voiny', *Kommunist Vooruzhennykh Sil* 1969:19; V. Shelyag, 'Dva mirovozzreniya – dva vzglada na voinu', *Krasnaya Zvezda,* 7 February 1974.

[4] Pukhovskii, 1965, pp.70–2; G. Khvatkov, 'Problema voiny i mira i revolutsionnyi protsess', *Kommunist Vooruzhennykh Sil,* 1970:1. 'The social character of every modern war must be determined from the standpoint of the interests of the proletariat's socialist revolution and the national liberation revolutions of the oppressed peoples, from the position of the main driving forces of social progress – the world system of socialism, the international

working-class movement and the peoples' national liberation movement' (*Marxism–Leninism on War and Army,* p.66).

[5] 'Civil war – a kind of a just and of an unjust war according to what class wages it and for what aims:

(a)   a *just war* of the oppressed classes, for the overthrow of the governing class, or of the previously oppressed class for defence and confirmation of the political rule against the armed uprising of the exploiting classes. Civil war is waged by the proletariat only when it has been imposed by the bourgeoisie. . . .

(b) | an *unjust war,* waged by the exploiting classes for armed defeating of the revolutionary movement of the exploited classes or waged by the overthrown exploiting classes for restoration of the previously exploiting system' (*Deutsches Militärlexikon,* Deutscher Militärverlag, Berlin 1961, p.81).

[6] S. Golikov, 'Leninizm o voinakh spravedlivykh i nespravedlivykh' *Propagandist i Agitator,* 1949:1, pp.26–8; P. Khrustov, 'Marksizm–leninizm o voine i armii', *Propagandist i Agitator,* 1950:1; 'Programma po kursu 'Marksizm–Leninizm o voine i armii', in *Propagandist i Agitator,* 1956:18; Prokop'ev, 1965, pp.40–8; *Marxism–Leninism on War and Army,* pp.62 ff.

[7] Nikita Khrushchev, 'Za novye pobedy mirovogo kommunisticheskogo dvizheniya', *Kommunist,* 1961:1, p.17.

[8] *Marxism–Leninism on War and Army,* p.70; Rybkin, 1974, p.11; 'Voina', in *Sovietskaya Voennaya Entsiklopediya,* vol.1, 1976, pp.282–5.

[9] Rybkin, 1969, p.12; Filosofskoe nasledie V.I. Lenina. . . , pp.32–4; *Marksizm–Leninizm o voine i armii* (1968). In *Marxism–Leninism on War and Army,* wars between developing countries are pointed out as an additional type of war, and the general democratic struggle of popular masses against monopoly associations is included into the main lines taken by the social struggle. However, although the types of war are said to be determined by these main lines, no type corresponding to the democratic struggle was presented.

[10] *Marxism–Leninism on War and Army,* p.70.

[11] In *Voennaya Strategiya* the following are enumerated as just wars: a liberating and revolutionary defensive world war of the socialist system against the imperialist one; national liberation wars; and civil wars. As unjust: an aggressive, predatory world war of the imperialist camp against the socialist one; small wars, undertaken by the imperialists to suppress the national-liberation movements, to seize or hold on to colonies, and small local wars between the imperialist states (pp.221–2); in *Marxism–Leninism on War and Army* the terminology is somewhat different: unjust wars include the following kinds: imperialist interventions and aggressive wars against the socialist countries; civil wars of reactionary forces against the revolutionary classes within the country; colonial wars against the oppressed peoples or newly independent states; wars between imperialist powers or aggressive attacks by the imperialists or other capitalist countries (p.71). S. N. Kozlov, 1971, applies different terms for two kinds of the wars between capitalist states: interimperialist wars are called 'wars between imperialist powers for the repartition of spheres of influence and world

supremacy' which is similar to the traditional notion of wars for the redivision of the world; on the other hand, just wars include 'the liberation wars of people of bourgeois countries which have become victims of the imperialistic aggression for their state sovereignty' which relates to wars between imperialist powers and small capitalist countries.

[12] Many scholars emphasise, however, such a division. 'The Marxist–Leninist thesis about two kinds of wars – just and unjust – makes it possible to assess each concrete war correctly, to determine the role which it will be able to play or has already been playing in the development of society...' (Rybkin, 1974, p.10); 'the division of wars into just and unjust is of extremely great importance' (V. Izmaylov, 'Kharakter i osobennosti sovremmenykh voin', *Kommunist Vooruzhennykh Sil*, 1975. Cf. 'Voina', in *Sovetskaya Voennaya Entsiklopedia*, vol.1, 1976.

[13] 'Unjust, predatory wars waged by the reactionary exploiting circles hamper, check the progress of human society, because they increase the exploitation of the oppressed classes and nations, defend all which is old, obsolete, and reactionary, and smother all which is new, developing, revolutionary ... Contrary to this, just liberatory wars ... are progressive, and revolutionary, they destroy, old, harmful, and reactionary institutions, which hinder the free progress of nations, they liberate the oppressed mankind from the capitalist slavery, they liberate nations from the imperialist oppression and create conditions for independence and national development of nations of colonies and dependent countries' (F. Khrustov, 1950); 'A reactionary, predatory war cannot be a just war, and an unjust war runs counter to social progress' (Rybkin, 1974, p.10); reactionary, aggressive wars cannot be just, and unjust wars retard historical progress (*Marxism–Leninism on War and Army*, p.63).

[14] The unity of the two characteristics was questioned by some Soviet scholars in the middle of the Sixties, especially at a debate on criteria of appreciation of wars in the Institute of History of the Academy of Science of the USSR in April 1966. Almost all participants contended that the two criteria should have been applied separately; wars were pointed out the positive and negative socio-historical and ethical features of which had not been tied together e.g. in 1866 Prussia waged an unjust but at the same time progressive war against Austria, similarly Japan against Russia in 1905, the just defensive war of the Swiss against the Austrian conquerors in fact checked their progress etc. The studies and pronouncements were published in *Istoricheskaya nauka i nekotorye problemy sovremennosti*, Stat'i i obsuzhdeniya, Moscow 1969 (esp. I.V. Bestuzhev 'K analizu kharaktera wneshnei politiki i voin. Kriterii progressivnosti i spravedlivosti'; utterances by M. Alperovich, I. S. Kremer, N. I. Tabunov, M. I. Geffer, A. M. Stanislavskaya).

[15] E.g. the national wars of the French Revolution were transformed into predatory wars for the extension of Napoleon's imperium.

[16] Metodologicheskie problemy voennoi teorii i praktiki, p.95; *Marxism–Leninism on War and Army*, Ch.II; V. Kozlov, 1969; T. Kondratkov,

'Sotsialnyi kharakter sovremennoi voiny', *Kommunist Vooruzhennykh Sil,* 1972:21.

[17] Kondratkov, 1972.

[18] Kende contends that the division of wars into just and unjust is unscientific (1972, p.24).

[19] Yermolenko, 1973, p.107.

[20] Rybkin, 1974; in an earlier article (1973 b), although he emphasises the importance of the division of wars into just and unjust, the main argument is made with regard to the typology of wars based on the socio-political criteria, and corresponding to the basic contradictions of our epoch.

In *Sovetskaya Voennaya Entsiklopediya,* 1976, two typologies are presented; a more general division of wars into just and unjust, at the same time progressive and reactionary, both characteristics of wars determined by their political content, is followed by a more detailed and concrete historical typology, where types of war in particular socio-economic formations are defined according to their political character – classes and states with their political aims ('Voina', vol. 1, pp.282–3).

[21] K. S. Bochkarev, 'O kharaktere i tipakh voin sovremennoi epokhi', *Kommunist Vooruzhennykh Sil,* June 1965; E. Khomenko, 'Wars: Their Character and Type', *Soviet Military Review,* 1965:9.

[22] Y. Rybkin, Yaderno-raketnaya voina i politika, Tsentralnyi klub im. Frunze, Moscow 1966; I. A. Grudinin, 'Po voprosu o sushchestve voiny', *Krasnaya Zvezda,* 12 July 1966.

The latter writes: 'The use by us of the nuclear missile weapon would be a forced act and would be the most just means, which would permit the bringing down on the aggressor the superior power of the same sort of weapons with which he will attempt to annihilate our country and other socialist countries.'

[23] 'It is not possible to accomplish all this solely by the use of nuclear weapons. Other types of weapons will also be needed, as well as the most diverse types of military equipment. In particular, in a future war one may expect the employment of chemical and bacteriological weapons. . .' (*Voennaya strategiya,* p.294).

[24] Talenski, Nikolski and others were criticised in Rybkin 1974, p.26.

[25] Rybkin, 1970, 1973; *Filosofskoe nasledie V.I. Lenina. . .*A 3 (V.I. Lenin o sotsialnykh posledstviyakh voiny); Kondratkov, 1972.

[26] 'Lenin always said that there is a close connection between the legitimacy and justness of wars and their progressiveness' (*Marxism–Leninism on War and Army,* p.63); in another study the term 'moral-legitimate assessment' of the political contents of war, of its aims, means and methods, is applied (*Filosofskoe nasledie V.I. Lenina. . . p.69*); the expression 'moral-legitimate characteristics' has also been used (Kondratkov, 1972, p.11).

[27] '. . . on the basis of dialectical analysis of political relations Marxism–Leninism produced a clear definition of the two opposite concepts of just peace and unjust peace. If political relations between states are based on

enslavement and outright robbery, class and national oppression, the seizure of other people's territory and infringement of the independence of states then the concept of peace that fully embodies all these essential features must be defined as an imperialist, unjust peace' (*Problems of War and Peace,* p.72).

[28] 'Our philosophy of peace is a philosophy of the social progress ... peace may be just and democratic, or it may be based on unjustice and oppression' (F. Burlatskii, 'Dialektika sovremennogo mira', *Izvestiya,* 21 November 1973). The two concepts of peace were discussed by the Polish political scientists, and the proceedings of the symposium were published in *Studia filozoficzne,* 1972:5; cf. Alfred Bönisch 'Zur Funktion bürgerlichen Friedensforschung im ideo-logischen Kampf', *IPW Berichte,* 1973:8; Peace, like war, is conditioned by socio-economic factors, therefore peace with and among capitalist societies is qualitatively quite different from a socialist peace. The former is only an imperialist pause between wars (imperialistische Atmepause), while the latter is an immanent feature of the social system (Bönisch, p.6).

[29] Chen Yi in an interview with John Dicon, September 1963, quoted in Arthur Huck, *The Security of China,* Chatto and Windus, London 1970, for the Institute for Strategic Studies, pp.51–2.

[30] Kardelj, 1961, ch. 'On Just and Unjust Wars'.

[31] By asserting that any war waged by a socialist country is just because socialism is progressive and capitalism reactionary 'the theory of just and unjust wars is brought to the ridiculous conclusion that any war which *I* wage is a just war. . .' A socialist country cannot judge itself what is socialist and what is just: 'There is an old Yugoslav proverb deriving from Ottoman Turkish times which says "The Kadi accuses and the Kadi judges" ' (Kardelj, pp.103, 105).

12   The future of war in the nuclear missile era

In recent years, the principal concern of Soviet scholars has been to decide in what way the change in the international situation brought about by the changes in the socio-political picture of the world and by the advent of nuclear military technology has affected the nature of war and its future. In their re-examination of the traditional Soviet position they have focused on three questions:

(a)   Is war still a continuation of policy?

(b)   Is it a practicable and advisable means of achieving political goals?

(c)   Do the answers to these questions have any bearing on the theory of the inevitability of war, which has played so profound a role in the Marxist–Leninist theory of war?

In discussing these three problems Soviet scholars have tended to restrict their comments to particular types of war rather than to relate them to war in general. [1]

The first problem, war as a continuation of policy, has already been treated above in chapter nine. To recapitulate that discussion briefly, Soviet scholars assert that this basic tenet in the theory of war remains valid. [2] The main form of the class conflict is the struggle between the forces of capitalistic imperialism on the one hand and the progressive forces of socialism and national liberation on the other. This antagonism might take the form of a war between the two socio-political systems — in other words, a world war between capitalist and socialist states [3] — limited or local wars instigated by imperialism, or revolutionary wars for social or national liberation.

The instrumentality of war

If the first question can be answered in the affirmative without difficulty, the second question – is war a practicable and desirable means of conducting policy? – has posed some problems. For war to be a useful instrument of policy, there must be some chance that it will lead to victory and the achievement of the desired political goals. Even if war is useful according to this criterion, it may still be an inadvisable means of pursuing policy if the same goals could be achieved by peaceful means at lower cost, or if total costs, direct and indirect, of using war are greater that the resultant gain. In such a situation, victory and achievement of goals would lose their meaning. According to Soviet scholars, the appearance of nuclear weapons, which immensely increased the costs of war, and the growth of the socialist system which created conditions for achieving revolutionary goals by peaceful means, have had a bearing on war in both these respects, although the impact has been different for each of the main types.

Nuclear rocket war between systems

In the Soviet view, it would be sheer suicide if the imperialist countries were to launch a nuclear rocket war against the socialist system, for they would be bound to lose; such a war would therefore have no instrumental value for them. On the other hand, the socialist camp could theoretically use such a war to achieve the ultimate victory over imperialism. [4] This conclusion is essentially a restatement of one of the fundamental predictions of Marxism, that socialism will ultimately triumph over capitalism. In a life-and-death struggle between the two, socialism must therefore win. Or again, it can be seen as a reassertion of the primacy of policy over war. Political control over the general course and on all material and moral aspects, and especially over military operations would be difficult in a nuclear rocket war. However, the difficulties would be experienced much more acutely on the imperialist side where the coordination between political, economic, and military efforts would be much less perfectly accomplished than in the socialist system. [5] Moreover, since any war the imperialist countries might launch would be, in the Soviet assessment, reactionary in character, resistance amongst the popular masses would grow and make it more difficult for the governments to assert control over their societies and to back up their armed forces. [6]

One or two authors have on occasion expressed doubts about whether the socialist countries could obtain victory in a nuclear war, but such views have always been sharply criticised. [7]

The Soviet scholars thus contend that if the imperialists were to unleash an intersystem nuclear rocket war, the socialist countries would win. However, they constantly repeat both officially and unofficially that such a war would be highly undesirable. The victory of progressive forces would lose much of its value because of the immense destruction it would cause. [8] Such arguments are expressed with special firmness when Soviet authors criticise what they regard as the governing Western view that nuclear war is 'thinkable'. [9] They are given an even sharper expression in the critique of the Chinese view on nuclear rocket war: 'the justness of aims and actions of socialist countries in a certain war can never serve as a motive of justification for unleashing it, as it is interpreted by the Chinese leaders, pursuing adventurous and provocative aims.' [10] In emphasising the certainty of victory of the socialist camp and thereby the usefulness and purposefulness of war, the Chinese are said to neglect the colossal damage war would bring.

Local and limited wars

In Soviet terminology wars less than total, unleashed by the Western powers, are quite another matter. While the new military technology has radically diminished the usability of a total war between the two systems, its impact on local and limited wars which are by definition less than total, has been less significant.

It is very difficult to make general assertions about the instrumentality of local and limited wars. [11] Each instance is a unique mixture of political and military factor. Soviet scholars do point out that such war is more practicable if there are good prospects of a quick victory than otherwise; the longer it lasts, the more internal and external factors become involved, and the greater become the costs in relation to potential gains. Moreover, as a local war drags out in time, the risk of it escalating into a major international war increases. Such a development would make such a war useless as an instrument of policy.

These risks are particularly acute because of the imbalance that usually pertains in the military effort made by the opposing sides in different periods of war. When an imperialist power engages in local limited war, the attacked country may mobilise a massive defensive effort; limited intervention is countered by a total defence. The situation can become more complicated since such movements for national independence are usually supported by socialist powers. Once the limited means originally adopted prove inadequate for achieving the political objective of the aggressor state, it may lose sight of its initial purpose. The war may thus escalate militarily without political control and may become deprived of any instrumentality. [12]

Soviet scholars usually cite the Korean and Vietnamese wars as examples of local and limited wars that lacked any instrumentality for the intervening imperialist power. Sometimes they are even called 'counter-instrumental'.

Revolutionary war

The instrumentality of revolutionary war has never been questioned in the Soviet study of war, and since the possibility of achieving the revolution without revolutionary war was until the 1950s never seriously entertained, the question of the desirability of revolutionary war was not raised either. In the re-examination of the nature of war that was undertaken after the adoption of the policy of peaceful co-existence, the doctrine of the inevitability of revolutionary war was for the first time officially denied. This made possible, and even necessary, a discussion about whether war was the best means of realising the revolution.

The place of armed violence in accomplishing the revolution can be divided into three interrelated questions. The most important of these is whether revolutionary war is necessary for victory. The second problem, perhaps less fundamental but still important, is whether a great international war must occur to generate the conditions necessary for revolution. The third problem is whether socialist states should encourage and give military support to armed uprisings led by revolutionary forces in other countries.

The traditional position Traditionally, the first question was answered in the affirmative, and categorically so. The socialistic transformation of society without an armed struggle was inconceivable. The possibility that the bourgeoisie might capitulate without armed resistance, or that if it were defeated it would refrain from counter-revolutionary war was simply not considered.

232

As to the second problem, although Soviet theorists have never stated that international war is necessary for the success of revolution, it can nevertheless be inferred from their description of historical revolutions that such an idea was at the back of their minds. They point out that the First World War created conditions which were extremely favourable for the October Revolution: it profoundly weakened the capitalist system by aggravating all its inherent contradictions and contributed to the growth of class consciousness in the proletariat. [13] Moreover, such favourable conditions did not recur until the end of the Second World War. It has never been concluded, however, that the socialist revolutions would not have succeeded had these wars never occurred.

Neither did Soviet scholars directly consider the third problem. They presumed in a general way that once the capitalistic system began to collapse, the proletarian revolution would quickly envelop all, or almost all, capitalist countries. The theory of permanent revolution, according to which it was the duty of the victorious proletarians in the first socialist country to support revolutionary uprisings in other countries by military means, can be seen as a literal interpretation of these ideas. Lenin's view was different however, although in some cases (Georgia 1921) such a policy was followed. Lenin concluded that the world revolutionary process would last a whole historical period, during the initial phase of which the revolution would succeed in only a small number of countries, perhaps only a single one. Since the involvement of the Soviet Union in any revolutionary events abroad would have been connected with great risks, the policy followed during the interwar period of building socialism in one country was derived from Lenin's theory. Afterwards, in spite of the fact that the Soviet Army actively helped communist forces to gain power in several European countries at the close of World War II, the official position, that troops should not be sent to support revolution in other countries, remained unchanged. Soviet officials repeatedly emphasised the importance internal conditions had in determining the success of the East European revolutions; the role of the Red Army, whose main task was to liberate these countries from Nazi control, was presented as determining only in the sense that it prevented counter-revolutionary uprisings.

Arms and revolution in the era of peaceful coexistence The need for a policy of peaceful coexistence in the conditions prevailing after the second World War led to a re-examination of both the official position and the tacit assumptions held with respect to these three problems. The identification of revolution with revolutionary war is denied. According to the new doctrine, the only condition necessary for the transition from capitalism to socialism is the establishment of the dictatorship of the proletariat. Revolutions without armed uprisings are therefore possible. [14] The necessity to resort to armed violence arises only if the bourgeoisie use it first; in other words, armed force is necessary only as a response if counter-revolutionary violence should be applied by thee exploiting classes. [15] Owing to the shift in the world balance of power between the revolutionary and counter-revolutionary forces, it could no longer be taken for granted that the bourgeoisie would cling to power with all the means at its disposal; it might well prefer to yield without armed resistance.

The suggestion that a connection exists between revolution and a preceding interstate war was also openly challenged. Interstate war might act as a catalyst by sharpening the internal contradictions and contributing to the emergence of a revolutionary situation, but it would be incorrect to regard it as a necessary prerequisite for revolution. [16] According to the official view, the general world crisis of the capitalistic system had now entered its third stage. Whereas both of the first two stages have been introduced by a world war, the beginning of the present third stage of the crisis had not been marked by a major interstate war but by the worldwide successes of socialism and the national-liberation movement, which together had greatly reduced the sphere of imperialistic rule.

Thus, according to the new Soviet announcements, peaceful revolution is possible. It is moreover maintained that peaceful revolution is desirable for many reasons. In the first place, civil war could result in great material destruction and human suffering. [17] Moreover, there is a risk in the modern epoch that civil wars could escalate into interstate wars, which because of the world-wide competition between the two opposing social systems, should be avoided. In the third place, it is argued that the struggle against war and for world peace best favours the development of the revolutionary process in any particular country. Peace allows the socialist countries to consolidate their gains whereby they can with greater strength contribute to the world revolutionary process and the national-liberation movement. [18] By contrast, war would be positively harmful to the revolutionary process. The demand for resources required in war would slow down the speed of socialist construction, it would nullify the progress made by the working masses in capitalist countries, it would set back the economic development of those countries which have recently won their freedom from colonialism, and it would retard the process of national-liberation in the remaining colonies. [19] In sum, the Soviet scholars reject the idea of a close connection between revolution and war as artificial. At the same time, the official policy of not giving armed support to revolutionary uprisings in other countries has been reconfirmed.

Commentary on the official position However, as in all matters of doctrine, the official position on these points has been expressed in terms general, and ambiguous, enough to allow scholars scope for complementary remarks, qualifications, and even reservations.

Regarding the place of civil war in the revolutionary process, some have contended that in spite of official optimism about carrying out the revolution by peaceful means, the prospect of armed uprising cannot be excluded. For example, the main thesis of a study published in a military periodical on Communist revolutionary strategy and tactics is that the choice of means, including armed violence, for seizing power will depend on the situation. The division of revolutions into peaceful and non-peaceful is not very meaningful since in all revolutions, which are a form of taking over political control by force, a combination of both means is used. Quoting Lenin, the article stresses that the choice of military means will be conditioned by many factors of which the ability of the proletariat to fight and of the enemy to resist, are very important. [20] The implications of this statement would seem to be quite different from those of the more usual official formulation, that the proletariat uses armed violence only in response to military action initiated by the enemy.

As to the thesis that interstate wars are not a prerequisite to revolution, some scholars point out that it does not mean that there is no connection at all between such wars and revolution. Even if interstate war is unlikely, there remains the possibility that intercapitalistic war may break out owing to the law of uneven development of the imperialist system. [21] Since a complete theory of revolution must include all possibilities of exploiting the growing weaknesses of the capitalistic system, it must not ignore the favourable conditions for revolution such wars might provide.

Finally, it might be added that since the first strong statement of policy on peaceful coexistence was declared by Khrushchev, some shifts of emphasis in the official party line can perhaps be detected. It is now pointed out with somewhat greater frequency in official statements that peaceful coexistence does not contradict the possibility of revolutionary war or wars of national-liberation. Occasionally in semi-official publications the necessity of using armed violence in socialist transformations is even mentioned; for example the revolutionary events in Portugal have been met by articles in which the democratic revolution has been assessed as a prelude to socialist transformations, both accompanied by armed violence in various forms. Under conditions of imperialism the fight for democracy and socialism takes the form of two social wars that become altogether typical. Such occasional statements confirm that the Soviets treat peaceful coexistence as a kind of class struggle that does not exclude violent revolutions. [22]

Wars of national liberation

Next only to socialistic revolution, national revolution is the primary motor of social progress in Soviet thought, since Lenin's own time. It is therefore not surprising that the problems relating to the place of armed violence in national revolution in the nuclear era have been solved in much the same way as those pertaining to socialistic revolution. [23] Here the problem of national revolution, with or without war, and the question of Soviet support for national revolutions will be discussed.

National revolution and war in the nuclear era It is one of the basic principles of Soviet theory that the national independence of the oppressed and dependent peoples in the colonised parts of the world will be brought about by revolution. To be free from all kinds of national oppression, whether economic, social,

235

political, or cultural, and to pave the way for social progress, some violent action is needed. Present-day Soviet ideologists remain as loyal to this principle as their forerunners were.

But just as it was deemed artificial to connect social revolution with war after the re-examination in the mid-1950s, so too the identity between national revolution and war was denied. [24] Once the possibility of achieving national liberation by peaceful means was allowed, the theoretical problem became to decide under what circumstances armed violence or peaceful methods should be used. On this point Soviet scholars maintain that since much will depend on the socio-economic conditions pertaining in the particular country in question, it is very difficult to generalise about the mixture of methods appropriate in the given case. [25] With respect to the international factors however, it can be said that the instrumentality of peaceful methods has gradually increased in the same measure as the capitalistic system has become weakened by the successes of the progressive forces. At the same time, there is in the present historical period the added danger that if war were chosen as a means of achieving national-liberation, it might escalate into a major international conflict and threaten total war. Thus on balance there is now usually more reason to prefer peaceful methods to military ones in the struggle against imperialism and neo-imperialism.

Recently Soviet scholars have emphasised that the oppression exerted by the colonialists can be regarded as a form of permanent violence, which may be manifested politically or, even after political autonomy is formally won by the country, in the tight control the imperialist power maintains over economic and social activities. In this perspective wars both for achieving national independence and for defending it can be seen as a response to violence.

Socialistic support to national revolution The support of socialistic states for national liberation movements is as natural as it is for socialist revolutionary forces. As has already been pointed out, according to Soviet views it is for this very reason that a peaceful road to national independence can now more often be followed in the present situation. Should the local circumstances require that the struggle for national-liberation be conducted by means of armed violence, the nature of the support given by the socialistic countries would be limited by two constraints. In the first place, all revolutions are ultimately dependent upon the abilities of the local progressive forces; revolution cannot be exported by the socialist countries, Soviet scholars say. In the second place, all support must be compatible with the policy of peaceful coexistence, which is presented as pre-cluding open armed interference in the internal affairs of other countries.

Double revolutions Revolutions the political goals of which are both national liberation and social transformation are a relatively new phenomenon. They have been successful in some countries, for example Vietnam, North Korea, and Cuba. The basic assumption of the Soviet position is that the two kinds of

236

revolution are distinct and that socialistic objectives can be saved until after independence has been won. [26] Even if the two revolutions should on occasion occur simultaneously, one should not make the mistake of regarding them as one and the same revolution; nor should each revolution for liberation be considered a part of a socialistic revolution nor even as a direct prelude to one. The path from national revolution to socialistic revolution, which in the historical perspective is inevitable, may be of unequal length and have greater or lesser number of twists and turns from one country to another, depending on the nature of the local revolutionary conditions. One possibility is that a national-democratic revolution may be gradually transformed into a socialistic one without any break in revolutionary activity, or may be a component of it, but this is by no means the only possibility.

The emphasis on the distinction between the two kinds of revolution is weakened however, by frequent statements in military periodicals that civil wars may become converted into national-liberation wars if imperialists intervene, and that national-liberation wars may become transformed into civil wars if the internal class contradictions become sharpened; this may occur if the working classes want to go from antifeudal to anticapitalist transformations and the bourgeoisie turns against the people. Under war conditions the mutual impact of the two kinds of revolution may be much greater than in the course of relatively peaceful transformations. In all these cases, the necessity of active support for the fighting peoples from socialist countries is stressed.[27]

The extremes criticised

From the changes that have occurred in official Soviet announcements concerning the appropriateness of various kinds of war in the nuclear era, single scholars have drawn extreme conclusions: that war as such is obsolete, at least any nuclear war. The amount of leeway allowed for the public expression of views diverging from the official line can serve as evidence of change in Soviet thought, but, on the other hand, the sharp criticism with which these extreme views are met, indicates that the limits imposed on such divergence are narrow.

In the early 1960s a book by Nikolskii appeared. It was remarkable because it argued that all kinds of war have become obsolete in the modern epoch. Such a view had never before been expressed publicly in the Soviet Union. The author argued that the historical development of social progress since World War II has gradually eroded the basic roots of all wars, namely the capitalistic system. At the same time the destructive capacity of war has continually increased with the invention of new weapons. As a result of these two tendencies, the forces of war (imperialism) have now lost so much ground to the forces of peace (socialism) that their ability to prosecute a war is greatly reduced; moreover total war has ceased to be instrumental for all imperialists' goals.

Thus a total war would in a sense be an anti-war: since it could not resolve any international dispute it would be neither a continuation nor an instrument of policy; since it would be useless and enormously destructive, it would be a crime against mankind.

Nikolskii does not really substantiate his claim to show the obsolence of all kinds of war, but his main assertion collides with the official line regarding the instrumentality even of nuclear war for the socialist states. Nikolskii was not alone on this point. Before his retirement General Talenskii also voiced the opinion that nuclear war would not have any instrumental value for either of the opposing social systems. [28] All such views were of course criticised.

The Soviet critique of Chinese doctrine and pacifism

In some respects the modern Soviet position on problems of war resembles that held by moderate pacifists – the anti-militaristic stress, the denigration of most kinds of war as a means of achieving political objectives. On the other hand, one would also presume that there is some similarity between Soviet and Chinese Marxist views, even if Chinese doctrine is ostensibly more militant. Yet Soviet scholars are anxious to make clear the integrity of their own position by pointing out the errors of what they call these two extremes. The virulence of the tone in official Soviet criticism of each of these two contending views can serve as an approximate indicator of the position of Soviet policy between them.

The instrumentality of war in Chinese doctrine It is a matter of fundamental doctrine for both the Soviets and the Chinese that a war between social systems would be instrumental for the progressive forces. Whereas the Soviets always add that the instrumentality is now much less owing to the great number of people who would be killed in the process, the Chinese have continued to assert this piece of doctrine almost without reservation. The Chinese discussion of the instrumentality of revolutionary wars for national independence or social transformation is equally faithful to orthodox doctrine. They do not give nearly as much attention as do the Soviets to peaceful alternatives to armed violence, nor to the problem of the desirability of war that arises once peaceful alternatives are allowed.

The militant tone in the Chinese position has been amplified by their scathing criticism of the Soviet position on these matters. They have accused the Soviet Union of trying to discourage revolutionary forces from engaging in armed struggle, which is, they say, the only way the revolution can succeed; of having put the interests of the so-called socialistic camp, which is to say the Soviet Union, before those of the revolution; and even of having betrayed the cause of the liberation of peoples from imperialist oppression with a policy of neo-imperialism.

It could be argued that the Chinese demonstrate in their actual behaviour a cautious moderation far removed from the aggressiveness of their declared policy. Be that as it may, the issue of immediate concern is the distinctions Soviet scholars draw between their position and the adventuristic Chinese policy.

According to the Soviet critique [29], the Chinese have abandoned the

Marxist conception of history based on the assumption that social activity is conditioned by the state of the material foundation. The objective conditions must be ripe for revolution before it can succeed; armed violence is then applied only to the extent made necessary by these same conditions. In its place the Chinese substitue a voluntarist-idealist theory of violence, according to which the course of social life, including the social and national struggle, is determined not by economic relations, but by coercion, in particular by means of armed struggle. [30] Mao's statements that 'the gun breeds power' and that 'the world can be remodelled only with the gun' are interpreted as apologia of force and the glorification of armed violence.

Because of this fixation with power, the Chinese embark on the mistaken policy of exporting revolution, of trying to foment revolutionary war even in countries not yet ripe for it. For the same reason, they ignore the costs that would be incurred in a total war. [31] Their position on total war is further distorted by the selfish expectation that they would emerge far less scathed than the Soviet Union, so the Soviet scholars say.

Pacifism Marxism has traditionally condemned pacifism for being both theoretically unsound and practically dangerous because it does not distinguish between just and unjust wars. Pacifists invariably err in their analysis of the basic cause of war. Since they fail to see it as a product of social contradictions, they are equally blind to its role as an instrument for furthering social progress.

The Soviet critique of pacifism has, however, concentrated on the political consequences implied in a total ban on war. It puts confusion into the minds of men by diverting their attention from the true causes of injustice and inequality; it thereby hinders the oppressed peoples from becoming conscious of their true situation; it furthermore deprives them of an essential instrument for their liberation; and it often lulls them into passivity with the worst sort of reactionary utopianism. Pacifism has therefore been attacked as one of the most dangerous enemies of forces aiming at national and social transformation. [32]

Recently, however, a number of prominent Western pacifists, who have all been denounced at one time or other, have gained some favour. For their active contributions against militarism, the nuclear arms race, and the outbreak of particular wars that seemed imminent, Max Born, Albert Einstein, Bertrand Russell, Jean Paul Sartre, Linus Pauling, Walter Millis, and Karl Jaspers are now considered as respectable allies in the struggle against imperialism and for peaceful coexistence. Even if the alliance is a tactical one – since Marxists hold that war cannot be completely eliminated until imperialism is finally defeated and that some kinds of war may be necessary to accomplish that goal – still it is a further indication of the Soviet view that a conscious effort must be made to prevent a war between systems. Despite fundamental differences with pacifists, a partial alliance on this common policy seems desirable. [33]

To sum up, the Soviet critique of Chinese doctrine for adventurism on the one hand and of pacifism for passivity on the other confirms that a middle position has been adopted in the nuclear era: armed violence should be neither extolled nor excluded, and its use should depend on its instrumentality in the prevailing circumstances.

The changing theory of the inevitability of war

The traditional Soviet view of war as a means of bringing about socialism might be summed up in the following two statements: since war is necessary for achieving socialism, it must be instrumental; moreover, since socialism is inevitable, then war must also be unavoidable. In the foregoing section, we have noted that by questioning the instrumentality of certain kinds of war in the postwar world, thus their practicability and desirability, Soviet scholars have in effect denied the necessity of war; in doing so they have removed part of the foundation on which the idea of the inevitability of war logically rested. Thus changing views about the instrumentality of war could not fail to affect ideas about the inevitability of war, which have played an important part in Soviet teaching, and are still occasionally used in current Soviet political doctrine and propaganda.

Development and change

There are two roots to this theory. According to the original Marxism, the only way to achieve revolutionary goals, whether in particular countries or the world at large, was by war. In the second place, Marx and Engels believed that war is inherent in the nature of capitalism, just as it was in all pre-capitalist class societies.

Lenin developed these assumptions into an elaborate theory concerning the inevitability of particular types of war. In his theory of imperialism he postulated that war between capitalistic countries is inevitable. In the imperialistic period, war would break out in order to redivide colonial empires. Colonial wars, which were more typical of the pre-imperialist period, would also occur from time to time. The internal contradictions of the colonial system would ultimately lead to wars by which the peoples of colonies and dependent countries would gain independence. And of course, Lenin considered revolutionary wars to be inevitable, although he foresaw that the conditions for armed uprising would vary from country to country, as a result of which the revolution would probably not break out in all capitalistic countries simultaneously as Marx and Engels had supposed, but rather in a few or a single country.

Before the October Revolution, the possibility of war between a capitalist and a socialist state was only treated indirectly by Lenin when he took up the duty of the successful revolutionaries to assist revolutionary movements in other countries, After the first socialist state was proclaimed however, he declared that war with capitalist states was inevitable; occasionally he indicated that such a war would result from an attack against the socialist country. [34]

Under Stalin the theory of the inevitability of war was upheld but the use made of it seems to have varied with the political conjuncture. At the beginning of this period, emphasis was placed on the inevitability of a counter-revolutionary war launched by capitalistic countries against the socialist motherland. A shift came in 1935, however, when to meet the growing threat of Nazi war preparations it was decided to cooperate with anti-fascist forces in Europe. It was stressed especially in the political propaganda that by untiring efforts the war that seemed imminent, as well as war between capitalistic countries, could be prevented. The thesis that war between the first socialistic state and capitalistic countries is inevitable was set in a long-term perspective. At the same time, the idea that revolutionary wars are inevitable was seldom mentioned.

It was not until the first postwar years that the whole set of ideas concerning the inevitability of various types of war was restated as a central element of Soviet ideology. [35] The theory seems to have reached its zenith at the climax of the Cold War. As the policy of peaceful coexistence gradually began to replace the policy conducted during the Cold War, the theory of the inevitability of war was referred to less and less frequently, and at the Twentieth Party Congress held in 1956, it was officially buried. In preparation for the congress Khrushchev reexamined Soviet doctrine to meet more adequately the conditions of the nuclear era and the problems of peaceful coexistence. Of the traditional view of the inevitability of war all that remained was the orthodox assumption that the economic and political system of imperialism is the ultimate source of all wars. It was expressly declared however, that all wars whatever their type could be avoided. [36] There was no fatal inevitability about the occurrence of wars between social systems or even of wars between capitalistic states; the possibility of peaceful transition from capitalism to socialism in particular countries was declared; the inevitability of war as a means to national liberation was denied. What is more, a theoretical innovation was introduced: instead of referring separately to the wars between the two antagonistic systems on the one hand and between the capitalistic states on the other, the principle of peaceful coexistence that wars can be avoided is formulated so that it applies to all states in general.

Some of the contrasts between the new position and the traditional one were afterwards toned down by emphasising the war-generating role of the imperialistic system, indicating the difficulties a peaceful seizure of power by the proletariat through parliamentary means would involve, [37] and by stressing that the Soviet Union and other socialist countries would support revolutionary civil wars and wars of national-liberation whenever and whereever they may break out.

Thus, at the same time, when the theory of the inevitability of war was irrevocably abandoned and peaceful coexistence was the order of the day, the threat of war was constantly repeated and the unchanging nature of imperialism as the source of all war continued to be emphasised. [38] Why, one might ask, were these contrasting views presented together? [39] To this question two answers might be given: one can try to give a theoretical explanation taking as the basis the Marxist–Leninist theory; the other answer consists of more speculative comments about the political factors that made this course of action expedient.

The theoretical explanation for the abandonment of the theory of the inevitability of war may lie in the changes that have occurred in the conditions of social life both internationally and within countries. Marxism–Leninism regards as ineluctable the progressive development of society, that is its evolution towards higher socio-economic and political formations, of which communism is the highest. Those social phenomena and actions which are necessary to carry this process forward are therefore inevitable. If progressive just wars were the only means of furthering progress, they would then be absolutely inevitable; to the extent that they serve social development at less social cost than alternative means, to that extent they are relatively unavoidable. Thus Marxism–Leninism does not equate inevitability with automatism: social phenomena cannot occur without activity springing from social consciousness, but such activity will inevitably take place if it is required by social progress.

Within these grand laws of social development operate at the same time many laws of a lesser order, including the laws governing reactionary socio-economic and political formations. The operation of these laws is, however, conditional upon the state of social development that has been reached. Thus because of these laws, reactionary and unjust wars directed against progressive forces are inevitable, but only so long as reactionary social forces are stronger than the progressive ones.

The conditions that had come to prevail by the mid-1950s were so radically different from those of any previous historical period that neither just nor unjust wars could be considered inevitable. The great forward strides that the socialistic system had taken at the expense of reactionary forces had had two effects: in the first place, it had become much more difficult for the reactionary forces to unleash unjust wars; secondly, possibilities had opened for progressive forces to execute social transformation without resorting to armed violence. At the same time, the development of nuclear weapons had so greatly increased the potential social costs of war that even just wars were now more expensive than non-military methods that had become instrumental as a means of bringing about social revolutionary changes. In sum, both kinds of war had ceased to be inevitable, even though they remain and will remain a theoretical possibility until capitalism is liquidated.

This interpretation of global conditions implies that a change in the guidelines for the correct behaviour of progressive forces required corresponding adjustment. The main goal of carrying on the permanent class struggle remained unchanged, but the resort to a just war as a means of achieving this goal should be made only if absolutely necessary. Within this general framework progressive forces should also struggle to crush any plans of unleashing unjust war.

So much for the hypothetical theoretical explanation. Several other reasons of a more political character may be suggested to explain the contrasting elements in the policy of peaceful coexistence.

Firstly, according to the new interpretation, war can be avoided only if a permanent struggle to prevent it is waged. One way of engaging in this struggle and thus of preventing war may simply be to emphasise the fact that the sources of war continue to exist.

Secondly, Soviet leaders may fear that in some indefinite future, under circumstances that cannot be defined despite the laws of socialistic development, capitalistic countries may be tempted to attack the socialistic system. Since they regard the military power of the Soviet Union as one of the best guarantees against such an attack, their emphasis on the permanent threat from imperialism may be interpreted either as an important premise of Soviet security policy or even as a justification for continuing to build up Soviet military capability.

A third possibility is that the call for resolute struggle against the unchanged enemy may be an intentional exaggeration to demonstrate for the Soviet allies the necessity of maintaining the socialistic military bloc under the leadership of the Soviet Union, to motivate a call for unity in Soviet society, and for the struggle against foreign ideological penetration, and so on. In other words, it is a theme that can be used by Cold War elements in leading political and military circles to advance their ideas.

The Chinese variant

From a common source in Marxism–Leninism, Soviet and Chinese theory on the inevitability of war have come now to differ on some important points.

Unlike the Soviets, the Chinese have never proclaimed an official policy of peaceful coexistence,[40] However, the contention that inter-system war is inevitable has seldom been mentioned in recent times, and when it has, it has seemed to refer to some indeterminate future.[41] One new element in the Chinese position is the often repeated statement that socialistic countries will not be the first to start an interstate war; such a war can only break out if it is unleashed by the imperialists. (This clarification of the Chinese position may be regarded as a response to Soviet and Yugoslavian charges that the Chinese contemplate the possibility of initiating a world war.) More recently a few statements have been made suggesting that the Chinese believe that some steps can be taken to prevent the outbreak of a world war.[42] It would seem then that the difference in substance is less than the polemics between the two would lead one to believe.

With regard to inter-imperialistic wars, the Chinese go beyond the orthodox position by including as an instance of this kind of war the conflict between the superpowers. Curiously enough, when the imperialistic character of the Soviet

Union in relation to countries in the Third World is pointed out, the Chinese emphasise the collaboration between the superpowers; [43] it is only when relations between the two superpowers are analysed that the long-run inevitability of an armed clash between them is mentioned. [44]

Unquestionably the greatest amount of attention is devoted to the inevitability of both revolutionary wars and wars of national liberation. The Chinese argue that since capitalism wages a relentless struggle against the oppressed masses and does not shirk at the use of violent means, the only conceivable outcome is the spread of open civil revolutionary war to an ever greater number of countries. [45] Apart from internal causes, [46] each revolutionary war is, moreover, part of the international world revolution. Since this revolution is an inevitable phenomenon according to the laws of social progress, then each of its parts must also occur. Thus in the Chinse view, imperialism may be defeated by a series of revolutionary wars instead of one decisive world war. Interstate war and war between the antagonistic social systems become secondary in importance to revolutionary wars.

The Yugoslavian variant

Although the clasical Marxist ideas constitute the ideological foundation of the Yugoslavian views on war and peace, and despite basic agreement that the capitalistic system, being the source of all wars, must be liquidated before wars can be eliminated, there are some differences between the Yugoslavian and both the Soviet and Chinese variants of the doctrine of the inevitability of war.

In the first place, the Yugoslavians maintain that the inevitability of war between social systems cannot be expressed as a law but rather as a relation between opposing forces. [47] Thus such a war was in some sense inevitable while the balance of power favoured those forces which desired it; such a situation existed in the epoch when imperialism was absolutely dominant and the anti-imperialistic forces were unable to prevent war. Now when the peace-loving forces, to which the policy of the socialistic states makes an important contribution, have come to the fore, war can be prevented. [48]

Owing to the corresponding weakening of the imperialistic system, war between the capitalistic countries is considered to be reduced to a theoretical minimum. [49] On the other hand, the possibility that a capitalistic country will attempt to conquer a weaker nation has theoretically increased, although this kind of war may be prevented through the vigilance of peace-loving forces even before the capitalistic system has become finally liquidated.

The Yugoslavians are perhaps most emphatic in their rejection of all suggestions that revolutionary war might be inevitable. They have suggested a variety of ways in which a revolutionary transition from capitalism to socialism might be made peacefully without any use of violence. As the working class becomes stronger, the capitalists will be compelled to make concessions. Therefore the proletariat can use parliamentary means of struggle to build up its political power and to realise gradually socialistic goals. It is even possible that non-Communist progressive forces may adopt socialistic measures as part of their programme. [50] The Yugoslavians are firmly convinced that the revolution can only succeed if it evolves naturally within the national context of each country. The internal conditions are always decisive for the choice of the right moment for the outbreak of revolution and the proper road to socialism. Any attempt to instigate or impose revolution from the outside under the pretext of supporting the working class or acting on behalf of the world revolution is bound to smother the true revolutionary spirit. Behind such an action would furthermore lie a fundamentally anti-socialistic and reactionary policy of great power hegemony.[51]

The nature of war: stable or changing?

Interwined in the Soviet discussion of the instrumentality and inevitability of war under conditions of the nuclear missile era has been an important issue, whether the essential nature of war has changed. [52] To the uninitiated observer the discussion may appear confusing owing to the Soviet tendency to use the term nature of war in two somewhat different senses.

The notions

In the broader of the two the nature or essence of war encompasses those general features which distinguish war as a political act and armed struggle from other socio-political and socio-historical phenomena. If used without qualification the term nature of war in Soviet studies is to be understood in this sense. [53] Simply because this definition is so broad, it is of little use in distinguishing between different concrete instances of war. For this purpose a more complex notion of the nature of war is needed. As was noted in a previous chapter, Soviet theorists conceive wars to have political, socio-historical, and ethical characteristics that reflect the socio-political features of the historical epoch in which they occur. When analysing the nature of a given war, they treat it as an inner combination and dialectic of features it shares with war in general, and other wars of its historical type, and of features peculiar to it alone. In other words, in such an analysis the notion of the nature of war comprises three notions: that of the nature of all wars, i.e. of war as a socio-historical phenomenon and a political act; of the nature of the particular type of war with which the given war is classified; and of the nature of the given war. [54]

When discussing the changeability of the nature of war, Soviet scholars have traditionally laid down that war is in all epochs a continuation of politics by military means and that its essential nature does not change. They have of course recognised that wars differ from epoch to epoch, but they have preferred to confine their remarks about the nature of war to the highest level of abstraction, and thereby to emphasise the timelessness of its essential nature.

The conditions in the nuclear missile era have persuaded several scholars, most of whom belong to some Western tradition, but one or two of whom are Soviet authors, to challenge the continued validity of the theses that war is a continuation of politics. This appears to be the stimulus accounting for the greater amount of attention Soviet study have devoted to clarifying the sense in which the nature of war may or may not be changeable. The general consensus seems to be that there is an essential core of meaning that has not changed but that certain basic elements have become modified.

There seem to be two main ways in which Soviet scholars have gone about dissecting the nature of war to make this distinction. One is to analyse the term in the context of political ends and military means. The other is to ask in what sense it can be said that each of the three parts the complex notion nature of war comprises; the general, the particular, and the individual can change.

With regard to the first type of distinction, one scholar has argued that while the basic relation between war and politics remains unaltered – war is still a continuation of policy, and policy is still primary – the content of policy, and the forms of warfare have evolved, whereby the form of the relation between politics and war have also changed.

Some scholars have put emphasis on the fact that policy is the most dynamic in the nature of war; what is progressive at one stage may become reactionary at the next, [55] for example, the policy of the bourgeoisie was once progressive but is now reactionary. Other scholars point to the military technical side of war as that aspect which is especially time-bound.[56]

Professor Rybkin elaborates on the idea that it is the form of the relation between politics and war that constitutes the transitory aspect of the nature of war. He indicates three ways in which this relation has changed in the nuclear era. The possibilities of starting war are more limited; war is thus less of a continuation of politics than it once was. Furthermore the possibility of achieving desirable political changes through war, the instrumentality of war, has become greatly restricted. Finally, the nature of modern military technology makes it increasingly difficult for politics to control warfare once the state of war has arisen.

Some scholars present the question of the changeability of war in a more abstract way. According to their argument, 'the essence of war is not only stable, but also in a state of motion (podvizhnyi)'. [57] The solution to this paradox is based on the premise that changes in the fundamental nature of war cannot be noticeable. An observer could not claim to have detected a sudden shift in the essential nature of war, because the same phenomenon cannot have two essences; a different essence implies a different phenomenon. Nevertheless, some gradual quantitative changes always occur within the framework of a given qualitative definition, but 'such change is insignificant with respect to the characteristic essence of the object'. [58] In time, the cumulative effect of these shifts may become important enough to allow one to speak of a qualitative change, but while the thing exists, the core of its nature remains essentially unchanged. [59]

This argument has also been put forward in other ways. Some assert that although some features of the nature of each phenomenon always change under real changing conditions, the phenomenon retains its identity (opredelonnost').

Thus the general nature of war does not change, as long as war is used as an instrument for the achievement of class interests; its manifestation in particular types of war, however, changes from one social formation to another, from epoch to epoch, and naturally there are differences between individual cases of war. [60] Both those who maintain that the nature of war is undergoing change and those who maintain that it is not, seek support for their respective positions in the writings of Lenin. The latter quote his opinion that the essence of each phenomenon is stable, the former his view that it is 'not only the manifestation of things that are transient and moving but the nature of things as well'.

The present Soviet position on the permanent and transient features of war can perhaps be summarised as follows:

Stable features

(a)   War is socio-political class action aiming at the achievement of socio-political class goals.
(b)   War is distinguished from other socio-political actions by the organised application of military means.
(c)   In the relation between politics and warfare, politics plays the determinant role.

Transient features

(a)   The policy governing a particular war will depend on the epoch in which the war occurs and on individual peculiarities.
(b)   The influence of politics on the outbreak and course of war will vary.
(c)   The way military violence is applied varies.
(d)   The instrumentality of war changes.

The Soviet interpretation of Clausewitz

During the re-examination of military doctrine in the conditions of the nuclear missile era, the discussion of such problems as the sense in which war could be considered a continuation of policy, or the meaning of the instrumentality, desirability, and inevitability of war, was often accompanied by comments concerning Clausewitz's views on these matters. Since there seems to be some correspondence between the development of Soviet political-military ideas concerning the nature of war and the shifts in Soviet interpretations of Clausewitzian formula, it may be of some use to review the latter. [61]

The traditional approach

Marxism–Leninism has incorporated many ideas that have originated in disparate philosophical traditions. As an integral part of Marxist–Leninist theory, they are sometimes interpreted without reference to their orginal source, and if reference is made to the origin of the idea, it is not the similarity but the difference between its Marxist–Leninist use and its original meaning that is emphasised. The adoption of the Clausewitzian formula that war is the continuation of policy by other means is a good case in point. Within the framework of Marxism-Leninism the meaning of the formula has been profoundly changed in two respects. In the first place, the principal terms, policy and war, were defined in relation to the fundamental Marxist concept of class. And secondly, in consequence of the broader concept of war, the formula was applied not only to wars between states, but also to internal war. Once these changes were made, it was possible for Soviet scholars to criticise Clausewitz for erring on these two points, and to dismiss him by branding him and his disciples as representatives of the reactionary classes.

Stalin's opinion of Clausewitz and its consequences

In February 1947, when Colonel Razin asked Stalin whether Clausewitz was still relevant, Stalin answered in the negative and gave three reasons for his opinion. [62] Firstly, Clausewitz only confirmed the familiar Marxist theses that there is a direct connection between war and politics. Secondly, his military doctrine, together with the whole German military ideology, was inconsistent with the interests of the Soviet Union and should therefore be criticised. Thirdly, as a representative of the hand-tool stage of military technology, Clausewitz could hardly be considered an authority in the modern machine-age of warfare.

The article in which Stalin gave his answer touched off a thorough critique of the political–military ideology and the philosophy of war on which Clausewitz based his formula of war as a political act.

Three main objections were raised against Clausewitz's philosophy of war. To begin with, the balance between his two concepts of war as an act of violence on the one hand and as a political act on the other was incorrect. In Clausewitz's view, the primary concept was absolute war, which in its pure form was by definition independent of political considerations. The political element of war was clearly secondary since its only role was to constrain the amount of violence used. Moreover the whole notion of absolute war was unacceptable to Soviet scholars. They interpreted it as 'the licentiousness of biological, animal instincts, massive physical violence unlimited by any norms or laws of conducting war'. [63] To regard absolute war as an ideal, as some sort of perfect state worth striving for, was an indication of Clausewitz's reactionary class ideology. The third objection concerns the second concept of war as a political act. Although this idea contains a grain of truth, it was distorted by an incorrect interpretation of what constitutes politics. In the idealistic interpretation of Clausewitz, politics and war are seen as the result of the voluntary actions of rulers. Marx had shown the folly of such a view by demonstrating that behaviour is rooted in the material base, that is in the system of production and the socio-economic structure. Owing to this philosophical approach, Clausewitz treats politics as a non-class phenomenon and confines himself to foreign policy. He thereby makes the mistake of assuming that the policy conducted by the ruler of a state embodies the interests of the whole nation.

Some of these errors can be explained by the fact that Clausewitz's philosophy of war is rooted in a reactionary ideology. He was, after all, a man who was an ideologist of Prussian militarism, a defender of the interests of the German Junkers, an enemy of the French Revolution and the democratic movement for uniting Germany, a proponent of the policy-state, and an apologist of predatory aggressive wars. By treating politics and war as matters of state, he was able to conceal the real class content of predatory wars. His concept of absolute war reflected the Prussian longing to annex its neighbours; the concept of real war described the frustration of an aggressor who fears that he is still too weak to carry out the great annexations of which he dreams and is compelled to follow a policy of caution. [64]

The Soviet critics therefore conclude that the Clausewitzian formula of war as a political act, taken in the context of his whole philosophy and ideological background, could be rejected as unscientific. His idea about the connection between war and policy was only partly true, and it was not until Marxism–Leninism gave it a correct interpretation that it became scientific and true. Marx and Engels were the first to reveal the social character of war, and Lenin and Stalin were the first to indicate that war had a class-political character. Following this official critique, Clausewitz's work fell into oblivion in the Soviet Union. Most often when the political nature of war was discussed in some work on political or military matters, no mention was made of Clausewitz.[65] When his name appeared, his whole theory was dismissed as unscientific and idealistic.

The reappraisal after Stalin

In the course of the re-examination of all political-military doctrine that took place after Stalin's death, Clausewitz's ideas on war and politics were initially met with somewhat more favour. [66] The deficiencies of his position – his failure to treat war and politics as class phenomena, his exclusion of the role of popular masses in historical and political processes – were pointed out, but more of value was found than had been in the traditional Soviet assessment. For example, it was to his merit that Clausewitz took into account the influence of social relations on political events and thus did not regard politics as the activity of the state exclusively. He also recognised that the destruction of enemy forces was not the only supreme aim of military operations; for him military action was only one of several means of achieving political aims.

What is more, the philosophy underlying Clausewitz's position was placed in a more favourable light. It was pointed out, in the East German literature but not without Soviet approval, that Clausewitz had reached his conclusions about the relation between war and politics through an application of the dialectical method to problems of war; this was 'a revolutionary act which made it possible to identify war as a social phenomenon and to analyse effectively the laws of the origin, roots, and course of war'. [67] The elements of idealism in his philosophy were said to be unrigourously applied and internally broken; absolute war should not be considered an existence in itself because according to Clausewitz, the truth of such logical constructions was proportional to the extent to which they reflected reality. It was suggested that despite the idealistic tone of his argument, Clausewitz actually adopted on some points a materialistic position. Instead of labelling Clausewitz's philosophical views anti-Marxist or reactionary as had traditionally happened before, [68] some scholars summed up their general assessment as indecisive. Clausewitz's writing was called in the East German literature an outstanding work on political-military questions and the foundation of the new military science and military art.

The return to criticism

Such glowing praise of Clausewitz can only be found in some studies from the mid-1950s. Although the general revision of Clausewitzian philosophy of war was never renounced, the Soviet scholars attitude has over the years gradually returned to the more traditional position. Occasionally he has been named the creator of the philsophy of war, and the first who gave a 'clear and penetrating definition of the nature of war and of the linkage between war and politics', [69] for which reason his contribution should be considered progressive rather than reactionary. But all the old criticisms have again been taken up and they tend to dominate the general tone of Soviet assessment. [70]

In recent years it has become usual to accentuate the differences between the Clausewitzian and Leninist interpretations of the the formula that war is the continuation of politics by other means. [71] The discussion of the subject in the history of the Soviet Armed Forces, written by the Minister of Defence, should serve as an adequate example. [72] There three differences between the Clausewitzian and Soviet interpetations are clearly presented: first, war as a continuation of foreign policy alone as opposed to war as the continuation of both domestic and foreign policy in their inseparable unity; second, politics as the embodiment of the highest intelligence of the state as opposed to politics as a class instrument; and third, idealistic philosophy, distorting the nature of politics and war and obscuring the roots and character of particular wars as opposed to the materialistic interpretation of war and the class system that generates wars. [73]

Moreover, in the modern critique of the philosophy of war considered to be most representative of Western thought, namely political realism, the Clausewitzian theory which is said to be adopted by realists is often to be tarred with the same brush, particularly with respect to its exaggeration of the role of armed violence and to its moral indifference. [74]

At the same time as the Clausewitzian contribution to the study of war has been soft-pedalled, that of Marxism–Leninism has been played up. Sometimes the thesis that war is a continuation of politics is described as a Marxist–Leninist principle without any reference to Clausewitz. [75] On other occasions Clausewitz is mentioned as the author of the principle, but the fundamental changes made by Lenin are underlined. [76] In the most zealous effort in this direction, it has even been suggested that some Russian theoriests defined the link between war and politics before Clausewitz did. [77]

Peace and peaceful coexistence

As in the West, there has been a tendency in Soviet political military thought to devote more attention to a detailed analysis of peace, and in particular to the question of how political goals, whose achievement through war is no longer possible, may now be realised by peaceful means.

Peaceful coexistence as a form of class struggle

In Marxist–Leninist theory there is no *a priori* reason why some given goal could be reached only through warfare and not by peaceful means. Like war, peace may be just or unjust depending on the nature of the political goals that characterise it. Since politics is an expression of the interests of the governing classes, the periods between wars in which armed violence is not used during the imperialistic epoch can be considered unjust, but peace may also be a state of affairs in which socialistic policy is realised, in which case it would be a just peace.

Peaceful coexistence is a set of ideas about the aims of such a policy and about how it can be achieved.[78] The cornerstone of the construction is the assumption that the strength of the socialist camp is sufficient to prevail over imperialism and to overpower forces working for an imperialistic peace or an imperialistic war. In this task socialism is aided by the forces struggling for national liberation and by revolutionary movements within the capitalistic countries.

Peaceful coexistence is conceived as a form of the international class struggle for the ultimate victory of socialism waged without the use of arms,[79] (on some occasions it is presented as an alternative to nuclear rocket war or world war, on others as an alternative to all wars between states). It combines an ideological struggle in the international arena with political class struggle within particular capitalistic countries and economic competition between the two social systems; the object of this struggle is to demonstrate the superiority of the socialistic system. It is a basic premise of the policy that such competition can take place without war and in observation of the sovereign and equal rights of nations, their right to self-determination, and the inviolability of the territory of other states. At the same time, peaceful coexistence calls for economic and cultural cooperation in fields in which this can take place to the profit of all countries involved.

The prevention of war

The declared aim of the political-military doctrine corresponding to the official general policy of peaceful coexistence has been to prevent war. To this end, the Soviet Union has put forward a number of proposals which indicated the limits to its non-involvement in war. [80] The Soviet Union would retaliate with all means at its disposal (even nuclear rocket war) to defend its territory and that of its allies from any act of military aggression. The imperialistic powers have furthermore been warned against intervention in local wars and especially against invading small countries; such actions could lead to major wars. More recently this veiled threat of Soviet counter-intervention has taken the form of a warning that imperialistic intervention would result in defeat for the intervening forces.

Thus the policy of peaceful coexistence does not prevent the Soviet Union from defending itself against armed attack, nor a nation dispossessed of its territory by force from using force to recover it. The exploited masses also retain their right to fight for social liberation. The latter exception provided the grounds on which the Soviet Union excluded internal wars (revolutionary wars and wars of national-liberation) from the doctrine of war prevention.

The Chinese and Yugoslavian variants

Within the so-called socialist system, the Soviet concept of peaceful coexistence has been criticised from two quarters.

On the one hand, the Chinese contend that such a concept is both unrealistic and harmful. It is unrealistic because the national and social contradictions of the modern world lead inevitably to war, if not to world-wide war. Peaceful co-existence is therefore nothing more than an unarmed truce unable to prevent local wars. But what is more, the policy is harmful because it amounts to a plot of the superpowers to divide the world into spheres of influence without regard to the interests of the small and the undeveloped nations, and to suppress revolutionary movements and movements for national liberation.

On the other hand, the Yugoslavians have presented a theory of peaceful coexistence that in effect is more peaceful than the Soviet theory in two respects. The first is the emphasis placed on the need to prevent all kinds of war, whether initiated by the imperialists or by the socialistic countries and thus not only inter-state wars but also armed interventions launched to support some social or national revolution. The Yugoslavians maintain that a war is not just, simply because it is waged by a socialistic country; on the contrary such a war is damaging for the cause of peace and revolution.

The second difference is that the Yugoslavians interpret peaceful coexistence to mean coexistence between states, not between blocs of states; they consider bloc-building to increase hostility and to create conditions for future wars. True cooperation, which favours economic development and fosters national independence, can only take place on a bilateral basis. Being tied to a bloc only hinders a state from such active coexistence. The alternative they propose is that each state follow an independent policy, a policy of non-alignment, which would best serve its interests. In this way bloc division would be overcome, international relations would be democratised, and since acute international problems could then be solved, peace would be preserved.

Notes

[1] The official papers, as well as the scholarly literature on the subject, are too vast to be listed here. For the exposition of the theory, see: *Voennaya strategiya*, 1968; Rybkin, 1959, 1973, 1974; Prokop'ev, 1965, Pukhovskii, 1965; Kozlov, 1971; Azovtsev, 1971, Grudinin, 1971, *Filosofskoe nasledie V. I. Lenina* . . . 1972; *Marxism–Leninism on War and Army*, 1972; *Problems of War and Peace*, 1972; Seleznev, 1974, Tyushkevich, 1975.

[2] 'It is well known that the essential nature of war as a continuation of politics does not change with changing the technology and armament.' (*Voennaya strategiya*, p.25).

[3] 'As regards its essence, such a war would also be a continuation of the politics of classes and states by violent means. . . . The social, class content of nuclear missile war and its aims will be determined by politics. The new world war will be, on one side, the continuation, weapon, and instrument of criminal-imperialist policies being implemented with nuclear missiles. On the other side, it will be the lawful and just counteraction to aggression, the natural right and sacred duty of progressive mankind to destroy imperialism, its bitterest enemy, the source of destructive wars' (*Marxism–Leninism on War and Army*, pp.28–9). In similar words: *Filosofskoe nasledie V. I. Lenina* . . ., pp.36–7; T. Kondratkov, 'War as a Continuation of Policy', *Soviet Military Review*, 1974:2.

[4] 'A thermonuclear war launched by imperialism would be qualitatively different from wars in the past or modern wars using conventional weapons

above all in its concrete political content, its political aims and its class character. It would resolve not individual, limited political issues, but an issue of historic importance, affecting the fate of all mankind.' (*Problems of War and Peace*, pp.96-7). Factors assuring victory of the socialist states are presented as follows: 'In the new war, if it should be allowed to happen, victory will be with the countries of the world socialist system which are defending progressive, ascending tendencies in social development, have at their command all the latest kinds of weapons, and enjoy the support of the working people of all countries. The balance of forces between the two systems, the logic of history, its objective laws, prescribing that the new in social development is invincible – all this predicts such an outcome. The might of the Soviet state, of the entire socialist community, which possesses the economic moral-political, and military-technical preconditions for utterly routing any aggressor, substantiates this view. Other factors and forces which will inevitably spring into action as soon as war breaks out must also not be thrown off the scales; they will include decisive anti-imperialist actions by the people, political, diplomatic, international legal, ideological and other actions against those responsible for unleashing a nuclear adventure.' (*Marxism–Leninism on War and Army*, p.30). In another place the authors state: 'Undoubtedly, a new world war, should it be unleashed by the imperialists, will bury the capitalist system' (p.78). '. . . a thermonuclear war is not an ordinary war. Because of the terrible devastation and annihilation it involves, on a scale absolutely without precedent, it cannot serve as a reliable means of acieving any political aims; *it would be suicidal for the aggressors*'(it. added) N. Sushko, *The Essence of War*, 1965:7, p.11). Cf. *Filosofskoe nasledie V. I. Lenina* . . . p.24.

[5] One of the main themes in Rybkin, 1965.

[6] 'If the imperialists unleash a new world war, the toilers will no longer tolerate a system which subjects people to devastating wars. They will mercilessly and irrevocably sweep capitalism from the face of the earth. . . . The more vigorously and resolutely they oppose the actions of the aggressor, the less damage will be inflicted on world civilization.' (*Filosofskoe nasledie V. I. Lenina* . . . p.25).

[7] 'To affirm that victory in a nuclear war is impossible is not only theoretically incorrect but politically dangerous. . . . *A priori* denial of the possibility of victory is harmful, because it leads to moral disarmament, disbelief in victory, fatalism and passivity' (Rybkin 1965).

[8] 'A thermonuclear war would kill hundreds of millions of people, lay waste entire countries, inflict irretrievable losses to material and spiritual culture. Mankind would be thrown back for many decades.' (*Marxism–Leninism on War and Army*, p.73). 'Such a war can cause substantial detriment to the development of world civilization, inhibiting the advance of the revolutionary process' (*Filosofskoe nasledie V. I. Lenina* . . . pp.23-4).

[9] E.g. McNamara's speech in Ann Arbor, 1962, was immediately sharply criticized by Marshal Sokolovskii in 'Strategiya samoubiitsev', *Krasnay*

Zvezda, 19 July 1962, and later by Gen. N. Sushko and Col. T. Kondratkov in 'Voina i politika v yadernyi vek', *Kommunist Voorushennykh Sil*, 1964:2.

Cf. G. Gerasimov, 'Twist of Military Thought', *International Affairs*, (Moscow), March 1963; V. Pechorkin, 'About "Acceptable War"', ibid. For a review of the criticism; see Lider, 1969, the analysis of Kahn's doctrine.

[10] K. Stepanov, Y. Rybkin, 'O kharaktere i tipakh voin sovremennoi epokhi', *Voennaya Mysl*, 1968:2, p.69.

[11] G. Malinovskii, 'Lokalnye voiny v zone natsionalno- osvoboditelnogo dvizheniya', *Voenno-Istoricheskii Zhurnal*, 1974:5; V. Mochalov, *Bolshaya lozh o malykh voinakh,* Voenizdat, Moscow 1965; his 'What Lies Behind the Theory of "Limited Wars"', *Soviet Military Review*, 1969:8; I. Shabrov, 'Lokalnye voiny i ikh mesto v globalnoi strategii imperializma', I–II, *Voenno-Istoricheskii Zhurnal*, 1975:3,4.

[12] The doctrine of flexible response and its theory of limited wars is 'inconsistent from the purely military standpoint: staking on provoking "local" conflicts, this doctrine does not ensure the imperialists the main thing – victory'. And: 'The new world balance of forces makes any "local" war difficult and futile for the aggressor, because the victims of the aggression receive aid from the socialist countries' (Mochalov, 1969, p.56). Malinovskii, 1974 contends that the analysis of the local wars unleashed by the imperialists (Algeria, Yemen, Bangladesh, Vietnam) substantiates the view that aggressors cannot achieve their political aims because of the changes in the correlation of world forces (pp.97–8).

[13] 'The historical evolution until now has always confirmed [the thesis] that the revolutionary overthrow of capitalism was at every time interrelated with world wars. Both the first and the second world wars were very strong accelerators of the revolutionary explosions' (*Osnovy Marksizma–Leninizma*, Moscow 1969, p.519, transl. *The Fundamentals of Marxism–Leninism*, Foreign Languages Publishing House, Moscow 1960).

[14] 'The revolutionary struggle for the dictatorship of the proletariat includes the exertion of open political coercion on the exploiters, but does not necessarily involve armed struggle. . . . The proletariat attempts to use primarily peaceful means for the revolutionary changes of the political system. Only when it has exhausted all peaceful means and encounters fierce resistance on the part of the reactionary classes, is it compelled to take to arms, to take up armed struggle (*Marxism–Leninism on War and Army*, p.79); revolution is a law of social progress, armed uprising, however, is dependent on circumstances (Pukhovskii, 1965, p.80); 'When, in what circumstances, did Lenin argue for resolving class or national contradictions by war, by a continuation of politics by violence? (P. Trifonenkov, 'Voenno-teoreticheskoe nasledie V. I. Lenina i sovremennost', *Voenno-Istoricheskii Zhurnal*, 1967:11).

[15] Both the seizure of power by an armed uprising and its armed defence are responses to the violence applied by the enemy. 'In both cases this is a coerced violence, an answer by the working class.' (Stepanov, Rybkin, 1968, p.70).

'Naturally, the capitalists do not renounce their class privileges and their political rule voluntarily. They fiercely resist the revolution. The intensity of the class struggle, its forms and methods of violence during the transition to socialism do not depend so much on the proletariat as on the resistance offered by the exploiters, on whether or not the bourgeoisie resorts to armed violence.' (*Marxism–Leninism on War and Army*, p.79).

[16] I. M. Ivanov, *Mirnoe sosuschestvovanie i krizis vneshne-politicheskoi ideologii imperializma*, Moscow 1965 (Ch. III 2 'Kleve-tnicheskoe otozhdestvlenie sotsialisticheskoi revolutsii c voinoi'); V. Tikhonenko, 'O sootnoshenii voiny i sotsialisticheskoi revolutsii', *Kommunist Vooruzhennykh Sil*, 1967:21, p.82; G. Khvatkov, 'Problema voiny i mira i revolutsionnyi protsess', *Kommunist Vooruzhennykh Sil*, 1970:1, pp.28–35.

A new problem was presented at the Twenty-second Party Congress: 'Communists never thought and do not think that the way to the revolution necessarily leads through wars between states. Socialist revolution is not necessarily connected with war. Although world wars unleashed by the imperialists resulted in socialist revolutions, revolutions are fully possible without wars. The great aims of the working class can be accomplished without a world war. The conditions for it are nowadays more favourable than ever before.' (Materialy XXII S'ezda KPSS, p.348).

[17] *The Fundamentals of Marxism–Leninism*, 1960.

[18] A. E. Kunina, B. I. Marushkin, *Mif o mirolubii SShA*, Moscow 1960; A. A. Kirillov, *Predotvrashchenie voiny – vazhneishaya problema sovremennosti*, Moscow 1962; V. I. Zamkovoi, *Kritika burzhuaznykh teorii neizbezhnosti novoi mirovoi voiny*, Moscow 1965; 'Peace and socialism are indivisible: socialism creates the socio-economic basis for the peaceful co-operation of peoples, and peace promotes the development of the world revolutionary process and the triumph of socialism in all countries' (*Marxism–Leninism on War and Army*, p. 78). 'Peaceful coexistence is a special form of class struggle. Not only does it not exclude liberatory people-democratic and proletarian revolutions, but creates more favourable conditions for them. It aims at the greatest mobilization of forces for strengthening and increasing the world socialist system, for liquidating the remnants of colonialism, for strengthening the independence of the young countries and their development toward democracy, for the growth and influence of communist and workers' parties . . .' (M. S. Kapitsa, *KNR: dva desyatiletiva – dve politiki*, Moscow 1969, p.274).

[19] Therefore: 'Communists reject the theory of pushing revolution by war. The victory of revolution in each particular capitalist country needs certain definite prerequisites of objective and subjective character' (Kapitsa, ibid. p.279).

[20] The forms of compellence are determined by the degree of the development of a given revolutionary class, by particular circumstances (e.g. the heritage of a long and reactionary war), and by the forms of the resistance of the bourgeoisie and of the lower middle-classes (L. Slepov, 'Sovremennaya epokha

i strategiva i taktika kommunistov', *Kommunist Vooruzhennykh Sil*, 1970: 2, p.20).

[21] 'Sovremennaya epokha i mirovoi revolutsionnyi protsess', A. M. Kovalev, ed., Izdatelstvo Moskovskogo Universiteta, Moscow 1970, pp.251–2. The conclusion is that wars do not generate revolutions but they in some respects favour their outbreak. Cf. A. P. Butenko, *Voina i revolutsiya*, Gospolitizdat, Moscow 1961, pp.10 ff.

[22] 'It goes without saying that there can be no peaceful coexistence where matters concern the internal processes of the class and national-liberation struggle in the capitalist countries or in the colonies. Peaceful coexistence is not applicable to the relations between oppressors and the oppressed, between colonists and the victims of the colonial oppression.' (Leonid Brezhnev, *Report of the Central Committee to the XXIII Congress of the Communist Party of the Soviet Union*, 1966.)

[23] V. L. Tyagunenko, *Voiny i kolonii*, Voenizdat, Moscow 1957; K. H. Brutents, *Protiv ideologii sovremennogo kolonializma*, Socekgiz, Moscow 1961; E. D. Modrzinskaya, *Ideologiya sovremennogo kolonializma*, IBL, Moscow, 1961; G. I. Mirskii, *Armiya i politika v stranakh Azii i Afriki*, Izd, 'Nauka', Moscow 1970; V. Tyagunenko, 'Nekotorye problemy natsionalno-osvoboditelnykh revolutsii v svete leninizma', *Mirovaya Ekonomika i Mezhdunarodnye Otnosheniya*, 1970:5; A. Iskanderov, 'Kommunisty i natsionalno-osvoboditelnoe dvizhenie', *Voprosy Istorii KPSS*, 1970:2; D. Kunayev, 'Lenin i natsionalno-osvoboditelnoe dvizhenie', *Kommunist*, 1969:17; V. Lyusinov, 'O formakh revolutsionnogo perekhoda k sotsializmu v razvivayushchikhsya stranakh', *Mirovaya Ekonomika i Mezhdunarodnye Otnosheniya*, 1973:2; G. Kim, P. Shastitko, 'Nekotorye voprosy sovremennykh natsionalno – osvoboditelnykh revolutsii v Azii i Afrike', *Voprosy Istorii*, 1973:8, *Marxism–Leninism on War and Army*, Ch.II.4: 'Wars Between the Colonialists and the Peoples Fighting for their Independence'; Y. Zhukov, 'The Rise of the National Liberation Movement After the Second World War', *International Affairs*, 1975:6; S. Bagdasarov, 'Krushenie kolonialnoi sistemy imperializma', *Kommunist Vooruzhennykh Sil*, 1975:15 (a study material in the armed forces); Y. Sumbatyan, 'Razryadka naprazhennosti i natsionalno-osvoboditelnoe dvizhenie', *Kommunist Vooruzhennykh Sil*, 1975:24.

[24] 'When capitalism was the only or the dominant force in the world arena, the peoples of the colonies and dependencies were restricted in their choice of the form and means of liberation movement, and at the time armed struggle was the surest way to national and social freedom. The victory of the Great October Socialist Revolution and the emergence of the socialist world system which followed the defeat of nazi Germany and militarist Japan, added to the arsenal of the national-liberation movement, and enabled the oppressed peoples to advance to independence along both military and non-military roads' (Y. Dolgopolov, 'National Liberation and Armed Struggle,' *SMR*, 1968:9, p.46).

[25] 'It would be wrong to counterpose the armed forms of struggle to the

peaceful ones, and vice versa. Lenin wrote that, in order to accomplish its tasks, the revolutionary class must be able "to master *all* forms or aspects of social activity without exception", it "must be prepared for the most rapid and brusque replacement of one form by another". The question is to utilise precisely those forms and methods of the liberation movement which correspond most fully to the given situation.' (Dolgopolov, ibid, p.47).

[26] 'The national-liberation movement, in spite of the great diversity of its forms, is a part of the world revolutionary process, but in the class content and in its nearest aims it is not a part of the world revolutionary socialist revolution. The proletarian movement and the national-liberation movement are two parts of the world revolutionary process, with limitless possibilities of cooperation.' (Kim, Shastikov, op.cit.).

[27] 'The direct international duty is to help (to a national-liberation war – JL) by military force if necessary, and also to help the victorious proletariat to prevent a counterrevolution. The Soviet Union is well known as fulfilling this duty honourably.' (V. Rutkov, Leninskoe uchenie ob imperializme i sovremennyi imperialism, *Kommunist Vooruzhennykh Sil*, 1969:13).

[28] 'In our times no illusion can be more dangerous than the view that a thermonuclear war may be an instrument of policy, that any political aims can be achieved by the use of nuclear weapons . . .' (N. Talenskii, Razumiya o minuvshei voine, *Mezhdunarodnaya Zhizn'* 1965:5, p.23); 'The disappearance of the possibility to win a world thermonuclear war, as a means of achieving political aims of states, and the negation of all military categories of war in a thermonuclear one indicate that a world thermonuclear war is not a war, but a self-denial of war' (N. M. Nikolskii, 1964, p.381). Cf. V. I. Zamkovoi, *Kritika burzhuaznykh teorii neizbezhnosti novoi mirovoi voiny*, Moscow 1965; P. N. Fedoseyev, 'Dialektika sovremennogo obshchestvennogo razvitiya', *Vestnik Akademii Nauk SSSR*, 1965:9.

[29] Some examples of the literature on the subject: M. S. Kapitsa, KNR: *dva desyatiletiya, dve politiki* Izd. Politicheskoi Literatury, Moscow 1969; *Kritika teoreticheskikh kontseptsii Mao Tse-duna*, Glavnaya Redatsiya Sotsialno-Ekonomicheskoi Literatury, Izd, 'Mysl', Moscow 1970; V. Zubarev, 'anti-marksitskii, avanturnicheskii kharakter voennykh kontseptsii maoistov', *Voennaya Mysl*, 1969:5; *Filosofskoe nasledie V. I. Lenina . . .* Ch.3 § 3.

[30] 'Rehashing the bourgeois theory of violence, the Maoists claim that war plays the role of principal motive force of the historical progress' (*Filosofskoe nasledie V. I. Lenina . . .*, p.75). M. A. Suslov accused the Maoists of contending that war is the principal means of transition to socialism (speech at the Central Committee Plenum, 14 February 1964, in *Pravda*, 3 April 1964). 'Maoists absolutize armed violence' (*Metodologicheskie problemy voennoi teorii i praktiki*, p.86).

'The idealistic theory of violence which Mao Tse-tung's group in China has adopted as a guide is of particularly great danger. It leads to a militarist bourgeois nationalist viewpoint, in keeping with which it is not class struggle or realistic estimate of all conditions contributing to the rising of socialist revolution, but war that is the decisive means for resolving the social contradictions' (K. E. Kalaushina, '*V. I. Lenin ob obshchestvenno-klassovoi prirode militarizma*' *Voprosy Istorii KPSS*, 1973:1).

[31] Mao Tse-tung's writings, Ling Piao's articles and the article 'Long Live Leninism' (Foreign Languages Press, Peking 1960, pp.20–2) have frequently been quoted and criticised. 'They consider that world thermonuclear war is inevitable and, attempting to hurry it along, they evidently suppose that the Chinese people will have the best chance since they are the most populous people on the earth. In case of the destruction of the majority of the peoples of the earth . . . there would remain in their opinion, the epoch of world domination by people of the yellow race. The Peking leaders have already come to terms with the idea of dividing people by race, by the color of their skins, rather than by class and social characteristics' (I. Yermashev, 'Pekinskii variant "totalnoi strategii" ', *Voennaya Mysl'*, 1963:10).

[32] 'Reactionary-utopian pacifists constitute a large group among those who hold this view. They state that "war has paralyzed itself", and that at present peaceful coexistence of peoples is guaranteed by the "nuclear equilibrium". . . . Many bourgeois and petty-bourgeois "wooers of peace", writers and philosophers of decadence, belong to the same group. The imperialist aggressors are not above exploiting their passive and contemplative position in order to lull the vigilance of the peace-loving forces' (Rybkin, 1965); abstract pacifist slogans aim at spiritually disarming the toilers (*Filosofskoe nasledie V. I. Lenina . . .* p.4). Marxist–Leninist theory proves 'the invalidity of the abstract pacifism, of the ideology of a part of small and middle bourgeoisie and of bourgeois intelligentsia, ideology which in fact gives a free hand to the militarism, and objectively is its servant'. (Dmitriev, 1975, p.16).

[33] Problems of war and peace, ch.16–18; *Filosofskoe nasledie V. I. Lenina . . .*, pp.74 ff; K. S. Bochkarev, O kharaktere i tipakh voin sovremennoi epokhi, *Kommunist Vooruzhennykh Sil*, June 1965. Georgi Zadorhzhnyi, *Peaceful Coexistence*, Progress Publishers, Moscow 1968; Mieczyslaw Michalik, *Moralność a wojna*, Wyd. MON, Warsaw 1973, ch. 'Absolutyzm moralny i negacja wojny'.

[34] '. . . the existence of the Soviet Republic side by side with the imperialist states for a long time is unthinkable. One or the other must triumph in the end. And before that end supervenes, a series of frightful collisions between the Soviet Republic and the bourgeois states will be inevitable.' (V. I. Lenin, *Sochineniya*, vol.29, p.133).

[35] I. V. Stalin, *Ekonomicheskie problemy sotsializma v SSSR,* Moscow 1952; Y. Usenko, *Prichiny imperialisticheskikh voin*, Gosizdat Politicheskoi Literatury, Moscow 1953.

[36] Materials of the XXth and XXIst Party Congresses; Khrushchev's Report delivered at the Fourth Session of the Supreme Soviet of the USSR, 14 January 1960, published as: *Disarmament: the Way to a Sure Peace and Friendship*

Between Peoples; R.Ya. Malinovskii, on many occasions (e.g. Report to the XXII Congress of the Communist Party of the Soviet Union by the Minister of Defence, 23 October 1961).

[37] While in a non-parliamentary system violence is the only means of seizing power, in a parliamentary system the struggle for power is also far from being peaceful. Transformation of the parliament into the instrument of popular masses, cannot be attained by election. Parliament ought to help to smash the military-bureaucratic machine, to deprive the ruling class of the army, police, and bureaucracy, to establish a new people's state. If all these conditions are not fulfilled, civil war is unavoidable (A. Belyakov, F. Burlyatskii, 'Leninskaya teoriya sotsialisticheskoi revolyutsii i sovremennost', *Kommunist*, 1960:13, p.20).

[38] The main theme of many studies, pamphlets and articles in *Kommunist Vooruzhennykh Sil, Mezhdunarodnaya Zhizn*' and *Kommunist* in the 1960s and 1970s.

[39] The main theme of Zadorozhnyi, 1968, esp. of the chapter 'Peace as a continuation of policy'.

[40] Mao's writings, Lin Piao's writings, esp. his programmatic article 'Long Live Victory of the People's War' (Peking New China News Agency, International Service in English, 2 September 1965). For a detailed bibliography, see Yin, 1971.

[41] 'Facts have proved that the aggressive nature of imperialism has not changed. . . . The danger is not yet over that imperialism will launch a new and unprecedently destructive war'.

'The Chinese Communist Party Resolution on the Moscow Conference', adopted on 18 January, 1961, by the Ninth Plenary Session of the Eight Central Committee, after G. F. Hudson, Richard Lowenthal and Roderick MacFarquhar, *The Sino-Soviet Dispute*, Doc. and anal. F. Praeger, New York 1963.

[42] '. . . owing to the fundamental change in the international balance of class forces a new world war can be preventd by the joint efforts of the powerful forces of our era – the socialist camp, the international working class, the national-liberation movement and all peace-loving countries and peoples. Peace can be effectively safeguarded provided there is reliance on the struggle of the masses of the people and provided a broad united front is established and expanded against the policies of aggression and war of the imperialists headed by the United States. The Communist Party and the people of China have always regarded the safeguarding of the world peace, the realization of peaceful co-existence and the prevention of another world war as their most urgent tasks in the international struggle' (ibid, pp.222–3).

[43] Yu Chao-Li in a programmatic article in *Red Flag* (1 April 1960) stated, that real contradictions were not those between the superpowers, but between governing imperialist groups and their own peoples, between imperialist countries and colonies (semi-colonies) and among the imperialist countries

themselves (after *Peking Review*, no.15, 21 April 1960, pp.17–24).

[44] '. . . the Soviet state monopoly capitalism, like imperialism, is subject to the objective laws of imperialism . . . the present struggle between the two nuclear superpowers — the United States and the Soviet Union — for world hegemony is the continuation of the history of contending for hegemony by the imperialist powers' (*Red Flag*, after *The Time of India*, 4 October 1973); and 'The nature of imperialism determines that while frequently colluding, the imperialist countries have no way of reconciling their conflicts in contending for world hegemony' (Shih Chun 'Again On Studying World History', *Peking Review*, 1972:24, p.11). See by the same author: 'U.S. Soviet Scramble for Hegemony in South Asian Subcontinent and Indian Ocean' (ibid, 1972:2); 'Superpowers's Contention for Hegemony in the Mediterranean' (1972:15); 'Why It Is Necessary to Study World History' (1972:21), 'On Studying Some History About the Imperialism' (1972:25,26).

[45] 'Whoever recognizes the class struggle cannot fail to recognize civil wars which in every class of society constitute the natural and under certain conditions inevitable continuation, development, and intensification of the class struggle. All the great revolutions prove it. To repudiate civil war, or to forget about it, would mean sinking into extreme opportunism and renouncing the socialist revolution' (Yu Chao-Li, *Peking Review*, 1960:15, p.20). Lenin's following statement was often quoted: 'no serious Marxist will believe it possible to make the transition from capitalism to socialism without a civil war' (*Long Live Leninism*, Foreign Languages Press, Peking 1960, p.36).

[46] 'Revolution is the affair of the peoples themselves in the various countries. The communists have always been against the export of revolution. They also resolutely oppose imperialist export of counterrevolution, against imperialist interference in the internal affairs of the people of various countries which have risen in revolution' (CCCR Resolution on the Moscow Conference, loc. cit., p.222).

[47] J. B. Tito, *Selected Speeches and Articles, 1941–61*, Zagreb 1963. Kardelj, 1961.

[48] 'Lenin did not hold war to be a fatal inevitability so long as there was a vestige of imperialism in the world, but always viewed the matter on the basis of the relationship of forces' (Kardelj, 1961, p.38).

[49] Kardelj, ibid, p.59

[50] Kardelj, 'On the Inevitability of Armed Revolution', in 1961, pp.80–97.

[51] Kardelj, 'The Policy of Coexistence and Marxism', pp.62 ff.

[52] *Marxism–Leninism on War and Army*, Ch. One; *Filosofskoe nasledie V. I. Lenina* . . . Ch. II; N. Sushko, 'The Essence of War', *SMR*, 1965:7; Rybkin, 1965; T. Kondratkov, 'The Substance of War', *SMR*, 1968:6; his: 'War as a Continuation of Policy', *SMR* 1974:2; Grudinin, 1971; Seleznev, 1974.

[53] 'The substance of war, as of any other thing, object, or process lies in its inner, deep-rooted, general basis that mirrors its specific features and its qualitative distinction from other social phenomena.' (Kondratkov, 1968, p.11);

Essence of war is its internal basis (osnova), which determines its character and the direction of its development (Grudinin, 1971, p.153); essence (nature) is the aggregate of the deepest, stable properties and relations of a thing (Seleznev, 1974, p.11).

[54] 'Using the category of essence, it is necessary to see within it the inner dialectic of the general, particular, and individual . . . the terms "essence of war as a socio-historical phenomenon", "essence of civil or any other types of wars" and "essence of a given, specific war" are non identical; they contain the dialectic of the general, particular and individual. Therefore, a study of any war should consist in moving upward from the abstract to the concrete, that is from a general (abstract) concept of war as a continuation of politic to a revelation of the specific and individual features of the given war . . .' (*Filosofskoe nasledie V. I. Lenina* . . ., pp.34–6). Cf. Tyushkevich, 1975, ch.II. Cf. note 34 in ch.9.

[55] Rybkin, 1965. He states: 'Theoretically there can be no doubt that the nature of war can change'. Grudinin, who earlier denied the possibility of essential changes in the nature of war (see notes 58, 60) admitted in 1971, that some changes are conceivable, and that the political aims are the most dynamic element in the nature of war (pp.159ff.) Cf. Rybkin, 1973.

'The mutability of the essence of war further consists in the fact that . . . the nature of relationship of the age, the political aims of the warring classes and states, as a rule differ substantially from preceeding ones, altering the specific and individual features of the essence of a given war in comparison with past wars' (*Filosofskoe nasledie V. I. Lenina* . . ., p.37).

[56] This is stated in many studies dealing with the impact of nuclear weapons on the nature of war.

[57] Grudinin, 1971, p.159; 'First, the interrelation between politics, the political content, and armed force is a stable one. This law all wars have in common. . . . Secondly, the interrelation between politics and war is changeable, because both elements involved in this relation are subject to change. That is why the essence of war is not immutable. . . . The interrelation between politics and war is both constant and changeable . . .' (*Marxism–Leninism on War and Army*, pp.21–2).

[58] I. A. Grudinin, 'K voprosu o sushchestve voiny', *Krasnaya Zvezda* 12 July 1966.

[59] Seleznev, 1974, p.15. 'It is important to bear in mind . . . that even in a period when a given substance remains within the bounds of measure and a phenomenon preserves its given qualitative definiteness, it is also changing, moving, developing. But in this period its changes are of a partial character. Only individual, although at times very important, aspects of a phenomenon can change substantially' (*Filosofskoe nasledie V. I. Lenina* . . . p.36); '. . . until the thing exists, it has a relatively stable internal basis in all stages of its development' (Grudinin, 1971, p.159).

[60] It has also been pointed out that war in its most abstract sense will last only as long as the phenomenon it defines, and therefore: 'The variability of the essence of war consists primarily in the fact that war arose as a result of a qualita-

tive transformation of armed conflicts of primitive society . . . and will end its existence with the elimination of antagonistic sociopolitical relations' (*Filosofskoe nasledie V. I. Lenina* . . . p.37); While the nature of war as instrument of politics remains unchanged each individual instance of war accomplished particular political aims of concrete classes with the use of historically conditioned means of armed violence; this is the changeable side of the nature of war (Dmitriev, 1975, p.13).

[61] *Problems of War and Peace*, ch.III; *Filosofskoe nasledie V. I. Lenina* . . ., Ch.III:1; *Marxism–Leninism on War and Army*, Ch. One:1.

[62] *Bolshevik*, 1947:2

[63] L. Leshchninskii, Bankrotstvo voennoi ideologii germanskikh imperialistov, V. Voenizdat, Moscow 1957 (after the Polish transl. 'Krach ideologii wojennej imperializmu niemieckiego', WydMON, Warsaw 1953, p.96).

[64] S. G. Golikov, 'Marksizm—Leninizm o voinakh spraviedlivykh i nespravedlivykh', *Propagandist i Agitator*, 1949:1, p.25; 'Voina', *Bolshaya Sovetskaya Entsiklopediya*, 1951.

'The Clausewitzian interpetation of politics concealed the class content of predatory wars, waged by the German princes and barons'.

[65] E.g. in F. Khrustov, 'Imperializm – istochnik agressii i zakhvatnicheskikh voin', in *Ideologi imperialisticheskoi burzhuazii – propovedniki agressii i voiny*, Gos. Izd. Politicheskoi Literatury, Moscow 1952, pp.24–6.

[66] See: Carl von Clausewitz, *Vom Kriege*, Eingeleitet von Prof. Dr Ernst Engelberg und Generalmajor A.D.Dr Otto Korfes, Verlag des Ministeriums für Nationale Verteidigung, Berlin 1957. In both introductions, Clausewitzian ideas were, in principle, highly appreciated.

[67] Korfes, 'Clausewitz' Werk "Vom Kriege" und seine Nachwirkung', ibid. p.LXIII, Engelberg, 'Carl von Clausewitz in seiner Zeit', ibid., p.V, LX.

[68] 'Er begab sich immer wieder auf materialistische Positionen', ibid.p.LXV. LXV.

[69] *O burzhuaznoi voennoi nauke*, Voenizdat, Moscow 1961, p. 52. Clausewitz established ('ustanovil') the linkage between war and politics, his ideas were a great step forward in the theory of relationship between war and politics (Strokov, 1965, pp. 258, 265); Krupnov, 1963, p.12.

[70] E.g. Rybkin repeatedly contended that Clausewitz didn't examine the roots and nature of politics in connection with war, but confined the analysis to the influence of the political goals on war. He didn't understand the class character of war and politics, and he separated them from their economic basis and thereby distorted the social character of war.

[71] V. Shelyag, T. Kondratkov, Leninskii analiz sushchnosti voiny i nesosto/yatelnost' ego kritiki, *Kommunist Vooruzhenneykh Sil*, 1970:12; 'Lenin didn't repeat Clausewitz's formula, but really gave a scientific definition of war; radically changing the social means of this formula' (*Metodologicheskie problemy voennoi teorii i praktiki*, p.76); there is a fundamental difference between Marxist–Leninist and Clausewitzian views on the essence of war.

Clausewitz 'propounded a false, idealistic view of politics, which he called the mind of the personified state. Besides, Clausewitz understood by politics only foreign policy, and ignored the fact that war is first and foremost a continuation of domestic policy . . . Clausewitz completely ignored the fact that politics is conditioned by deep causes rooted in the economic system of society' (*Marxism–Leninism on War and Army*, pp.7–8); 'The fact that Marxism–Leninism accepts the words of the formula that war is the continuation of politics by violent means should not be allowed to mask the totally different content with which materialists and idealists imbue it. . . . In fact . . . Marxism–Leninism brought about a complete revolution in views of war and politics, providing the first scientific interpretation of them, banishing idealism, metaphysics and sophistry' (*Problems of War and Peace*, p.81); 'Lenin . . . created on the basis of dialectical and historical materialism a totally new doctrine on war, which differs radically from the teachings of Clausewitz and other bourgeois theorists' (*Filosofskoe nasledie V. I. Lenina* . . . p.55).

[72] Grechko, 1974, p.291.

[73] V. Shelyag has presented an extremly sharp critique of the Clausewitzian idealistic, antiscientific and methodologically weak military philosophy in: 'Pod lzhivym flagom "renesansa"' *Krasnaya Zvezda*, 3 July 1975. Clausewitz is quoted as stating that his work *On War* is a relatively formless mass, which ought to be rewritten by a greater mind.

[74] '. . . Clausewitz exaggerated the role of armed coercion in the implementation of policy and spoke in favor of its unrestricted utilization in the so-called "absolute war"' (*Filosofskoe nasledie V. I. Lenina* . . . p. 60).

[75] E.g. Rybkin 1973(b). Cf. *O sovetskoi voennoi nauke*, 2nd ed., Voenizdat, Moscow 1964, p.57; A. A. Kirillov, *Predotvrashchenie voiny – vazhneishaya problema sovremennosti,* Socekgiz 1962, pp.5 ff.; G.A. Deborin *O kharaktere vroroi mirovoi voiny*, Voenizdat, Moscow 1960, pp.8 ff.

[76] Strokov, 1965, pp.258–9; cf. fn. 71.

[77] E.g. I. G. Burtsev was presented as the theorist who pointed out the link between war and politics before Clausewitz did ('Mysli o teorii voennykh znanii', *Voennyi Zhurnal,* vol. I, Petersburg 1819, p.55. After: *O sovetskoi voennoi nauke,* 1964, p.162).

[78] Some items of the vast literature on the subject: Zadorozhny, 1968; D. Volkogonov, 'Mirnoe sosushchestvovanie – alternativa voiny', *Kommunist Voorushennykh Sil,* 1973:19; B. T. Bardin, 'How the Party Carries Forward Lenin's Idea of Peaceful Coexistence in Present-Day Conditions', *Voprosy Istorii KPSS,* 1973:12. after *Daily Review*, no.11 (1751). One of the main themes in *Problems of War and Peace.*

For an analysis of conditions and factors which according to Soviet views make it possible to prevent war, see: Tyushkevich, 1975, ch. V. Cf. *Sovetskaya Voennaya Entsiklopediya,* 'Marksizm–Leninizm o probleme voiny i mira', in 'Voina' (War), vol. 1, 1976, p.284.

[79] Peaceful coexistence 'in no way implies the possibility of relaxing the

possible struggle. On the contrary, we must be prepared for this struggle to become more intense and an even sharper form of confrontation between the two social systems' (L. Brezhnev, in *Pravda*, 28 June 1972) 'detente creates favourable external conditions for the class and national liberation struggle' (Y. Krasin, 'Detente and the class struggle', *Pravda*, 24 May 1976, after Pra-APN, 29 September 1976).

Peaceful coexistence cannot encompass relations between classes within the capitalist countries, nor relations between imperialism and the national-liberation movement. 'Peaceful coexistence is a specific form of the class struggle on the international arena' (Y. Sulimov, 'Nauchnyi kharakter vneshnei politiki KPSS', *Krasnaya Zvezda*, 20 December 1973). Peaceful coexistence is a peaceful form of the revolutionary class struggle (V. Stepanov, 'Sotsializm i mezhdunarodnye otnosheniya', *Kommunist,* 1973:16); 'The principle (of the peaceful coexistence-JL) doesn't mean giving up the struggle for social and national liberation. This is a special form of class struggle' (V. Trukhanovskii, 'Leninskii kurs vneshnei politiki KPSS', *Krasnaya Zvezda*, 17 February 1974). 'all forms of class struggle between the two systems – political, economic, and ideological – are closely interrelated' (F. Ryzhenko, *Pravda*, 22 August 1973).

Cf. Sh. Sanakoyev, 'Foreign Policy of Socialism: Sources and Theory', *International Affairs,* (Moscow), 1975:5; Y. Morozov, Klassovyi kharakter Sovetskoi vneshnei politiki, *Kommunist Vooruzhennykh Sil*, 1975:19 (esp. the section 'Mirnoe sosushchestvovanie – forma klassovoi bor'by'). In many studies, the following pronouncements of Lenin have been quoted: 'War is a continuation, by violent means, of the politics pursued by the ruling classes of the belligerent powers long before the outbreak of war. Peace is a continuation of the very same politics' (vol. 23, p.163) and, 'war is the continuation of policies of peace and peace the continuation of the policies of war' (*Polnoe sobranie sochinenii,* vol. 23, p.192). His saying, that the imperialist peace is a 'breathing space' between imperialist wars is also frequently quoted (vol. 22, p.274).

[80] I have reviewed the Soviet disarmament and arms control propositions up to 1965 in *Zachód a rozbrojenie,* ZAP, Poznań-Warsaw, 1966.

Part III

CONFRONTATION OF VIEWS

13   Approaches to approaches

If the presentation of Western and Soviet ideas about the nature of war has shown nothing else, it has surely indicated the complexity of the subject. There may therefore be legitimate reasons – interests, purpose, method – for a scholar to organise the material on war in some particular way or to limit himself to some small part of it: on the basis of its dimensions – political, economic, social, biological, etc.; its roots; its social functions; its strategic aspects; or one of its many other features and functions. At the same time, each such new perspective tends to add to the complexity of the total view.

In this study I have tried to include as much material as possible since my subject is the whole nature of war. Yet I too have been obliged to fit the welter of ideas about war into a scheme that reflects my interest in what I regard as the main difference between particular Western concepts of war. Therefore I have attempted to combine the review of all approaches to war with a comparison of the two approaches that seem to be the most representative of the contesting ideologies. Since this decision was in some measure arbitrary, it might be worthwhile to examine briefly how the various approaches to the study of war are grouped by other scholars and where political realism and Soviet Marxism–Leninism fit into their particular schemes. In this way the faults and merits of my own treatment may be more clearly appreciated.

Classifying approaches to war.

There seem to be five main ways in which it has been proposed that theories of war as a form of social behaviour might be classified:
(a)   according to their ideological origins;
(b)   according to their philosophical premises;
(c)   according to the social level at which they presume war to have its roots;
(d)   according to the dimension of social activity (economic, political, social, etc.) to which they claim war chiefly belongs;
(e)   according to the purpose of the theory to explain the nature of war or to measure it empirically.

Before turning to these classifications, [1] it should be noted that by limiting the discussion to war as a social phenomenon, we have in effect taken it for granted that war always is a social activity. Though this is an assumption very widely held by scholars of war, it is not universally held. There are a few approaches outside this frame of reference, for example those which regard as war any violent contact between distinct entities in the Universe, or those which view war as hostile interaction between members of the animal kingdom. [2]

Philosophical classifications

Apart from the classification of theories according to their ideological origins, which will be saved until last, the most general lassification of approaches is that according to their philosophical outlook. One such proposal divides approaches according to their theory of causation into deterministic, indeterministic, and teleological.

Characteristic of deterministic approaches is that they treat war and all other social phenomena as determined by laws or regularities. Such theories can be further subdivided according to the determining force, whether material, spiritual, or metaphysical, or according to the determining power of the force, ranging from absolute in a fatalistic approach to less strict theories in which the working of the determining force is interpreted more generally and room is left for the play of other factors in forming the details of social activity.

The indeterministic or voluntaristic approaches view war as a deliberate act taken by men of their own free will upon their appreciation of the situation.

The teleological approaches consider the world and man's existence as part of a plan in which certain forces work toward some ultimate end designed by God, nature, or history itself. All social events are interpreted in the light of the movement towards this end. Setbacks are often accounted for by conceiving the process of development as cyclical: out of the short-term periods of birth, growth, maturity, and decline, through which civilisations, nations, or countries are obliged to pass, is discerned some long-term betterment in terms of spiritual, religious, technological, or other values. War too is considered to have a role to play in carrying forward mankind towards a higher stage of progress. [3]

Rapoport has made the interesting suggestion that all concepts of war can be placed in one of three groups: political, eschatological, or cataclysmic.[4] In the first category belong philosophies that view war as a rational political act. By eschatological is meant a theory in which war is used to achieve some mission, such as the dominance of a race or religion or a socio-political system, the establishment of world order without injustice and war, and so on. And a cataclysmic philosophy considers war to be a catastrophe befalling a nation or humanity as a whole, one for which no human agent can be assigned responsibility.

Roots

Different approaches to war can also be grouped according to the social level they deem war to have its main roots. A distinction is often made between theories that find the origin of wars in the nature of man and those which trace it to the nature of society (or, more abstractly, between microcosmic and macrocosmic theories). [5] Although war is a political and institutionalised form of social conflict in the overwhelming majority of both kinds of approaches, their different understanding of the roots of war have important consequences. This can perhaps most clearly be seen in their prescriptions to eliminate war. A microcosmic theory might claim that this goal could only be achieved by perfecting men or by eliminating evil leaders. According to macrocosmic theories, the solution perhaps lies in changing the social structure, or in asserting democratic control.

A threefold grouping in which a differentiation is made at the macrocosmic level between the state and the international system, is also quite common. [6] It is used, for instance, by Kenneth Waltz, who in his theoretical analysis of man, the state, and war concludes that the anarchy of the international systems can be regarded as a permissive cause of war. [7]

The international system may have a more direct role to play in generating war in other theories as those which attempt to determine the propensity for war that an international system has by relating it to the nature of the polarity of power within it, to the ideological division of the world or to the economic disparities between parts of the world.

Dimensions

Perhaps the most frequent division of approaches, and the one here adopted in Part I, is based on the social activity in which the various theories claim to find the primary root of war. [8] Sometimes a dimension corresponds to one particular level, as the psychological dimension applies at the level of the individual, but others, like the economic dimension, are not specific to any one level.

For the most part theories of war recognise that whatever the primary roots may be deemed to be, war is at least in an immediate sense an political act. Their different interpretations of war thus imply separate understandings of the mechanism by which war is generated by its main source through policy. In other words, each approach contains a different interpretation of what constitutes policy and political activity; for example, a manifestation and instrument of the class struggle, or a search for domination over other people, or a search for better conditions of life through geographic expansion. Indeed, the nature of politics is so tied up with the nature of war in these interpretations that might well call such theories – on the model of sociopolitical and geopolitical – psychopolitical, biopolitical, national-political, and so on.

Since it often proves difficult to place approaches to the study of war in unidimensional categories, the possibility of multi-dimensional classifications arises. [9] In some cases it may be truer to the approach to regard it as multidimensional at the same time as the special determining role of one of the dimensions is noted.

Methodological classifications

A remarkable feature of Western social science since the Second World War has been the attention devoted to making social phenomena amenable to statistical analysis. Out of this preoccupation with quantitative methods has emerged the somewhat dubious distinction between qualitative and quantitative analysis. A scholar using the former would be more concerned with a theoretical analysis of war, whereas a colleague using one of the latter methods would be more concerned with measuring war empirically. To the extent that a scholar allows his criteria of empirical measurement, the size of armies, the duration of hostilities, the amount of life lost, etc., to determine his concept of war, to that extent the distinction between qualitative and quantitative approaches may unfortunately be valid. When such techniques are used to throw light upon theoretical problems and are thus confined to the role of instruments, the idea of a quantitative approach seems meaningless.

Ideological classifications

The classification of concepts of war according to their ideological origin is the only one applied in the Soviet Union in accordance with Marxist–Leninist ideology. Since only two primary ideologies are recognised, that of the fighting proletariat and that of the bourgeoisie, only two theories of war are conceivable, the Marxist–Leninist and the bourgeois. Thus whatever the similarities and differences between the many Western theories, they comprise so many variants of one type of theory in the Marxist–Leninist view.

There are some non-Marxist schools and scholars who also make use of a division of theories of war on an ideological basis. When comparing Communist ideology with non-Communist (liberal, Western, free-World etc.) ideology, Western scholars treat it as an attribute of the Soviet Union and the other Communist states and additionally of international Communism and Communist organisations in various countries as well. In the Marxist–Leninist interpretation, however, ideology is an attribute of class alone. Each class has its own ideology through which it expresses and defends its vital interests and essential aspirations. Imperialist ideology (anti-Communism) is the ideology of monopoly capital, whereas Communist ideology (Marxism–Leninism) expresses the vital interests of the working class and of all working people. Thus Soviet and Western scholars may mean quite different things even when they employ ostensibly the same classification.

The classification of political realism and Soviet Marxism–Leninism

Soviet scholars regard political realism as the school in the non-Marxist camp that is most representative of bourgeois ideology. It is in any event that part of Western ideology which incorporates the political goals towards which Western leaders strive.

In the West, the dimensional classification is the one into which it is generally considered that political realism most naturally fits; it is very much a political theory of war. For the most part, political realism focuses on the state as the main actor, although some concepts like the balance of power, may also suggest

272

the system-level of analysis. Amongst those who use the ideological classification, political realism embodies the Western ideals of freedom, peace, democracy, and the sovereign equality of nations. This is a much more disputed claim, however: some political realists themselves would argue that their analysis shows how states really use their power regardless of the political goals this power serves, and other Western scholars would argue that the very amorality of the position of political realism completely disqualifies it as the bearer of Western ideals. As to methodology, most of the famous realists have been associated with the traditional qualitative approach to the discipline, but many of the younger generation use so-called scientific quantitative methods.

To place political realism into the philosophical classifications is also very difficult because of the very general character of its main ideas and because of the variety of positions taken by particular scholars. The theory is certainly not teleological, but some ideas have a deterministic ring – the course of international affairs is seen as the outcome of the struggle between national interests, the balance of power is a mechanism which in many theories seems to operate independently of the nation-states – and the emphasis on prescription carries a strong indeterministic overtone, that statesmen can shape the future to their will by the conscious application of power. It is probably this aspect of political realism that has led Rapoport to regard political realism as a neo-Clausewitzian theory and to include it in his political category.

While Western scholars thus classify political realism as indeterministic or deterministic, they treat Marxism–Leninism as a deterministic theory that has definite teleological features. Rapoport argues that in terms of his conceptual scheme the Soviet theory of war has undergone such great transformation that it must be treated as two theories: eschatalogical in its orthodox form, since Lenin expected that the world proletariat would convert an imperialistic war into a victorious class war against imperialism and establish a classless and warless world order, but it became cataclysmic during the interwar years when war was seen as a disaster that threatened to befall the Soviet Union. The increasing polarisation of the world into ideologically opposed camps reinforced the Soviet fear that war might break out with devastating results unless active steps were taken to guard against it. Rapoport adds that others who believe that world revolution is still the ultimate goal of Communist leaders might prefer to regard modern Soviet philosophy of war as messianic or eschatological, or, if they discount the ideological determinants of Soviet foreign policy and regard it as determined by state interests only, they may find the political classification more appropriate.

There seems to be general agreement amongst Western scholars that Marxism–Leninism is a macrocosmic theory, one which finds the roots of war in the social structure of states. It is usually considered a political or economic theory of war, [10] but sometimes a combination of these two dimensions. [11] None of these classifications are completely unobjectionable. Since in the Marxist–Leninist approach politics and economics are inextricably intertwined.

it would be a distortion to classify the theory exclusively according to either one of these dimensions. It may be valid to classify it as a macrocosmic theory, yet the differences between Marxism–Leninism and other macrocosmic theories regarding the determinants of the internal structure of states and the origins of war are more significant than this one common feature.

The description of Marxism–Leninism as a hyper-deterministic theory seems to be erroneous, Marxism–Leninism does stress the determining role of the material conditions, but it disassociates itself completely from a fatalistic determinism and emphasises the importance of the role of the conscious activity of men in deciding to wage war. Finally, against Rapoport's proposal it can be objected that although Marxist–Leninist ideology does set itself the mission to rid the world of injustice and war by means of revolution, and although it is true that Marxism–Leninism allows the possibility that this political activity may take the form of war, nevertheless, it has never been part of the Soviet concept of war that war be used to purge the world of war. In the second place, to consider the modern Soviet concept of war as cataclysmic suffers from the deficiency that in this view, the threat of war is not ominous because its roots are unknown or unknowable, but because of its consequences. Soviets claim to know that all wars spring from a social system based on class exploitation and are therefore confident that they have a plan of action that will prevent war from breaking out. The Soviet view of the own theory of war is in one way straightforward: it is the true scientific theory of war in contrast to the bourgeois theories. [12] Sometimes it is also called social or sociological to indicate that it is part of the Marxist–Leninist theory of society. [13] Internal war is regarded as the highest form of the class conflict, and international war is the projection of this struggle into the international arena.

Mutual assessment and critique

Since according to Marxism–Leninism ideology is the only important criterion for distinguishing between theories of war, the Soviet critique of non-Marxist approaches treats them as different instances of the same basic view. This point of departure also gives the Soviet appreciation of Western views a characteristicly aggressive style. Nevertheless, an effort is also made to meet each type of theory on its own grounds. Moreover, the evolution of the Soviet critique throws some light on the shifts in the Soviet theory of war.

One might suppose that the Western critique of Marxism–Leninism would be comparatively fragmented owing to the great variety of approaches. Despite this, however, a Western critique is not difficult to put together. Before turning to the next chapter for a detailed comparison of the political aspects of war, I believe that it is therefore both possible and appropriate to examine the critique that Western and Soviet scholars direct at the general position of each other.

In the late 1940s a Soviet article or study on the bourgeois theory of war usually consisted of a review of some non-Marxist concepts of war together with a biting condemnation of them as unscientific and reactionary. [14] Typically it would begin with the ethical approach, described as a view that exalted war to a noble principle and scorned peace; the stagnation of peacetime bred moral decay and led consequently to the downfall of nations. This approach was branded genocidal and included with Nazism as the most reactionary variant of bourgeoisie. Next came the biological approach, which, it was held, claimed that war was a natural necessity of the life-cycle of nations; the purpose of such false laws could only be to justify and canonise the predatory wars of the bourgeoisie. The theory that war is a manifestation of the struggle between races was frequently the third item on the list. It was presented as a thinly weiled proposal to destroy the lesser races. [15] Equally anti-human were 'Malthusianism' or 'neo-Malthusianism'. In the Soviet description, these theories of war amounted to recommendations that the superfluous population of the world be killed off in order to redress the balance between the number of people and the limited means of subsistence. Western scholars developed geo-political theories in order to justify predatory wars on the basis of geographic factors. Finally it was maintained that the purpose of their 'cosmopolitical' concepts of war was to justify its use to establish world government and liquidate the sovereignty of nation-states.

None of these theories could be regarded as scientific, since they were all derived from a faulty philosophical and sociological base. They conceived of war as something eternal, its outbreak being accidental or dependent on the will and decision of God or great statesmen. Thus they completely failed to see war as a social phenomenon closely related with other social phenomena and dependent on the material basis of social life, and they ignored or denied the existence of laws of social evolution.

In sum, the critics held, none of these theories had any scientific value. United by their common ideological basis, they were merely rationalisations for the bellicose policy of imperialism. In particular, they served the long-range policy of the United States for world domination: the racial and geopolitical theories were said to justify the American policy of subjugating underdeveloped peoples, [16] and the cosmopolitical theories gave the United States an excuse to use war to create a world government under its control. [17] According to the Soviet critics, the imperialists hoped by making war seem inevitable with such pseudo-scientific theories to conceal the criminal character of their predatory wars, to divert the attention of the popular masses from the struggle against the capitalistic system, and to spread the conviction that it is impossible to eliminate war.

After Stalin's death the Soviet critique of Western theories of war very gradually became broader and more subtle. [18] By the end of the 1950s Western theories with psychological, sociological, and socio-political features had been recognised and incorporated.

In the subsequent decade, a more penetrating analysis of the philosophical differences was developed, and at the same time, political realism emerged the main target of Soviet criticism. [19] Since these additions could not be fit within the old framework, new ways of categorising Western approaches had to be found, and a great variety of proposals were put forward. For the most part a distinction was made between the theories traditionally taken as representative of bourgeois ideology, which were grouped together under the heading biological approach, and the new ones. [20] Once these psychological sociological and other theories were admitted, a refutation had to be worked out.

The main objection levelled against the bourgeois psychological theories was that they misinterpreted the motive force of social behaviour. In the Soviet interpretation they attempt to explain behaviour as the outcome of psychic processes when these were not themselves autonomous but in turn depended upon the complexity of the material conditions of social life. Thus for example, the psycho-analytical theory that aggressiveness can be explained as a product of frustration is incomplete; since frustration is itself not a primary force, but is generated by contradictions in the external circumstances, the aggressiveness which is to be explained by frustration must also be rooted in the external conditions. A second problem with psychological theories of war is that they assume what they observe in individuals to be equally applicable to social groups. The aggressiveness of the ruling classes cannot be equated with that of the average man, however, the Soviet critics pointed out. Behaviourism rightly posits that behavioural patterns are learned rather than instinctive, they said, but it had failed to see that the contradictions of class society are the main variables influencing the formation of the social psyche.

Sociological theories of violence were subjected to criticism as well. However, since the arguments received their fullest expression later on in the critique of political realism, they will be left for the moment.

The multi-dimensional approach was exposed to severe criticism [21] on the grounds that by suggesting there are many factors that cause and influence war, it hides the fact that the true source of war is the imperialistic system. Such an approach often reflects agnostic ideas towards war, which are a definite hindrance to a true understanding of the essence of war and its causes.

In books devoted to the philosophical foundations of the Western theories of war, the main line of attack was to show that all these theories belonged to one of two alternative types of idealistic philosophy. In one group, called objective-fatalistic, were included theories that believed man's actions to be guided by forces lying beyond society and history. From such an outlook, war may be the manifestation of the action of mystical laws or supernatural forces, or an expression of the will of some deity, or a form of punishment for the sin of mankind, or in another variant, an instrument by which the true faith achieves its conquest. Soviet critics emphasised that theories of this type do not condemn nuclear war, for if it were to break out, it would be because the supernatural force required it.

The second form of idealism Soviet scholars call subjective-voluntaristic philosophy. Theories of war representative of this branch regard war as a product of the inherent biological or psychological characteristics of the individual personality. According to this view, man is free to act according to his moral and spiritual nature. Society influences him only to the extent that the restrictions it imposes may or may not be respected by him. [22]

During the 1960s, the critique of political realism as the most representative of the bourgeois theories of war began to appear, and in the 1970s it occupied the first place in the criticism of non-Marxist approaches. [23]

In the Soviet presentation, two interconnected ideas are said to underlie this approach. The first is that violence is a primary force rooted in the nature of man. It is in principle the basic means by which men defend their rights and interests against encroachment from others. Within states, individuals seldom actually have recourse to violence owing to the order imposed by the sovereign through the use of legitimate violence. Furthermore, states behave like individuals, but the reality of international relations is in contrast to society anarchic. To protect their interests states must therefore use violence and in consequence must accumulate power to be in a position to win in any test of strength. Thus the second underlying premise of political realism posits that the motive force of international relations is the quest for power, particularly military power. A corollary of this position is that war is natural and inescapable, and that it moreover is the most effective, the most decisive means by which a state can promote its interests.

Both of these tenets are sharply criticised in the Soviet appreciation. In the first place, violence is not an inherent feature of man but a product of socio-political and, ultimately, economic forces. Its use is therefore conditional upon the historical circumstances. It is an instrument which progressive and reactionary forces may or may not use in the class struggle, depending upon the historical stage of social progress reached.

Similarly, the notion that the search for power constitutes the motive force of history is fallacious. A state may follow a power policy, but only if its ruling class believes it to be in their socio-political and economic interest to do so. The same can be said of war as a means of policy. Consequently, the accumulation of military power has never reflected some autonomous historical force, but has been a product of economic and social relations. Moreover, even as a means of policy, the power of states has always been a combination of economic, political, scientific, technological, and military components. The relative importance of each has varied with the historical circumstances; military power has never held a position of pre-eminence on its own merits.

Even more sharply criticised has been the related idea that the search for power leads to a balance of power and that by regulating international conflicts war may serve to maintain and stabilise the balance of power. Six main objections have been raised.

(a)   Balance of power theories are used by states to justify their policy of domination. By claiming to play the role of balancer, which in the theory is necessary to secure equilibrium, states may interfere in the relations between other countries in order to acquire economic and political gains, and in particular to control world policy. The role of balancer is now played by the United States.

(b)   The balance of power theory has been used in the struggle between capitalism and socialism to justify building up a net of anti-socialistic military blocs and bases and to prepare for war against the socialistic countries. In this way, the United States has been able to extend the influence of its political-military doctrine and thereby to gain great influence over the military policy of its allies.

(c)   In balance of power theories, war is usually the way in which the balance is maintained. Nowadays such theories only serve to aggravate conflicts, to justify the arms race, and to excuse the most bellicose policy.

(d)   The balance of power theory is now used to justify the suppression of movements striving for national-liberation or social revolution on the grounds that changes in the internal structure of states may shift the world balance of power to the benefit of socialism.

(e)   The balance of power theory is unscientific because it overemphasises military power as the determinant of the total power of states and groups of states. It disregards such vital factors as the nature of a political system and the policy it conducts, the ideological consciousness of its masses, and its economic potentialities. It underestimates the world correlation of antagonistic social forces, and fails to take into account the laws of social development according to which the dynamics of the balance of power will change to the favour of socialism.

(f)   Finally, the balance of power theory involves a falsified picture of the international system. It replaces the struggle of the two principal antagonists for the future shape of world society with various artificial models of triangles, quadrangles, etc. of power: thus the real correlation of forces of progress and of forces of the past is distorted.

Although during the mid-1970s the Soviet critique of Western approaches had come to focus on political realism, the non-political approaches were still reviewed, and new variants were also taken up, for example, the theory that war occurs because man is unable to control the development of science and technology, which creates pressures for new sources of raw materials and new weaponry. [24] Increasingly, however, these theories were treated as supplements to political realism.

Since the range of such criticised theories has been increasing, new classifications have appeared; to this point, none of them has gained general recognition. [25]

The Western critique of the Soviet theory of war

In the West the Marxist-Leninist theory of war is treated either as one of plurality of contending approaches, none of which has any official status, or as the official position of the Soviet Union. In both approaches there are a great many different opinions about the merits and demerits of the Marxist–Leninist view. To present these in detail, interesting though it might be, would take us away from the main issue. In the following I shall therefore present some general criticisms held fairly widely by Western scholars and review only the more detailed comments of political realists.

It is commonly considered a fundamental error underlying the whole Marxist approach to assume that all social structures and social activity are rooted in the economic relations between men and their means of production. One of the strengths of Marxism is the way in which it integrates such social phenomena as the economy, the social structure, the political systems, and ideology into one consistent whole. But the attempt at causal reductionism, to explain changes in this complex set of factors on the basis of one of them, yields a model so over-simplified as to be unconvincing if not even misleading. [26]

There is no good reason for treating economic relations as an independent variable, and much to commend the view that their development have to some extent been influenced by political and social factors. If this be true, the political order cannot simply be a reflection of the economic structure, and social development must be a much more intricate interplay of various forces than the Marxist–Leninists would allow.

For these reasons it is generally held that classical Marxism has overestimated the continuity and intensity of the internal struggle. Some have gone so far as to suggest that the Marxist conception of society as divided into two irreconcilable classes amounts to a variation of Social Darwinism. [27] Be that as it may, most Western critics would argue that Marx's preoccupation with economic classes reflected the great rift that existed between workers and industrialists in Victorian England. They would point out that his opinion of society as being in a state of permanent struggle and disequilibrium that must end in revolutionary war has been proven by history to be exaggerated. There have been explosive conflicts between workers and their employers, that cannot be denied, but these have never led to socialistic revolution. On the contrary, the intensity of the struggle has been reduced in many countries once labour has become organised and ways have been found of solving conflicts with management without using violence. [28]

As to the Marxist view that society is in a constant state of struggle, but that the struggle will suddenly cease once the revolution is achieved, they would point out that society can also be sub-divided into other groups – religious, political, cultural, etc. – none of which need be congruent with economic classes nor completely dependent upon them. The competition between various groups may contribute as much to an understanding of the development of a society to which they belong as the conflict between classes. Moreover, such competition is intrinsic in the nature of society: it cannot be eliminated by revolution. Nor is it desirable to do so, for as long as it is kept within bounds, it provides the dynamic by which social progress proceeds.

The theory of war offered by the Soviets is therefore both too narrow and too crude, Western scholars would argue. It is too narrow because it ignores the effects of other factors that independently of economic relations contribute to social conflicts. [29] As a result, the interpretation given to many wars whose cause is not obviously economic, i.e. most of interstate wars, seems to distort the facts to make them fit the theory. It is too crude because even in wars with plausibly economic origins the simplification imposed by the concept of class prevents any explanation of capturing the subtleties of reality.

These shortcomings are revealed in the Soviet thesis that imperialism is the source of all modern interstate wars. Apart from the criticism of the theory of revolutionary war, objections are raised against each of the three types of interstate war as they are presented in Soviet theory – war between capitalistic states, war between social systems, and colonial wars.

The preoccupation with economic class struggle has blinded Soviets to the changes that have occurred in the way capitalistic states see their national interests. It may be true that the two World Wars were fought because of the expansionistic policy of some countries; a policy in which economic considerations may have had some part but were by no means predominant. [30] The action taken by the capitalistic allies to defend themselves from Nazi Germany was not essentially different from that taken by their socialistic ally; economic interests are marginal when national survival is at stake. Moreover, all states that suffered the Second World War learned the great costs that modern military technology demands, costs that have spiralled to unfathomable heights with the subsequent development of nuclear missiles. Capitalistic states have therefore come to find it more in keeping with their national interests to co-operate for common benefits and to solve their differences through diplomatic channels. The risk of war between capitalistic states has thereby almost disappeared.

For similar reasons, the capitalistic countries have no interest in provoking a war with the socialistic bloc, so long as the socialistic countries respect the conventions of international law and refrain from direct or covert aggression. On the contrary, it is more in keeping with the economic interests of the capitalistic countries to foster trade and peaceful relations with the socialistic countries and to avoid war.

As to colonial war, there may be something in what the theory of imperialism says about the economic motive for obtaining colonies, at least in the heyday of colonial empires. What the theory has difficulty in accounting for is the fact that some colonial powers have freely granted their colonies independence because the costs of administration and defence were so high, while others have fought to keep their colonies, regardless of the cost, for reasons of national prestige. It would seem that the relation between colony and metropolitan power is more than a matter of simple economics. Moreover, the Marxist–Leninist hypothesis that the colonial power would be economically ruined after losing its colonies is difficult to substantiate. Certainly France has seemed to become economically more fit each time it yielded to the demands of one of its colonies for independence.

Finally, it is a weakness of the Marxist–Leninist theory, the Western critics would argue, that it takes no account of war between socialistic states. In trying to explain the Soviet interventions in Hungary and Czechoslovakia and the border clashes between Soviet and Chinese troops, Soviet scholars must resort to far-fetched rationalisations, when it has been quite obvious that the Soviets were protecting their state interests.

One of the reasons why the Soviet theory of war between states has proved lacking in explanatory power is that the worsening of the class struggle within capitalistic states, which was postulated to be the source of the imperialistic drive, has not come about. Capitalistic states have not only increased their total production and the productivity of their citizens; they have also adopted policies to redistribute income more equitably and to provide more secure living conditions for all. With the great expansion of the service sector this has required, the industrial sector of the economy has greatly decreased in relative importance and in the more developed capitalistic countries is now no greater than the service sector. [31] The Marxist–Leninist theory has in the process lost much of its significance as a diagnosis of modern social problems.

The weakness of the Soviet theory can also be shown by considering that in none of the great revolutions in history, the English, American, French, Russian, and Chinese, was the economic cleavage of society the main cause. [32]

In their critique of the Soviet theory of war, Western scholars often adduce evidence from the foreign policy of the Soviet Union itself. They argue that its behaviour corresponds much better to the model of political realism, in which national interests receive first priority, than to the Marxist–Leninist ideas of class policy. For example, the actions of the new Soviet state in annexing some territories after the Revolution can be interpreted as a continuation of the Tsarist imperial policy. [33] Like other great powers before it, it came after the Second World War to interpret its national interests in global terms. Thus it has taken upon itself the task of balancing the American (and more recently the Chinese) influence in various parts of the world. This is the reason why the countries in the Third World that receive Soviet support are by no means always the bearers of social progress. It also seems to be the explanation that best accounts for Soviet naval expansion, amongst other things.

Moreover, the defence policy of the Soviet Union is to a large extent based on geopolitical considerations. [34] So great is the Soviet fear of invasion from the

West that it is apparently insufficient that the Soviet heartland be protected by friendly buffer states: the Kremlin must also be able to exert decisive control over the policies adopted by the governments in these socialistic states. In this respect as well it appears that state interests have greater priority than international class collaboration.

Some political realists state that the true object of Soviet foreign policy is to establish the Soviet Union as the world centre of socialism and the international revolutionary movement. Thus even the international class struggle acquires geopolitical dimensions.

Finally, the Soviets are charged with using the concept of a 'dynamic correlation of forces' as an ideological instrument of their expansionist policy. The concept holds that in our world there exists a balance of power between forces struggling for socialism and forces defending capitalism, which continuously changes in favour of socialism. The Soviet Union as the main component of the forces of socialism has the right to support all other detachments of this camp and in turn ought to be supported by them. Such ideas are used to justify military and political support for revolutionary and national-liberation movements, and instigation of internal troubles in capitalist countries. They hide the ulterior motives of the Soviet Union: to expand its political influence, to speed the military build-up to continue the Cold War.

Some comments

I have applied in this study a two-fold division in presenting ideas about the nature of war. First, I have distinguished between Soviet (Marxist–Leninist) and Western non-Marxist views, because my object has been to compare these two views. In addition, I have sub-divided the non-Marxist category into a number of approaches in order to choose from the multiformity of theories that comprise it the one or two approaches that share some very general ideas with the Soviet theory of war. One might perhaps say that the first division denotes differences, the second similarities; and an element of both is necessary if comparison is to be made.

The Western approach singled out for special discussion in Part I was political realism. In the foregoing, political realism was regarded by Soviet theorists as the governing theory in the West, while Western scholars acknowledged it to be the most well-established of the political approaches. The place of the Soviet theory to other theories of war was in turn commonly viewed to be with political theories of war, especially theories with a socio-economic element. At the same time, both approaches are broad enough to regard war as a multi-dimensional phenomenon, even if the main focus is on the political dimension: this at least is my own understanding of the two.

From the review of the criticisms each approach has levelled at the other as an approach, nothing has come to light that would affect this assessment. Although each side finds the other too narrow, that it is uni-dimensional or uni-causal and that it ignores other significant dimensions and causes of war, this criticism seems to be exaggerated, based solely as it is on an assessment of some extreme views or a simplistic interpretation of the other approach.

The Soviet complaint that political realism is preoccupied with military power and ignores completely the domestic sources of state policy does not seem justified, for example. Political realists usually include economic, political, psychological, and other factors in their concept of power. A recent tendency within this school has been to include domestic factors in assessing the causes of war, and even to include internal wars with their obviously internal sources in the general analysis of war.

On the other hand, the critics of the Soviet approach mistake, consciously or unconsciously, the tenet that all social existence and activity including war is rooted in the economic base, which is thereby the ultimate source of all social conflicts, for a hypothesis that only economic factors can be the direct cause of all war. Marxism–Leninism asserts the former but not the latter. The misunderstanding of the political realists on this point reveals their own inclination to be satisfied to describe the direct causes of war.[35] To none of these do they ascribe special significance *a priori*. Together these causes comprise the national interest, which is thus a general term for all direct causes rather than a more fundamental cause on which they depend. Marxism–Leninism goes beyond direct causes and asks what caused them. Therefore, while it might be possible to question whether economic forces really are the roots of war, or the only roots, it is invalid to reject the theory on the grounds that not all wars have economic causes. At the same time, the position of the political realists leaves them open to the charge that since they only relate the obvious in general terms, they provide only a shallow understanding of why wars arise. In consequence, they consistently underestimate the impact of deep underlying currents, as economic forces.

A second point of contention concerns the importance of geopolitical ideas in each other's theory. Political realists maintain that Soviet scholars despite their disavowal of geopolitics actually apply geopolitical ideas when assessing their strategic situation; there is therefore a geopolitical element in the Soviet concept of war. The Soviets in turn denounce political realism for overemphasising geopolitical location as a war-generating factor. The Soviet charge is sometimes put too extravagantly, since this factor is in principle only one of several in the theory,[36] and in turn, political realists do go too far in attributing geopolitical ideas to the Soviets.[37] Their error does point to the fact that geographical location must enter into the Soviet concept of war, as into official concepts of other states, even if it is hardly ever mentioned in theoretical studies on the subject.

Finally, a few words about the mutual accusation that the whole concept of war in the adversary's doctrine is ideologically determined and unscientific, that it is merely an instrument of propaganda in the political and ideological

283

struggle. Such opinions seem to be based on some misunderstanding of the relation between ideology and science. Ideology is a system of mutually inter-related ideas, views, and teachings about society and the laws of its development. The concept of some social phenomenon reflects reality but interprets it in accordance with the main ideas on which the whole ideology is based; there is thus both a scientific and ideological component in all social concepts. The scientific value of the respective ideas on war depends both on the degree to which they reflect the real processes of war and on the degree to which the basic ideological ideas reflect the laws that actually govern society. Therefore, the ideological character of a concept is not *per se* a factor determining the scientific or unscientific character of the theory.

Notes

[1] Other criteria for classifying approaches are also applied, e.g. the sociological theories which underly them. In the classification by Wilbert E. Moore (*Social Change,* Englewood Cliffs, Prentice-Hall, N.J., 1963) four types are listed: evolutionary theories, Marxism, functionalism, and conflict theory (political realism can be here classified with the conflict theory); in the classification presented by Richard P. Appelbaum (*Theories of Social Change,* Markham Publ. Co., Chicago 1971), which includes evolutionary theory, equilibrium theory, conflict theory, and 'rise and fall' theories, both political realism and Soviet theory can be classified with conflict theory.

[2] Wright presents five periods of war – animal, primitive, civilised, modern, and recent war; animal war is considered to occur even today (1968, p.455). Clarke in his analysis of the science of war and peace presents some views asserting the occurrence of 'animal warfare' (1971, pp.196–200).

[3] Many classifications have been presented, based partially on the philosophical assessment; e.g. Sorokin, 1928, presents three groups of theories: war as a manifestation of a 'universal law of struggle' (struggle theories), as manifestation of human instincts (instinct theories), and theories which emphasise particular factors (economic, diplomatic, political, etc.); Bernard, 1944, writes about theories of war as produced by external forces (God, natural law, physical cataclysm), human nature, 'cultural lag' theory and others. Timasheff, 1965, distinguishes groups of theories based on the fatalist approach, great-man theory, decision theory, biological, psychological, demographic and economic theories, the morbid national culture theory. Cf. Berenice A. Carroll, Clinton F. Fink, 'Theories of War Causation: A Matrix for Analysis', in Nettleship, et al., eds, 1975, pp.55–71.

[4] Rapoport, 1968 (1971), pp.13 ff. In an earlier study (1966) he divided conflicts into strategic and systemic (cataclysmic) categories; the former obviously corresponds to the political concept of war.

[5] Dougherty, Pfaltzgraff, 1971, chs. 7 and 8. An example of the emphasis on

the 'microcosmic' roots of war: '. . . war manifestly has deep roots in human personality, whether natural or acquired; the political, economic and diplomatic "causes", which historians have been want to argue over, stop short of telling the whole story' (Marwick, 1974, p.4); Bruce M. Russett, 1974, presents a macro-cosmic view of international politics, including war as its instrument.

[6] Haas, 1965, writes about systemic, societal, and psychological approaches. The second category is subdivided into functional approach (state or governing elites consciously employ war when they find it necessary or useful) and prerequisites approach (types of political and economic systems as pre-requisites to war) in Northedge, Donelan 1971, the following terms are proposed: systemic, statecraft and sociological approaches.

[7] Waltz, 1975, pp.230 ff. In a variant Deutsch and Senghaas present five levels of analysis, where different scholars see the main source of wars: inter-national system as a whole (its character and functioning), regional political or social subsystems, interests of the national states, national subsystems (social classes, political institutions, or interest groups), and individuals (their charac-teristics and behaviour) (1971, pp.26–9).

[8] Some other classifications in which the concepts of the nature of war are closely tied up with the interpretation of their causes: Bernard, 1944, in one of many kinds of classification that he presents distinguishes between biological, psychological, economic, political, religious, etc., theories (the 'departmenta-lized' classification). Wright, 1968, presents six approaches to the study of war – political, economic, technological, legal, psychological and sociological. These are not called 'approaches', however, and each of them is differently presented ('the political value of war', 'the economic causes of war', 'war as a psychological problem', etc.) (1968, pp.460–5).

Alcock, 1972, divided the 'conventional' theories of war into: arms race theory, the profit motive theory, the nationalist theory, theories based on the so-called innate aggression, population pressure, differences of race, language and religion; Brodie, 1974, lists the historical approach to war, economic theories of war causation, psychological theories and political theories (pp.276–340); Deutsch, 1974, when analysing theories of imperialist expansion, which seemingly are also theories concerning a great number of international wars, presents four 'old' 'packages of folk theories': biological-instinctive, demo-graphic-Malthusian, geographic-strategic and psychological-cultural. He also presents the conservative, the liberal and the Marxist theories as sociological-economic class of approaches. M. Estelle Smith writes about six approaches to the genesis of conflicts of wars: sociological, psychological, ecological, bio-logical, economic, and ideological ('Cultural Variability in the Structuring of Violence', in Nettleship et al., 1975, p.600).

[9] 'If the foregoing analyses are correct, war has roots that are simultaneously biological, psychological, and social.' (Aron, 1966, p.355).

[10] Dougherty, Pfaltzgraff, 1971, ch.6. 'Economic Theories of Imperialism and War'.

[11] Yin, 1971, proposes to treat the Soviet approach to war as a mixed economic-political, since Soviets consider war between capitalist countries as having economic roots, and war between capitalist and socialist systems as having political roots.

[12] The bourgeois military theory can discover neither roots and essence, nor general laws of war, because it is based on a wrong, idealistic philosophy and methodology, this can be accomplished only by the Soviet military science (*O burzhuaznoi voennoi nauke,* Voenizdat, Moscow 1961; Introduction, pp.3–5): (after the critique of bourgeois theories of the genesis and nature of war): 'Only the Marxist–Leninist science presents a scientific definition of war' (*O sovetskoi voennoi nauke,* 2nd ed. Voenizdat, Moscow 1964, p.6).

Science was in no position to discover laws of war before the appearance of Marxist theory, it was only empirically based, and couldn't use the true scientific approach (Savkin, Ch.1); cf. *Metodologicheskie problemy voennoi teorii i praktiki,* p.11; *Voennaya strategiya,* p.65.

[13] 'The Marxist–Leninist dialectic is opposed to abstract views on the essence of war and insists on a social, class definition of individual wars' (*Problems of War and Peace,* p.66); *Filosofskoe nasledie V.I. Lenina . . .* ch.II, cf. Kondratkov, 1972; he considers the investigation of the social class character of war and main task in the analysis of the nature of war.

[14] I. Usenko, *Prichiny imperialisticheskikh voin,* Moscow, 1953, pp.4–6; *Agressivnaya ideologiya* i *politika amerikanskogo imperializma*, Moscow 1950; *Ideologi imperialisticheskoi burzhuazii – propovedniki agressii i voiny,* Akademia Nauk SSSR, Institut Filosofii, Moscow 1952; F. Khrustov, 'Marksizm–Leninizm o voine i armii', *Propagandist i Agitator,* 1950:6, pp.51–2; V. Gorynin, 'Marksizm–Leninizm o voine i armii', *Propagandist i Agitator,* 1954:1, p.49; N. Petrov, 'Reaktsionnaya suschchnost' i agressivnyi kharakter sovremennoi voennoi ideologii amerikanskogo imperializma', *Propagandist i Agitator,* 1952:14; V. Vasilenko, 'Reaktsionnaya sushchnost' i agressivnyi kharakter imperialisticheskoi voennoi ideologii', *Propagandist i Agitator,* 1953:17.

In my early writings I reviewed the Western theories of war from the standpoint of Soviet politico-military thought (*Czynniki zwyciestwa,* Wyd MON, Warsaw 1952; *Wybrane zagadnienia. wojskowo-polityczne, Materialy dla studiów wojskowych,* Wyd MON, Warsaw 1958, pp.31–44).

[15] Soviet scholars often compare the present American and British racialism with the German Nazi one. While the German racialists preached the superiority of the Aryan race, the American and British ones assert the superiority of Anglo-Saxons. While the German racialists alleged that the Aryan race was assigned the mission to rule over all other peoples, called by them the 'slave race', the American and British racialists state that they have a right to rule over the 'inferior peoples'. And, finally, while the former prepared World War II and organised the destruction of 'lower races', the latter now prepare a war to subjugate peoples of other countries.

Many Soviet authors quoted an interview by I. S. Stalin, in which he made the above comparison ('Interviu tovarishcha Stalina s korrespondentom 'Pravdy' otnositelno rechi g. Cherchilla', *Pravda,* 14 March 1946; *Propagandist i Agitator,* 1946:5, p.2).

[16] The modern geopolitical theories which hold that Northern America and not Eurasia should be considered as the 'world island' and should rule over the world, are said to be used for the justification of the aggressive policy of the USA.

(P. Fedoseev, 'Amerikanskie geopolitiki – propovedniki agressii', in *Ideologi imperialisticheskoi burzhuazii – propovedniki agressii i voiny,* 1952); geopolitical theories are combined with racialist ideas, the latter being transformed into the American state ideology (O. Sayapina, 'Rasizm i kosmopolitizm – reaktsionnoe oruzhie razzhiganiya voennoi isterii', in *Ideologi . . .,* p.125).

[17] Cosmopolitanism propounds the rejection of national sovereignty and aims at paralysing the resistance of nations that are objects of the American expansion. Combined with the anti-Communist propaganda and slogans of a world government it provides an important means of preparing a new world war (M. Rubinshtein, 'Burzhuaznyi kosmopolitizm – ideologiya imperialistichesckikh agressorov', in *Agressivnaya ideologiya i politika amerihkansogo imperializma,* 1950).

Cf. M.A. Suslov, *Zashchita mira i bor'ba s podzhigatelyami voiny,* Moscow 1949, p.13. Report to the conference of Cominform.

[18] Initially, however, no great change in the content of the critique could be noted. See for instance *Marksizm–leninizm o voine i armii,* Biblioteka ofitsera, Voenizdat, Moscow 1957.

[19] S.A. Tushkevich, *Neobkhodimost' i sluchainost' v voine,* Voenizdat, Moscow 1962; *Marksizm–leninizm o voine i armii,* 2–5 eds., Voenizdat 1963–1968; Prokop'ev, 1965; Nikolskii, 1964, ch. 'Nekotorye burzhuaznye teorii proiskhozhdeniya voin'; V.I. Zamkovoi, *Kritika burzhuaznykh teorii neizbezhnosti novoi mirovoi voiny,* Moscow 1965; I.M. Ivanova, *Mirnoe so sushchestvovanie i krizis vneshnepoliticheskoi ideologii imperializma SShA.* Izd. Mezhdunarodnye Otnoshenia, Moscow 1965.

Problemy voiny i mira: Metodologicheskie problemy voennoi teorii i praktiki; Aleksander Yakovlev, *Ideologiya amerikanskoi 'imperii',* Izd. 'Mysl', Moscow 1967, his *Pax Americana,* Izd. 'Molodaya Gvardiya' Moscow 1969; N. Ponomarev, 'Krizis burzhuaznykh teorii voiny i mira', *Kommunist Vooruzhennykh Sil,* 1964:16; *Kritika burzhuaznykh teorii i vzgladov po voennym voprosam,* Voenizdat, Moscow 1970.

[20] F.N. Fedoseev, 'O prichinakh voin v sovremennuyu epokhu', in *Istoricheskii materializm i sotsialnaya filozofiya sovremennoi burzhuazii,* Izd. Sotsialno-Ekonomicheskoi Literatury, Moscow 1960, pp.550–69. For a comprehensive critique of the philosophical ideas underlying Western theories of the origin and nature of war, see: B. Shabad, *Politicheskaya filosofiya sovre-*

mennogo imperializma, Izd. Mezhdunarodnye otnoshenia 1st ed., Moscow 1963, 2 ed., 1966; his *Imperializm i burzhuaznaya sotsialno-politicheskaya mysl.* Izd. 'Mezhdunarodnye otnosheniya', Moscow 1969.

[21] The Soviet critics often quote writings of Quincy Wright, Bernard 1944, Bouthoul, 1951, 1953.

[22] Cf. Tyushkevich, 1975; *Filosofskoe nasledie V.I. Lenina; Marxism–Leninism on War and Army,* Ch. One:5, Seleznev, 1974.

[23] V.M. Kulakov, *Ideologiya agressii,* Voenizdat, Moscow 1970, ch.2. 'Ideino-teoreticheskie istochniki amerikanskoi voennoi ideologii'; pp.54–87; A. Karenin, *Filosofiya politicheskogo nasiliya,* Izd. Mezhdunarodnye Otnosheniya, Moscow 1971; *Voennya Sila i Mezhdunarodnye Otnosheniya,* 1972; *Problems of War and Peace,* ch.X; *Marxism–Leninism on War and Army,* Ch.One:, A. Kunina, 'A Critique of Bourgeois Theories of the Development of International Relations', *International Affairs,* Moscow 1973:2; A. Volkogonov, 'Mirnoe sosushchestvovanie – alternativa voiny' *Kommunist Vooruzhennykh Sil,* 1973:19; *Filosofskoe nasledie V.I. Lenina* ... pp.98–9.

[24] Rybkin, 1973, pp.21–3.

[25] E.g. a division into four groups has been presented: theories that identify as the main source of wars: (a) human nature (biological and psychological theories); (b) shortcomings of the social order (political realism, philosophical-historical, socio-cultural and other theories); (c) divine will; (d) functional causes (geopolitical, demographical, etc.) (Barsegov, Khairow, 1974).

> The third group includes concepts according to which the will of God is the main and cardinal cause of wars. It includes numerous 'theories' proclaiming war either as a result of the general sinfulness of people, or a means of establishing the 'true' religion, or a 'punishment of God' which may lead to an 'end of the world'.... The utter scientific untenability and practical harmfulness of religious concepts of the origination of wars, fatalistic by their very nature, is obvious.... Neo-Malthusian concepts hold that population growth (being increasingly out of proportion with the means of existence) inevitably leads to war, creates a potential and very real danger of the 'aggression of starving countries' against 'rich' countries with a low birth rate.

Such theories not only mistake the causes of wars, but also point out alleged functions of wars as solving the problem of overpopulation by destroying large numbers of people.

[26] Lewis A. Coser, 'Karl Marx and Contemporary Sociology', in Coser, 1967: his: 'Social Conflict and the Theory of Social Change', ibid; Dahrendorf, 1959; Dougherty, Pfaltzgraff, 1971, ch.6 'Economic Theories of Imperialism and War'; Michael Haas, 1965. Lambeth writes: '... ideology dictates that Soviet doctrine follow Lenin's classic thesis on imperialism which maintains that war is solely derivative of economic causes' (1973, p.211); Rapoport contends that besides the allocation of resources, two basic other sources of social conflict are

discernible, namely the struggle for power (over a position which confers on the possessor the power of decision) and the need for autonomy (1974, Ch.17, 'Some Aspects of Endogenous Conflict').

[27] One of the basic contentions in Bramson, Goethals, 1964; Introduction to Part II.

[28] Coser, Karl Marx . . . p.25.

[29] What is more, many economists have considered competitive free trade which is the governing principle of capitalistic society to be a guarantee of peace and not a major cause of wars, at least in our times. Wright summarises his review of the economic causes of war with the contention that wars have not arisen from economic competition (1968, pp.461–3).

[30] Raymond Aron contends that the Marxist view of sources of modern wars is wrong: '. . . the actual relationship is most often the reverse of that accepted by the current theory of imperialism: the economic interests are only a pretext or a rationalisation, whereas the profounder cause lies in the nation's will to power' (*The Century of Total War,* Bacon Press, Boston 1955, p.59).

[31] Therefore 'class conflicts have been minimized', some scholars observe (Daniel Bell, *The End of Ideology,* Collier, New York, 1962). The above book was sharply criticised in many Soviet studies and articles (e.g. L. N. Moskvichov, *The End of Ideology Theory,* Progress Publishers, Moscow 1974).

[32] Perez Zagoria, 'Theories of Revolution in Contemporary Historiography', *Political Science Quarterly* Columbia University, March 1973, pp.32ff.

[33] E.g. at a conference of the Strategic Studies Center of the Stanford Research Institute, May 1973, several scholars 'stressed that the Soviet Union pursued a grand strategy which was derived from Russian tradition, and which orchestrated political, diplomatic, military, and all other courses of action, and coordinated all instruments of power toward its more enduring goals in which the control of Europe featured prominently' (Joshua, Rabin, 'American – Allied Relations in Transition: A Synthesis of the Conference Discussions', in Foster, Beaufre, Joshua, 1973, p.3).

[34] Garthoff, 1954, p.9; Arnold 1961, p.42; Blasius II, 1966, p.354; Strausz-Hupé, 1962, pp.31–2; Boris Meissner, Gotthold Rhode, eds, *Grundfragen sowjetischer Aussenpolitik,* Stuttgart 1970.

B. Moore, *Soviet Politics – The Dilemma of Power*, Harvard Univ. Press, Cambridge, Mass., 1950; L. J. Halle, *The Cold War as History,* Chatto and Windus, London 1967:

See Robert G. Wesson's presentation of the Soviet policy, which he considers expansionist, motivated by geopolitical reasons (the tendency to conquer and defend territory) as well as by ethnic tensions and tradition of despotic rule *(The Russian Dilemma: A Political and Geopolitical View,* Rutgers University Press, New Brunswick, N.J., 1974).

[35] The study by Blainey on the causes of wars, 1973, seems to be a good example. He analyses in detail factors that influence the decision to unleash war

('the disagreement of nations on their relative strength') and calls them 'causes'.

[36] The Soviet critique of the geopolitical approach is directed towards the most extreme form, and its most extreme application: 'The *geopolitical* concept regards war as the result of a country's geographical location, its natural conditions, climate, etc. (Cohen, 1964; Huntington 1961; Mackinder); '. . . these imperialist theories closely intertwine with the so-called theory of *Lebensraum* in its German, British, Japanese, American and other variants outlined in the works of Ratzel, (Kjellen). Mackinder, Haushofer, Fitzgerald, Spykman, Huntington, Weigert and other pillars of the geopolitical "science". The geo-political "science" of the seizure of foreign territories, naturally, was used and is being used by the most aggressive, most reactionary forces of imperialism. It was officially used by fascism' (Barsegov, Khairov, 1973, p.75).

[37] Many political realists present geopolitical ideas in such a way that their approval of geopolitics can be supposed. For example, representatives of the 'forward strategy' school, William R. Kintner and Robert L. Pfaltzgraff, presenting Robert Strausz-Hupé's views, and especially his book on geopolitics, write: 'Geography influences the foreign policy of a nation in many ways. . . . Much of foreign policy consists of the effort of a state to gain control of, or to prevent others from controlling territory'. Strausz-Hupé appealed for strong US support for Britain and France, because German politics was based on geo-political theories; the Germans aimed at dominating the 'heartland' which in turn could permit them the domination of all Eurasia, and finally the remaining areas of the world (*Strategy and Values: Selected Writings of Robert Strausz-Hupé,* Lexington Books, Lexington, Mass., 1974, p.81).

14 War and politics

Both of the approaches under comparison summarise their position with the formula that war is the continuation of politics by other means. Crucial to an understanding of what each approach means by war are therefore the terms politics and continuation.

However, the two approaches are not simply two different scientific theories. They are rooted in different ideologies, two incompatible views of man and society, of social relations and progress, and of the role of war. Moreover, from these more general ideas are derived the basic political positions of the two antagonistic socio-political world systems, the relations between which will be decisive for the future of mankind. An understanding of the different ways in which the two approaches relate war and politics to each other must therefore begin with an appreciation of the more fundamental cleavage in first principles.

The ideologies

In the Marxian perspective, all ideas are derived from the material base, and since the so-called class society is divided into two antagonistic classes, each of them has an ideology, a characteristic system of ideas about man, society, and progress. The ruling class imposes its ideology upon the society, but the other ideology also does exist. Both ideologies reflect the division of society into economic classes – in it the powerful are the powerful and the weak are the weak – but they evaluate the causes and merits of such social order, and perspectives of its change quite differently. The governing class tries to present it as a natural order of things, in which each class can not only recognise itself, but also accept its position. The prevailing ideology is thus an instrument by which the governing class maintains the *status quo*. Since both the governing and the governed classes act in accordance with the conception they have of their place in the social order, to the extent that the governed class accepts the prevailing ideology, it remains passive and cannot fight for its true interests.

Marx wished to transform radically the society he observed and therefore composed an ideology that was intentionally contrary to what he perceived to be the ideology of the capitalist class. His object was to present society from the perspective of the suppressed proletarian class. He believed that if the working class could see its situation in the light of its true interests instead of accepting its place in the ideology of the governing bourgeoisie, it would also be in a position to act to free itself from servitude. The Marxian ideology is thus a guide to transforming society as well as a description of it. As Marx himself wrote, 'the philosophers have only *interpreted* the world, in various ways, the point, however, is to change it'.

The first philosophical pillar on which Marxism rests is thus materialism. The second is the dialectic; its application to the history of the society, combined with the idea of the primace of the material (i.e. economic) base in the social life, constitutes historical materialism, which is the Marxist theory of sociology.

According to the Marxian dialectic, very briefly, society progresses through a two-stages process: contradictions inherent in all social forms in class societies, and first of all the contradiction between antagonistic classes, become accentuated to the breaking point; the resolution of this conflict takes the form of annihilating the old and obsolete, and emergence of the new and progressive; all positive elements of the old are absorbed, however, and the new whole constitutes a higher level of social organisation.

The dialectic and materialism, when applied to the social history, complement each other so well that together they provide a view of social life almost unrivalled in bredth and power. Materialism saves the dialectical view of change from sterile tautology by presenting the economic contradictions as primary and driving progress forward. On the other hand, the dialectic infuses Marxism with a spirit of confident optimism that a better world lies within man's grasp. For the point is not merely to change the world, but to change it in the right direction. Materialism provides the driving force, dialectic points to the goal, a goal which in the logic of the argument will be reached some day. The dialectic thus merges knowledge with value: the end-state is both inevitable and ultimately desirable.

It is difficult in such a short space to do justice to the basic conceptual apparatus of Marxism and the more to its Soviet transformation, it is almost impossible to draw a comparable sketch of the Western intellectual framework, at least in non-Marxian terms. It may perhaps suffice to note a few points of contrast.

There is no scientific social theory in the West that claims the explanatory scope of Marxism–Leninism, and it could hardly be claimed that there exist a prevailing systematic concept of society. Soviet scholars regard non-Marxist views as a single ideology, one theme with many variations, on the grounds that they all reflect, directly or indirectly, and to various extents, the interests of the governing class. That assumption is put in question in the West, however, and the coexistence of many disparate schools of thought is explained by the fact, that reality can be seen from many perspectives, none of which need be more or less true than another.

The most widely held views seem to belong to the positivist tradition. Observed phenomena are explained as the effect of certain causes according to some universal laws. The discovery of laws is an empirical process, and the truth of any one law is dependent upon the validity of the actual observations on which it is based. Since laws ought also to cover cases as yet unknown, they cannot be completely verified. The causal pattern of the empirical science, while very fruitful in connection with the natural sciences, made it difficult both to explain motives of social behaviour (which were often treated as causes), and to predict the direction of social change. While in the Marxist–Leninist perspective, the analysis of the past and contemporary reality allows for predictions about the future of the society, the positivist tradition contributed to a strong trait of scepticism both with regard to what can be known about the past and to the possibility of predicting the future. The result is that progress is seen in another light.

According to historical materialism, all the traditional problems of political thought, justice, equality, freedom etc., can only be solved in the classless society, and there is a surety that they will be solved. The primary object of social acton is therefore to reach the classless society (or to destroy class society, which amounts to the same thing) as quickly as possible. Social change is a revolutionary process.

In the non-Marxian traditions, political ends are not so structurally concrete nor so future-oriented. Such political goals as freedom, equality, and justice are conceived as relative values that are found to a greater or lesser extent in all social forms. There cannot be any social structure that is absolutely just, or in which men are absolutely free and equal. If there were, it would of course be in everyone's true interest to create such a society right away. But since the future cannot be known, neither can it be reached by any short cuts. What is possible is to have an idea of a better world and to strive to reach it. The non-Marxist idea of social change, like the Marxist, therefore implies conscious activity towards a goal. There is no ultimate goal, however. Social development is an unending process of doing what is possible to improve existing society. Since it is only possible to take one step at a time the process is an evolutionary one. In time society will be radically different from what it once was, however, a revolution will have taken place.

To place political realism in this setting, it is closely tied to the positivistic tradition. A partial theory of international relations, it does not try to explain the origin of states, nor does it offer a recipe for the good life, the political ends toward which men (or states) ought to strive. Perhaps some of the appeal of the realists' position is related to the failure of their intellectual opponents, the utopians, to show how schemes for a better world could be put into practice. States exist, and they must survive to achieve their political ends. Whatever these may be, these are the 'givens' of political realism. The necessity to survive dictates certain patterns of behaviour, which political realists try to describe and explain. The mode of explanation is thus causal rather than teleological. In sum, the lack of the vision of a better future world and the emphasis on the evolutionary process of change in which the basic values of the society will be saved, open the door to the charge of *status quo* bias in the realists' political thought.

Politics and war

Politics and the state

Both approaches under comparison agree that politics is the activity by which social groups attempt to realise their main goals and protect their fundamental interests. They differ, however, over which social group is the main actor in politics, and thus over the interests and goals at the centre of political activity.

The differences are rooted in different views of the society. In the political realist perspective, society is an organism in relative equilibrium, united both by cooperation and coercion, where armed coercion is applied only in extreme cases when the existing order is threatened. Soviet theorists view presocialist society to be divided into antagonistic classes, whose basic interests are incompatible. The ruling class may resort to armed violence if its rule is threatened, and the suppressed class may accomplish revolutionary change of the existing system only using political violence, usually including armed uprising. Concomitantly, in political realism theories it is usually the society as a whole, or in other words the nation whose interests are paramount in issues of peace and war, whereas according to Marxism-Leninism economic classes are the main political actors both within states and in relation between them. The state in non-Marxist theories acts on behalf of the nation as a whole. Since the population recognises the state as its agent, it is prepared to submit voluntarily to its authority, and coercion is applied only to maintain law and order. Consequently, the state protects the interests of the whole nation, both inside and outside the bounds of its sovereign competence. In Marxism-Leninism the state is an instrument of the ruling class who maintain their position of power by using the authority of the state to suppress other classes with violence or other more subtle methods. The state therefore acts in the interest of the ruling class; only exceptionally does it pursue truly national aims. Concomitantly, in non-Marxist approaches international politics consists of a competition between national aims, whereas in the Marxist–Leninist approach it is competition of the interests of different ruling classes.

The difference between Marxist–Leninist and non-Marxist approaches concerning the interests that are pursued in politics goes beyond the distinction between national interests and class interests: the very concept of interests is also different. Political realists define the national interest in such general terms that power, security, independence, position amongst other nations and similar factors are included. Marxist–Leninists maintain that the economic interests of classes underlie all other interests and that political competition is a projection and concentrated expression of the basic economic division of society. They do not deny the existence and importance of such national interests as security, independence and the like, but they emphasise that economic class interests are fundamental and primary in the policy pursued by all classes and all class states.[1] Politics is the struggle of classes for state power, either to preserve it or to seize it.

One consequence of this difference is that in the view of political realists the policy of a state is divisible. Their main focus in on foreign policy, which they assume is based on essentially the same considerations in all states whatever the nature of their internal policy and institutions. Domestic factors may affect foreign policy, but such influence is marginal, acting more as a constraint than a driving force. War, which is an outcome of foreign policy, is therefore fought for national goals and not, in the first instance, for the goals of some part of the nation. In Marxism–Leninism on the other hand, the policy of the state is a seamless whole. The foreign and domestic policy of a state are not two separate, if interrelated, policies, but two complementary expressions of the same policy. For this reason Marxist–Leninists point out that a study of the general nature of war or of particular types of war or even of individual wars must relate war to the aggregate policies of the classes and states involved. The goals for which wars are fought reflect the political goals towards which the ruling class aims in general, whether internally or externally, by violent means or peaceful.

Each of the confronted approaches maintains that the other purposely distorts the truth. Soviet scholars claim that non-Marxist concepts of state and political authority are deceptive inasmuch as the civil rights of citizens, including the right to vote in free elections for the party of one's choice, which the apologists stress in their vindication of the capitalistic system, are simply formalities without true significance. [2] The true base of political power is economic power; the rest is mere window-dressing. Similarly, Soviet scholars maintain that the preoccupation in Western approaches with foreign policy is part of a deliberate effort to mask the true nature of social conflicts and to divert attention from their internal sources. [3]

Western critics of Soviet theories retort that the government of the Soviet Union cannot be said to act on behalf of the true interests of its people, simply because the Communist Party, which maintains itself in power through dictatorial methods, claims that it does so. Marxist–Leninist rhetoric distorts reality to disguise the Soviet ambitions for power.

The international system

Two concepts Marxism–Leninism views international relations as an arena in which intersecting and partly overlapping confrontations take place. First, there is a struggle for the interests of classes ruling in particular countries. Since states are instruments of these classes, the struggle is carried on between states. Secondly, there is the world-wide struggle between two main forces of the contemporary world – the world working class and the world bourgeoisie. This struggle is waged on many fronts and takes many forms, as described in chapter 9. The core of the theory is that all forms of confrontation spring from class relations, which constitute the principle factor of international relations.

The picture presented by political realists is based on quite a different assumption: international relations constitute the struggle for state (i.e.

national) interests. A growing importance is attributed to the world-wide struggle of the two antagonistic systems, but the ideological struggle between them is said to be subordinated to a considerable extent to the competition of state interests, the states, as always, searching for power. The two theories consequently predict different futures. In the Soviet assessment, the international struggle will disappear together with capitalist states; in the political realist conception, the structure of the international system as a community of independent states with conflicting interests will remain for a long time, if not indefinitely.

Balance of power versus the correlation of forces. Both of the approaches under comparison devote much attention to the problem of the balance of power as a feature of international system. [4]

(a)   A difference may be observed even in the wording. Instead of the traditional term balance of power, the Soviet scholars usually use the term correlation of forces (of power) or correlation and deployment of forces.

(b)   In the political realist conception the balance of power refers to the balance between the superpowers, the main opposing blocs, and is complemented by the balance of power between the other great powers, each of which fights for the best position in the general configuration. The importance of the local (regional) balance is also mentioned. In the official Soviet view, the balance between the superpowers is replaced by the balance between the two antagonistic systems and all states, not only the great ones or members of the antagonistic coalitions, are included in the balance; other forces are also included.

(c)   From another perspective, while political realists use the idea of balance of power with reference to states, in the Soviet concept, it is the balance between world classes, between the world working class and world bourgeoisie that is considered. Therefore national contingents of the international classes are seen as components of their global force, and in addition to states, political movements of international character are also included, the world communist movement being the most influential.

(d)   In the Soviet concept, much more than in the political realist one, the multi-dimensional character of the balance of power is emphasised, especially its ideological-political components (e.g. the political awareness of masses is regarded as a component of the anti-imperialist forces).

(e)   The character of the socio-political system plays an especially important role, Soviet scholars say. It includes the scope of the economic potential and the solidity of the economic structure of the society, the scope of cooperation between its different classes and strata, the amount of support granted by the population to the government, the political and ideological unity and strength of the society, efficiency of the state power, and the amount of cohesion among the members of the system.

(f)   In the political realist approach balance of power is interpreted as maintaining the political *status quo*; the contemporary balance favours the

296

preservation of independence of nations and of peace as well as the social and economic development.

In the Soviet concept, the dynamic character of the balance of power (correlation of forces) is strongly emphasised; the balance is said to be constantly changing in favour of the world working class and its struggle for revolution.

In both approaches the concept of the balance of power is descriptive and normative: in the political realist view it is the stable character of the balance of power that is the main condition of peace and progress, whereas in the Soviet concept it is the dynamic character of the balance that is said to serve the cause of peace and progress.

When we turn to what is in practice the current policy of the superpowers, all of these theoretical differences seem to be somewhat less. In the first place, both sides regard the power of the two giants as the main item in the balance of forces, and both consider power to be mainly composed of military strength backed by a powerful economy. Secondly, in the West, the necessity for a permanent struggle to maintain the balance of power is now fully recognised, and the value of the ideological and social components of power is now appreciated more fully than ever before. At the same time the Soviets in fact regard their military power as the determining component of the so-called forces of peace, socialism, and independence of nations; all other contributions are treated as secondary and subsidiary. While the concepts of balance of power and correlation of forces are different, the policies of the superpowers connected with them are much more similar.

The actors in war

Corresponding to these different interpretations of the sense in which war is a political phenomenon, and of the arena in which it may be conducted, are separate conceptions of whose policy it is that is continued in war, that is who the main actors in war are. Not withstanding these doctrinal differences, the distinction between the two concepts of actor seems to have become less clear-cut as the respective theories of war have become elaborated during the past two decades.

The classical view of political realism is that wars are waged by states or coalitions of states for national goals. To the extent that internal wars are taken into consideration, the role of the state in acting against part of its society is considered that of the defender of the interests of the nation as a whole, these being the maintenance of law and order, the integrity of the society and of its economic and cultural progress.

The main actor in the traditional Marxist–Leninist teaching, which regards war as the highest form of class struggle, is of course class, even if it acts behind the scenes in most types of war. The principle focus of the traditional Marxist-Leninist analysis was on civil war, which was a direct expression of the class struggle and the principle means by which the whole social system was radically transformed. In such wars, the contradictory political goals of different economic classes meet head on; the ruling class uses the state apparatus as its instrument. In international wars, which because of their frequency, size and consequences, occupy an important place in the study of armed conflict, states are the ostensible actors; however, since the state is merely an instrument of the ruling class, in each society, such wars are also rooted in the class struggle and are its indirect expression. In wars of national-liberation, the nation fighting for independence is usually led by classes or by a coalition of classes that combine the interests of the nation with their own.

The distance between these two traditional views has lessened somewhat as a result of theoretical development on both sides. On the one hand, political realists now pay more attention to domestic sources of foreign policy. They are gradually departing from the view that the confrontation between the two blocs of states with different socio-economic systems is simply as result of the interplay for power in the international arena. They still deny that domestic conflicts are based on contradictions between economic classes, but by paying more attention to the analysis of the internal sources of conflicts and wars, they have narrowed to some degree the gulf separating the two approaches.

On the other hand, Soviet scholars emphasise more frequently, and discuss in more detail this idea that the sense in which class contradictions should be understood as a motive for war is not self-evident. The class struggle is fought more or less directly on different internal and international fronts: between antagonistic classes within particular countries; between the antagonistic systems; and across state boundaries, between on the one hand the working class in all countries united in an international revolutionary movement, and on the other, international capital, which supports bourgeois rule. More indirectly the class struggle is reflected in the contest between imperial powers and the nations that are or have been dependent upon them. The struggle for full national independence is in the course of historical progress interrelated with the struggle for social transformation of society. Moreover by weakening the imperialistic system as a whole, it contributes to the world-wide struggle between the opposing social systems. A further stage removed from the class struggle are wars between developed capitalistic countries; for the most part such wars are internecine struggles between ruling classes for their selfish interests. In some cases, the class character of war cannot be pointed out at all: some wars do not reflect the conflict between classes.

There seem to be indications that the Soviet approach is evolving towards an admission of states as well as classes as the proper actors in war, although all such suggestions are carefully hedged with reservations, and class terminology is still used to explain the role of state as a party to war. In some studies, and even some articles in textbooks, war is defined as a continuation of the policy of classes or states or both. In others, the actors mentioned include classes, states, nations, and popular majorities. A second modification is the thesis that the ruling class is stratified, the various strata may have different attitudes towards particular wars. Together these shifts suggest that Soviet scholars have come to consider class as the main source determining the position of the parties in war rather than as the main direct actor in it.

Typologies

These similarities and differences can be summarised in a comparative review of the types of war recognised by both sides.

The Western approach with its traditional view that war is a matter between states arising out of the inevitable conflict in their respective interests and the impossibility of their peaceful resolution in the anarchic realm of international affairs, regarded as war only those armed conflicts which are waged by legitimate authorities. This position has proved too formal to be useful as irregular wars have greatly increased in frequency. It has gradually given way to a threefold typology of internal, interstate, and mixed wars. [5] Since a potential war between the superpowers is recognised to be an interstate war with many special features, this typology comes close to the basic classification applied by Soviets ever since the October Revolution into interstate wars, civil wars, wars of national liberation, and wars between the socialistic and capitalistic states. However, the difference between the criteria at the base of the typologies is crucial: in Western definitions it is the outer attributes of the hostilities that signify, moreover, typologies based on military-technical characteristics have remained the most frequent; in Marxist–Leninist typologies the competing socio-political forces and their political aims are always the criterion.

Both approaches accuse each other of unscientific method. Marxist–Leninists are charged with making reality fit their metaphysics and therefore try to reduce all wars to one main type, or to apply one single criterion, when it is quite clear that some wars, [6] particularly internation ones, have nothing to do with class conflicts. Non-Marxists are said to be blind to the distortion that their class outlook brings to all their interpretations of war and all their typologies of war, the latter reduced to military-technical or other secondary characteristics, for ideological-political reasons. [7] Soviets accuse Western approaches of trying to make it more acceptable to put down movements striving for social change or national liberation by refusing to treat such conflicts as war. Western scholars have in turn declared that the Soviet scholars insist on relating all wars to the class struggle in order to make justifiable the active participation of Communists in support of the progressive side in all conflicts.

To sum up, the two approaches have become more similar in one or two respects despite the polemics. Political realists have taken greater interest in studying internal wars, but they consider such wars to have a different political dimension and to be fought by different actors from the interstate wars they have traditionally studied. On the other hand, the Soviet scholars have to some extent softened the class criterion such that its role in determining the political nature of war and in defining the parties to war is interpreted more flexibly. The scope of both approaches to war encompasses similar types of war, although the significant features by which the types are distinguished from each other is quite different. The final judgement must still be that the separate ideological postulates lying at the base of each approach are far too divisive for the gap between the two views to be much narrowed by these tendencies.

Problems of the continuation

There are four main problems relating to the sense in which war can be seen as a continuation of policy. These might be expressed as four questions: what is the particular policy of which war is a continuation? what is the distinction between peaceful and violent means? can political control be kept over military operations? and, how does the course of war affect policy?

The policy continued

As we have seen above, the opposing approaches have quite separate views regarding the essence of politics, which has profound implications for their concept of war. Apart from such fundamental differences, there are other points on which the two views differ in interpretation or emphasis regarding the policy that is actually continued in war.

In their traditional forms, Marxism–Leninism interpreted the idea of the continuation of policy very broadly while political realism usually treated it in a comparatively narrow sense. Western scholars have been inclined to emphasise the policy a government has on some usually international issue on which states take positions irreconcilable by peaceful means. Soviet scholars have seen war as the continuation of the whole politics of the ruling class rather than the particular policy it may have in a given question. Put another way, the Soviets have tended to stress long-term policy, political realists to stress short-term policy.

Modern developments in both theories have brought the two approaches somewhat nearer each other. In the West, much research has gone into analysing the postulate of political realists that states go to war when their leaders decide after rational deliberation that it is in the national interest to do so. The findings of these studies have alerted Western scholars to the pitfalls of assuming that the national interest is in some sense revealed to decision-makers, who have only to find the best means of pursuing it. What the national interest is for the state leaders depends on what they believe it is; and the interpretation they put on international developments depends on their perspective on life in general. Beliefs, perceptions, values, judgements – all spring from the social heritage and life experience of the individual. Thus the political reality he meets is not a piece of hard data but in large measure what he takes it to be. These socio-psychological refinements to Western theory correspond in some measure to the ideological dimensions on which the Marxist–Leninist view of the politics of the ruling class is based. Their introduction thus makes discussion of a much contested point more possible.

The Soviet scholars have on their part refined the maxim that all wars are the result of the class struggle by treating cause as a complex concept. The basic cause of war may be the economic division of society into antagonistic classes, but each war will also have more direct causes. In recent work more attention has been given to immediate causes than to underlying ones, partly because of the pragmatic need to know what the correct policy on particular issues should be if war is to be avoided. At the same time this new interest has increased the resemblance between Marxist–Leninist and non-Marxist theories.

Both approaches have also been developed in response to the new features of international affairs in the nuclear epoch. Owing to the risks of war, more attention has been given to accidental, irrational, and fortuitous factors in order to eliminate their influence as far as possible. Moreover, in recognition of the intensity of the political division between socialistic and non-socialistic states, the meaning attached to the policy that is continued in war is in some contexts extended beyond the policy of the direct participants in war. The ramifications on the policies of the two global camps, to which the warring parties usually belong or from which they receive support, are also taken into consideration.

Change of methods

The idea of continuation implies that a change of method occurs. The traditional statements of the confronted approaches were in general agreement that there are three main aspects to this change: firstly, armed violence becomes the main instrument of policy; secondly, the violence is delivered by organised armed forces; and thirdly, other methods are also used simultaneously although they are subordinated to the military operations. The Soviet approach has remained essentially unchanged. Political realists have however, introduced new ideas, some of which have reduced differences with the Soviet approach, others of which have rather enhanced them. Amongst the former is the redefinition of the conditions that must be fulfilled before armed forces might be recognised as a party in war. Whereas political realists previously refused to recognise hostilities as war unless a rather high level of organisation was achieved, many of them now consider to be sufficient indications that organised forces are in the process of being gathered, even if they have not yet appeared in the course of fighting. Presumably this modification has been made in recognition of the great number of modern wars in which irregular warfare has occurred, at least in the initial stages.

Amongst the new ideas that increase the difference between Soviet and Western concepts of war are those which suggest that the boundaries between peace and war have become obliterated (though not typical of the approach as a whole, they are sometimes presented by political realists). It has been suggested, for example, that the transition from peace to war be treated as a continuum in which hostile moves gradually change from non-military to military, since the point at which peace becomes war is often difficult to determine and is chosen arbitrarily. A second example is the proposition that with the increase of military threat in peacetime and the use of political methods in guerilla warfare the difference between peace and war is not so much qualitative, based on the use or non-use of military violence, but more one of degree.

Such ideas are completely unacceptable to Soviet scholars. They insist that there is an essential difference between peace and war, and that the mark of the qualitative distinction is the dominant method by which the political struggle is carried out, whether by peaceful or military methods. Any other suggestion is not only patently false, but also dangerously deceptive since it can be used to convince the members of a society that war has already started or that it is meaningless to try to prevent war.

Political and military priorities

The last two questions with which this section on the continuation was introduced are perhaps best treated as two aspects of the same problem, namely the extent to which military goals are incorporated in the political goals of a country. The way this comes about during peace is itself an interesting topic. In time of war, the political stature of military goals must inevitably rise, and the interesting question becomes how far they remain subordinate to supreme political (non-military) goals, which are thereby continued in war. Do military goals in fact take the upper hand once a country finds itself in a state of war? Is there a continuation of policy or two different policies?

Both schools of thought have in common the general theoretical premise that political considerations are paramount even during war: goals, strategy, and tactics are all subordinated to political interests. And both argue that this is as it should be.

Political realists sometimes appear to be ambivalent whether they mean that wars actually do remain under political control or whether they should be although they often are not. The thesis of many studies of the Second World War is, for instance, that the Western Allies allowed the military goal of unconditional surrender to override all other objectives, whereas the Soviet Union always had a clear idea of the political goals it wished to achieve by military means. As a result, the postwar arrangements were more favourable to Soviet than to Western interests. The Soviet success in this respect has usually been attributed to the iron hand of Stalin, but implied in this argument is a criticism of the decentralised system of decision-making traditional in the West. Moreover, the experience of the Western great powers in conflicts where the opponent has used guerilla tactics has hardly dispelled these doubts. All too often, the realists have argued, governments have failed to define clearly their political goals on the basis of the national interests and have allowed comparatively tangible military objectives to have priority by default.

On most of these points Soviet scholars would agree with their Western colleagues, but they would put a different interpretation on the facts. They would argue that the difficulties of the Western Allies in asserting political control over their military operations were caused by weaknesses inherent in the capitalistic system. The political efficiency of the centralised democracy in the Soviet Union demonstrated the superiority of a social system in which the leaders have the full backing of the people because they act in the true interests of all. Thus, these same tendencies which have been noted by political realists are taken as evidence confirming the Soviet belief in the crisis of the capitalistic system and the superiority of socialism.

When considering the prospects of a potential nuclear war both sides tend to project these views. Political realists wonder whether it would be possible for any belligerent in such a war to assert the primacy of political values. They indicate the difficulty if not impossibility of determining the political goals or of achieving victory in a nuclear rocket war because of the magnitude of the destructive forces unleashed. Moreover, the strategy of nuclear war would require that the hostilities unfold with such rapidity that political leaders would have no opportunity of exerting conscious and effective control to keep the war within the compass of their general political aims. Soviet scholars agree with most of these judgements, but in contrast to the policial realists, who treat these difficulties as inherent in the nature of nuclear war, the Soviet scholars discuss them only in connection with the capitalistic countries.

Both approaches also take up the problem of the impact the course of war would have on the definition (or rather redefinition) of political goals. In older theories of war, this feedback affect usually referred to the need to adjust policy according to the outcome of military operations as the war proceeded. In modern theories, much attention is also devoted to the social costs of war, regardless of whether the military operations lead to victory or defeat. Thus political realists underline two most dangerous effects. First, the aims of war would in all probability be redefined because the whole international situation would require a radical reassessment. Second, the enormous loss of life and material destruction would lead to extreme social crises, and perhaps even to the complete breakdown of the society with unpredictable consequences for its future. Soviet scholars also note that such a war would make more unmanageable, problems existing even in socialistic countries, but it would not affect the internal cohesion of their societies. The negative impact on capitalistic countries would in contrast be devastating: it would deepen all contradictions, intensify the class struggle and the struggle for national liberation, and accelerate the victory of both revolutionary movements. [8] Therefore, if the imperialists were to unleash a nuclear war, it would only hasten their ultimate defeat.

Finally, both approaches note that the military requisites of a nuclear war have already left their stamp on society by contributing to the growth in the role of the military in peacetime. Soviet scholars indicate that the military-industrial complex has gained an ever more dominant influence over both the internal and external politics of the largest capitalistic countries, particularly the United States. [9] The military have exploited the fact that preparations for a nuclear war require very large investments of time and resources. They have succeeded in welding an alliance with the powerful interests of monopoly capital, which reap enormous profits by manufacturing weapons of mass destruction. At the same time as military technology has become increasingly sophisticated, the military have become increasingly indispensable to political leaders because of their expert knowledge. The militaristic circles have thus been able to exert great political power.

Political realists are somewhat divided on this matter. Many would agree with the substance of the Soviet analysis of developments in the West, although 'hard-boiled' realists would reject it as a distortion. The Sovietologists amongst them have observed that the military have become more politically prominent within the Soviet Union as well. [10]

Notes

[1] 'As the Marxist–Leninist science has indisputably proved dictatorship of the economically strongest class is the nature of each state; it uses political power to protect the existing economic system and to suppress the resistance of its class antagonists. . . . Bourgeois democracy is, in fact, a dictatorship, a political rule of a minimal barch of exploiters over the overwhelming majority of the exploited' (E. B. Peskov, B. A. Shabad, Apologiya imperialisticheskogo gosudarstva', in *Istoricheskii materializm is sotsialnaya filosofiya souvremennoi burzhuazii,* p. 379, cf. note 20 in ch. 13).

[2] All Soviet publications concerning the science on state, politics, law, and economics deal with the problem in question and include critique of non-Marxist theories. A brief summary in 'Gosudarstvo' in *Pliticheskii slovar'*, Gospolitizdat, Moscow, many editions (e.g. 1958, pp.136–7); cf. *Istoricheskii materializm*, F. W. Konstantinov ed., Gospolitizdat, many editions, ch. 'Gosudarstvo i pravo'.

[3] In all Soviet studies on the relation between politics and war, the failure mentioned above of the non-Marxist interpretation of the problem is pointed out. '... the bourgeois ideologists artificially set up domestic policy in opposition to foreign policy and maintain that foreign policy decides domestic policy and thus attempt to prove that war is the product and continuation only of the former' (*Marxism–Leninism on War and Army*, p.24).

[4] G. Shaknazarov, 'O sootnoshenii sil v mire', *Kommunist*, 1974:3; A. Sergiyev, 'Leninism on the Correlation of Forces as a Factor of International

Relations', *International Affairs* (Moscow, 1975:5); S. Tyushkevich, 'Sootnoshenie sil v mire i faktory pre dotvrashcheniya voiny', *Kommunist Voorushennykh Sil,* 1974:10; *Voennaya sila i mezhdunarodnye otnosheniya,* Izd. Mezhdunarodnye Otnosheniya, Moscow 1972; A. Topornin, 'The Balance of Power Doctrine and Washington', *SShA: Ekonomika, Politika, Ideologiya,* 1970:11; V. Kelin, 'Dangerous conceptions', Izv.-APN, 25 February 1975; V. Zhurkin, 'Détente and International Conflicts', *International Affairs,* 1974:7, Michael Voslensky, 'The Correlation of Forces: the Soviet View', paper prepared for Peace Science Society,Conference in Zurich, 1975; Sh. Sanakoyev, 'The Problem of the Correlation of Forces in the Contemporary World,' *International Affairs,* 1974:11; V. Zhurkin, 'Détente and International Conflict', *International Affairs,* 1974:4, G. Shaknazarov, 'Deistvennye faktory mezhdunarodnykh otnoshennii', *Mezhdunarodnaya Zhizn, 1977:1.*

[5] Typologies including internal, international and colonial wars can be regarded as coming even closer to the Soviet typologies based on the socio-political criteria; some of such typologies have been mentioned in Ch. 4, note 8.

Similarly, classifications of internal armed violence based on the political motives for action can be seen as likening more the Soviet ones; some of such classifications have been listed in Ch. 4, note 10.

[6] Dahrendorf, 1959; Coser, 1956, 1967; Nicholson, 1970 states that there are many 'counter-examples' to the assertion that class conflict underlies each war, p.20, Cf. Ch. 16.

[7] Cf. Ch. 16. 'The classification of wars according to military-technical features only is typical of bourgeois military theoreticians. This is because it is unprofitable for them to reveal the class essence and the aggressive character of the military policies of imperialism. They therefore confine themselves to a "technical" classification of wars, ignoring their class-political content'. The authors contend that nuclear wars (total and limited), conventional wars (world and local) and local wars against the national-liberation movement are the three types of war taken into account in framing modern US strategy. (*Marxism–Leninism on War and Army,* p.71).

Rybkin, 1973b, writes that it is characteristic of bourgeois military ideologists that 'no matter what social science they represent, they usually separate themselves from the problem of classifying wars according to their socio-political content. They prefer to differentiate wars primarily according to scale – "world war", "local war", or "limited war". To these they also add the "cold war". . .'

Cf. the critique of the Western typologies of war in S. Tyushkevich, 'Razvitie ucheniya o voine i armii na opyte Velikoi Otechestvennoi voiny', *Kommunist Voorushennykh Sil,* 1975:22, p.15; R. Simonyan, 'Voiny glazami Pentagona,' *Krasnaya Zvezda,* 27 May 1976; *Sovetskaya Voennaya Entsiklopediya,* 'Sovremennye burzhuaznye teorii voiny' in 'Voina', vol. 1, 1976, p.283.

[8] War is a major crisis, and any crisis accelerates the disclosure and intensification of contradictions; nuclear war would generate an 'ultimate' crisis of imperialism and be disastrous for it. Nothing less than a complete collapse of the

system would be the result (*Marxism–Leninism on War and Army*, pp.2, 30 and others); *Filosofskoe nasledie V. I. Lenina . . .*, pp.24-5; *Metodologicheskie problemy voennoi teorii i praktiki*, ch.IV 2.

[9] S. M. Vishnev, *Sovremennyi militarizm i monopolii*, Izd. Akademii Nauk SSSR, Moscow 1952; S. A. Dalin, *Voennyi gosudarstvenno – monopolisticheskii kapitalizm v SShA*, Izd. Akademii Nauk SSSR, Moscow 1961; A. A. Kornienko, *K kritike sovremennykh teorii militarizatsii ekonomiki.* Voenizdat, Moscow 1960; P. A. Faramazyan, *SShA: militarizm i ekonomika*, Izd, 'Mysl', Moscow 1970; M. Kulakov, *Ideologiya agressii,* Voenizdat 1970, ch. I, ch. V; *Filosofskoe nasledie V. I. Lenina . . .*, ch. IV; *Problems of war and peace,* ch. 5; P. Zhilin, Y. Rybkin, 'Militarism in Contemporary International Relations', *International Affairs* (Moscow) 1973:10; Y. Rybkin, A. Migolatyev, 'The ideology of the Present-Day Militarism', *International Affairs*, 1974:7.

[10] 'The Military-Industrial Complex: USSR/USA', *Journal of International Affairs*, 1972:1, esp. Roman Kolkowicz, 'Strategic Elites and Politics of Superpower'; Thomas W. Wolfe, 'Soviet Interests in SALT: Institutional and Bureaucratic Considerations', in *Comparative Defence Policy*; Benjamin S. Lambeth, 'The Sources of Soviet Military Doctrine', ibid.: Klaus Mayer, 'Die sowjetischen Streitkräfte und Breschnew: Konsens oder Dissens?', *Beiträge zur Konfliktforschung*, 1974:1; Leon Goure, Foy D. Kohler, Mose L. Harvey, 'The Role of Nuclear Forces in Current Soviet Strategy', University of Miami 1974; David Holloway, *Military Component in Soviet Policy*, Paper presented at the III International Conference of Heissische Stiftung Friedens- und Konfliktforschung, Frankfurt a/M., December 1974.

15   The interpretation of Clausewitz

Despite the fundamental differences sketched in the foregoing chapter, both Soviet theory and political realism do assume that war is still a continuation of politics by means of armed violence; they are therefore conceptually related to the famous Clausewitzian dictum. Moreover, the whole development of the two approaches up to and including the modern debate on the nature and future of war has been accompanied by recurring reference to the traditional formula. A review of this discussion may cast some additional light on the differences between the two approaches.

In the foregoing chapter it was argued that such diametrically opposed ideologies as the Marxism–Leninism of the Soviet Union and the variety of non-Marxist ideologies prevailing in the West could only adopt the same formula by interpreting the principal terms in different ways. Therefore, although both agree that there exists a relation between politics and war whereby war is the subordinate instrument of politics, each views the nature of the actual tie between the phenomena in a different way.

Approaches, philosophies, and the formula

Part of the difference may perhaps also be traced to the fact that the two traditions have tried to answer two basically different questions in their approach to the study of war. Clausewitz and his followers continued a long-standing tradition of analysing war to find the most effective method of achieving victory. His point of departure was clearly military. In his view, the Napoleonic campaigns had demonstrated the need for a new concept of war and a new kind of strategy. His interest in politics derived from his belief that political aims in the post-Napoleonic world influenced military strategy more than ever before. A knowledge of policy would thus be helpful in winning war. The course he took in his reasoning was thereby from war to politics.

In contrast to the Clausewitzian approach, the aim of the Marxist analysis of war was to discover how to win the great political struggle between classes. Armed uprising seemed to be the only way the working class could gain power, although the possibility of exploiting the First World War for revolution soon became a central focus. In both cases, however, the primary object of Marxist analysis was to win a particular political struggle. War was studied as one of several factors that might contribute to the victory of the working class. The path of analysis was from politics to war.

To digress for a moment, there is a second and related respect in which the two traditions diverged right from the beginning. Clausewitz took the state-system and the fact that wars normally occur between states as given. He was interested in how states made use of war in the conduct of their foreign policy. In the Marxist tradition, however, what was given was the need to put an end to the inhumanity of capitalism through revolution. The main focus was thus on the role of war in domestic politics. This difference between the two approaches is discussed more fully in the next chapter.

Clausewitz thus began with war and considered politics to be a factor influencing and modifying (moderating) its course. He therefore found it logical to treat unmodified war as pure violence, as violence pushed to its utmost bounds. This he called absolute war. In his scheme it became the ideal form of war, the archetype in relation to which real wars, limited to some extent by policy, were to be compared and measured.

Marxism and then Soviet theory have, on the contrary, never adopted a concept of ideal war nor accepted that there can be any such thing. The only wars are real wars, all of which are a type of activity by means of which class goals may directly or indirectly be obtained. When wars are assessed it is with respect to their effectiveness in bringing closer the goal of long-term policy.

While the Clausewitzian run of thought might be reduced to the formula 'from absolute war to real war', the direction in the Marxist tradition might therefore be expressed as 'from absolute policy to real war'.

Clausewitz's heirs in the long tradition that regards war primarily as a military phenomenon expanded upon the idea of war as violence aiming at the destruction of the enemy. These militarists either ignored Clausewitz's position on the moderating function of politics on real wars or else considered the idea on primacy of politics relevant only to the outbreak but not the course of war.

Political realists have preferred to see Clausewitz in the light of the positivistic tradition to which they are anchored. They have argued that Clausewitz recognised that wars are in reality a form of political activity. His discussion of absolute war is conducted at a purely theoretical level, and is therefore of less practical value. It is to misconstrue his intent, they argue, to maintain that he advocated war as unlimited violence. According to this interpretation then, Clausewitz is the father of a line of thought in which war is in actual practice a political phenomenon.

Soviet scholars regard Clausewitz as an exponent of bourgeois, i.e. reactionary ideology, and are therefore unconcerned with the Western debate about the correct interpretation of his ideas. On the one hand, his preoccupation with violence is connected with a militaristic and expansionistic policy. On the other, his ideas on war as a political act distort the true character of politics by neglecting its class essence and relating it to power policy. In either interpretation, Clausewitz's ideas are unscientific.

The distinctions can also be conceived of in another, perhaps overly simplified way, if Clausewitz's contribution is divided into his philosophy of war on the one hand, politics on the other, and the formula by which he links the two together.

In general, the militaristic line of interpretation of Clausewitz stressed his military philosophy so much that even though they accepted his view of politics, they tended to ignore the political implications of the formula. The tradition of political realism adopted the formula and the political philosophy underlying it, but played down the significance of Clausewitz's philosophy of war. The modern neo-Clausewitzian variant of political realism has brought the political philosophy up to date with the new political and military-technological conditions. And, quite differently, while adopting the formula of war as a continuation and instrument of policy, Marxism–Leninism placed it in a completely new setting and rejected Clausewitz's concept of both war and politics.

Clausewitzian ideas in the mutual assessment

The Soviets look at the West

It is therefore a mark of criticism when Soviet scholars refer to the Clausewitzian influence in the Western theories of war. Nor is it sufficient to win approbation that non-Marxist scholars reject or modify Clausewitzian ideas: they must do so for the right reasons.

In the analysis of Clausewitz's influence on Western thought up to World War II, Soviet scholars mainly pointed out the military tradition that idolised war. [1] The Nazis themselves claimed to continue the heritage of Frederick the Great, Clausewitz, Moltke, and Ludendorff, and to apply the military principles of these men in their war campaigns. The Soviets criticised such ideas for the emphasis placed on the theory of absolute war and all its corollaries (e.g. that the main aim of war is to destroy the enemy forces), [2] and for the neglect of the primacy of politics over warfare. This error led logically to a view in which military strategy is independent of policy, or even, as in Ludendorff's writing, to the thesis that policy is completely subordinated to warfare. Western militarism was therefore incorrect on two counts: it was based on the false idea of absolute war inherited from Clausewitz; moreover, unlike Clausewitz, who despite his non-class political philosophy at least recognised the primacy of politics over warfare, the militarists committed the blunder of elevating warfare to an end in itself.

In the critique of Clausewitz's heritage in the Western thought after the Second World War, the references to Clausewitz in Soviet assessments of Western theories of war were more diversified, reflecting perhaps the Western debate on the interpretation of Clausewitz's ideas. Nevertheless, all three schools of thought, militarists, political realists, and pacifists, have been criticised.

Firstly, such ideas as massive retaliation, which dominated Western politico-military doctrines during the 1940s and 1950s but traces of which can still be found, are interpreted as applications of the Clausewitzian concept of absolute war. [3] These ideas err in underestimating the importance of policy as a moderating and controlling factor. Pointed out as advocates are not only leading military officers, who might be expected to carry on the militaristic tradition, but also those political realists, who during the period when the United States had a nuclear monopoly and then gradually lost it, supposed that total war might be applied as an instrument of policy.

Secondly, Soviet scholars criticise the main body of political realism for adopting the political philosophy underlying Clausewitz's formula of war as a political act. Their treatment of war as a normal instrument of foreign policy, which is conceived not as class policy but as state policy, must be regarded as unscientific. [4] It leads them to neglect internal politics as well as internal wars, and to depreciate the impact that changes in socio-economic structures and in military technology have had on the nature of international politics. [5] By restating the Clausewitzian assumption that the only valid criterion for assessing war is its usefulness in increasing the power of the state that uses it, political realists exploit a false amorality to conceal the immorality of the wars unleashed by imperialism. They have no moral qualms about nuclear war; it is merely their callous opportunism that moves them to condemn it.

Thirdly, Soviet scholars disagree with the type of opinion common amongst pacifists and held by some political realists as well, that the Clausewitzian formula conceiving of war as an instrument of policy is no longer valid. Such a view fosters the dangerous illusion that war is no longer possible. On the contrary, the imperialistic forces may well try to gain their ends by war. The only way they can be prevented is by actively struggling against their every attempt. The anti-imperialistic front is only weakened by such a complacent attitude as that the pacifists express. [6]

Reference to Clausewitz has thus been a vehicle through which Soviet scholars have criticised the main ideas on war held in the West in modern times. The thread linking these three types of criticism together, and joining them to the critique of Clausewitz, is the concept of politics in terms of class. This is the foundation on which Soviet scholars reject the Western notion of war and politics, (ideas shared by or derived from Clausewitz) as well as the absolutism of pacifists, who fail to see that war may still be an instrument of class policy.

The non-Marxists assess the Marxists–Leninists

When Western scholars take up Clausewitz in relation to Marxism–Leninism it is usually in connection with one of two themes. According to the first, Soviet theory is greatly indebted to Clausewitz for some of its basic concepts; the Marxist–Leninist notion of the relation between war and politics is interpreted as an outgrowth from Clausewitzian roots. [7] In the second, more critical theme, the militant tone in Soviet politics is attributed to its adaptation of Clausewitz's idea of absolute war.

The argument of those who interpret the Soviet concept of war as a variant of the Clausewitzian is fairly straightforward. It is known that both Engels and Lenin read and admired Clausewitz's study of war. [8] Lenin himself has declared that war is a continuation and instrument of policy by other means – virtually a direct loan from Clausewitz. [9] There can be no doubt there-fore of the Clausewitzian influence: the object of analysis is to define its extent. [10] Some scholars go so far as to claim, that almost each part of the Soviet theory of war borrowed some basic ideas from Clausewitz.

Something of the same type of argument is used by those critical of the militancy in Soviet politics. They maintain that historical materialism paints a picture of society so simplified and so utterly devoid of any faith in the ability of men to solve their differences through mutual understanding that it must be called a distortion. The only hope Soviets can offer is that when the misery of the exploited class has reached the point where it is no longer bearable, revolution will break out and deliver mankind not only from the capitalistic yoke but from all internal and international struggle and violence. Her Soviet theory shares with Clausewitz, or perhaps even borrows from him, a concept of violence as a great outburst of destructive force, which at the same time paves the way for the new balance or the new order.

This idealisation of violence including the use of armed force can be traced throughout Marxian thought, from Marx himself to Mao. Marx and Engels provided the basic idea of inevitable revolution or class war, but they confined its application to internal situations. Lenin expanded the scope of confrontation with his theory of imperialism, in which all interstate wars are interpreted as a reflection of the class struggle. Stalin further emphasised the antagonistic dichotomy between capitalism and socialism at the international level by declaring all revolutionary wars and all wars fought by the Soviet Union to be one and the same war. After the Second World War, this policy of total confrontation was practised in the strategy of the Cold War, in which a combina-tion of all military and non-military methods was used to realise long-range political goals. [11] The resultant state of neither-war-nor-peace can be regarded as a kind of warfare that corresponds with the Clausewitzian concept of war as political intercourse conducted with an 'admixture of military methods'. Some scholars assert that Mao has rounded off the adoption of the Clausewitzian theory of war by the socialist camp. [12] To the theory of civil war as a continuation of internal class struggle, and world war as a continuation of world class struggle he has added the theory of the peoples war as a struggle in the underdeveloped areas against the alliance between domestic reactionary forces and world imperialism. The theory of war between the 'rural areas' of the world and the 'cities' is said to be the logical extreme of the Marxist–Leninist interpretation of war as absolute violence, since it reduces all political conflicts into one great struggle, that can only be resolved through one great war.

Common to both these themes is the presumption that Marxists–Leninists have made Clausewitz's theory of war their own by making a few adjustments so that it better fits their theory of class politics. No criticism need be implied in this observation, although Soviets themselves are very sensitive to any suggestion that they have cribbed their theory of war from a Prussian militarist. For the most part, however, Western scholars who see a connection between Clausewitz's idea of absolute war and the Soviet concept of revolution reject the notion of violence pushed to its utmost bounds. They therefore use Clausewitz as a platform from which to attack the Soviet belief in the inevitability of violent revolution.

The meaning of the debate

When such various sins are attributed to the influence of one man's writing, one might be pardoned for suspecting that the admirers, as well as the critics themselves, are guilty of a certain amount of exaggeration. Have not Western scholars progressed in their political thinking beyond the state-centred balance of power concepts of Clausewitz? Does not the Soviet concept of evolution in the context of peaceful coexistence imply in non-Marxist terms a less violent form of militancy? Indeed, when all adjustments and modifications are made to take into account all the social and technological developments (however these are interpreted) that have taken place since Clausewitz wrote *On War* almost 150 years ago, can there really be anything of the Clausewitzian essence left beyond the idea that war is a continuation of politics?

Apart from the problem of deciding whether modern ideas that bear some resemblance to ideas put forward by Clausewitz are sufficiently similar to be called Clausewitzian – and that is what the name-calling is about – one might also ask what it signifies at all to call a family of concepts Clausewitzian. It would seem from the way some critics argue that they believe the object of their attack never would have come into being had it not been for that evil genius Clausewitz. For example, the Western version might run: if only Clausewitz had never written 'war is a continuation of politics by other means', then Lenin would never have expressed the idea himself. This argument is surely a false one; it presents the development of ideas in a simplified fashion. Clausewitz's *On War* was not a necessary prerequisite to Lenin's theory of interstate war; it was rather the case that Clausewitz gave expression to an idea that was also inherent in Marxism and then in Marxism–Leninism. [13] Reading *On War* may have stimulated Lenin to find the idea in Marx's conceptual system, but so well does it fit into the Marxian train of thought that it would have emerged sooner or later. Similarly, militarists, political realists, and pacifists may be indebted to Clausewitz in one sense, but their ideas are not causally dependent upon his.

These reflections point to a common place that seems nevertheless to be often overlooked. Certain ideas are called Clausewitzian, not because Clausewitz was the first to express them, nor because his presentation was complete, but because he succeeded in articulating many of the conclusions that could be, and were being, drawn by men of military and political science about the great socio-political changes occurring in nineteenth century Europe. Clausewitz was not the first to see that wars are fought for political aims, nor did he ever provide a comprehensive description of war as a political instrument. It is nevertheless to his credit that in one excellent book he presented both a very valuable analysis of the military theory of his times and the war/politics formula, even if he did not succeed in combining them. And his study of war has become a source of intellectual stimulation to several generations of scholars. Therefore, the Clause-witzian tradition neither begins nor ends with Clausewitz, even if *On War* is its focal point. There may be some historical interest in trying to determine what Clausewitz himself meant, but modern interpretations of Clausewitz inevitably say more about the interpreters' general ideas of man, society, and war than they do about Clausewitz. The debate on the Clausewitzian formula is therefore a useful tool for discovering the differences between the modern ideologies, and in particular, between the modern theories of war.

Notes

[1] *Sovremennaya imperialisticheskaya voennaya ideologiya*, Voenizdat, Moscow 1958; M. A. Milshtein, A. K. Slobodenko, *O burzhuaznoi voennoi nauke*, 2 ed., Voenizdat, Moscow 1961; *O sovetskoi voennoi nauke*, 2 ed., Voenizdat, Moscow 1964.

[2] 'The idea of unlimited armed violence, stemming from Clausewitz's assertion on the "absolute war" was basic in the German doctrine' (*Sovremennaya imperialisticheskaya voennaya ideologiya*, p.65).

[3] 'Statements by Clausewitz on the employment of unlimited force in war exerted some influence on the forming of the military theory of imperialism. The most shameless ideologies of American imperialism, grasping at erroneous, contradictory statements by Clausewitz, hyperbolize armed coercion, depicting it as the sole effective means of policy, its foundation; in their statements they endeavour to find arguments to justify the aggressive military-political doctrines and strategic concepts of imperialism' (*Filosofskoe nasledia V. I. Lenina*, p.43); Strokov contends that some Clausewitzian assertions are now applied by the American military writers and West German militarists; Smith, 1955, serves as an example (1965, p.257).

'The ideologists of US imperialism are particularly fond of applying Clausewitz's erroneous propositions for their selfish ends, notably his view on the unlimited use of armed violence in an "absolute war". General Dale O. Smith, for example, frankly said: "The roots of the policy of a massive retaliation go back a long way ... The Clausewitz conception of war emphasized massive attack, instantly, at the critical point of enemy strength". The US Professor H. Speier, an expert on international affairs, deliberately adapts his aggressive doctrine to some of Clausewitz's propositions. He writes that total war, which has in the past formed the foundation of the nazi doctrine and is now being made much of the American doctrine, is essentially unlimited war, or, to use Clausewitz's expression, "absolute war".' (*Marxism–Leninism on War and Army*, pp.26–7).

[4] Contemporary Western ideologists are accused of taking from Clausewitz the erroneous assertion, that politics is non-class, and wars are 'national' and not connected with the politics of the ruling classes (V. Shelyag, 'Pod lzhivym flagom "renesansa" ', *Krasnaya Zvezda*, 3 July 1975).

[5] 'In extolling Clausewitz and ignoring historical experience the ideologists of the reactionary bourgeoisie, especially those closely connected with the top brass of the aggressive NATO bloc, make it appear that no changes have taken place in the interrelation between politics and war. They justify the policy of nuclear blackmail, insist on keeping thermonuclear war in their political arsenal. . .' (*Marxism–Leninism on War and Army*, p.26).

[6] Comp. Ch. 12; '. . . many bourgeois ideologists currently adhere to a pacifist formula in their views on nuclear war, maintaining that the latter "has ceased to be a continuation of policy". . . Certain of these [politicians, theorists] have reiterated and continue to repeat that the Clausewitzian formula on war as a continuation of policy is not applicable in our time' (Rybkin, 1973:20 p.26).

[7] Earle, 1943; Hans Rothfels, Carl von Clausewitz, *Politik und Krieg*, Berlin 1920; Garthoff, 1954; Fuller, 1962; Gerhard Ritter, *Staatskunst und Kriegshandwerk*, Bd. I. Munich 1954; W. Hahlweg, 'Das Clausewitzbild einst und jetzt', Introduction to: *Vom Kriege* 17 ed., Dümmler Vg., Hanover-Hamburg-Munich 1966; Ruge, 1967; Byron Dexter, 'Clausewitz and the Soviet Strategy', *Foreign Affairs* Oct. 1950; Stefan T. Possony, *Jahrhundert des Aufruhrs*, Munich 1955, Ch.X: 'Clausewitz als Lehrmeister des Kommunismus'; C. Beyerhaus, 'Der ursprüngliche Clausewitz', *Wehrwissenschaftliche Rundschau*, 1953:3; Wilhelm Ritter von Schramm, 'Von der klassischen Kriegsphilosophie zur zeitgerechten Wehrauffassung', *Wehrwissenschaftliche Rundschau*, 1965:9; Dirk Blasius, 'Carl von Clausewitz und die Hauptdenker des Marxismus', I-II, *Wehrwissenschaftliche Rundschau* 1966:5–6; Gerhart Matthaus, 'Krieg ist Politik mit Blutvergiessen', *Wehrwissenschaftliche Rundschau*, 1967:7; Gerhard Ritter von Schramm 'Ist Clausewitz noch zeitgemäss, Nachwort zum *Vom Kriege*, Rowohlt Vg. 1963; his 'Clausewitz als politischer Klassiker', *Wehrkunde*, 1973:4; Bernard Brodie, 'On Clausewitz: A Passion for War', *World Politics*, Jan. 1973; Roger Parkinson, *Clausewitz, A Biography*, Stein and Day, New York 1971; Vincent J. Esposito, 'War as a Continuation of Politics', *Military Affairs*, 1954; Wallach 1972, ch.11: 'Engels: Der Gründer der marxistischen Kriegslehre', 12. 'Lenin: Der Verwirklicher'.

[8] In all West-German analyses the influence of Clausewitz on Marx and Engels is emphasised; Alastair Buchan observes that Marx and Engels were 'great admirers of Clausewitz' (1966, p.95).

[9] In West-German analyses study of Clausewitzian writings by Lenin is described in detail; in many American and British articles Lenin is called

'disciple of Clausewitz' (a typical example: Donald E. Davis and Walter S. G. Kohn, 'Lenin as Disciple of Clausewitz', *Military Review*, 1971:9).

[10] And the assessement was very different; it varied from the opinion that the influence was moderate (Karl Marx 'rejoiced at finding in such an eminent military authority substantiation for his own theory of the relationship between war and politics' – Esposito 1954, p.22) to the assertion that almost the whole analysis of politics, war, revolution, etc. performed by Lenin was based on Clausewitzian ideas (Blasius stated that Lenin was able to elaborate the strategy of the Communist party in war, revolution, and in the post-war politics of the first socialist state only because he had read Clausewitz and had found in his writings scientific tools that enabled him to develop the theory and to realise the policy). Fuller, after an analysis of Clausewitz' influence on the founders of Marxism, concluded that Engels, Marx, and Lenin were 'deeply indepted to Clausewitz' (1962, p. 208).

The main conclusion of Gen. Antonio Pellicia's analysis: 'Clausewitz and Soviet Politico-Military Strategy' is that: 'The Soviet concept of war without any doubt, is the same as Clausewitz's (*NATO's Fifteen Nations*, Dec. 1975-Jan. 1976, p.32)': he contends, however, that Soviets acknowledge only the absolute form of war. A well known Sovietologist, one of the founders of the 'forward strategy' school, Stefan T. Possony, in a chapter entitled 'Clausewitz as the teacher of Communism', writes that Clausewitz demonstrated to the Russian revolutionists how to seize power (*Jahrhundert des Aufruhrs*, p.60); he also contends that Clausewitz was the only military prophet of Communism up to the end of World War II.

Even a more moderate Sovietologist writes that 'it is common knowledge . . . that much of Soviet military doctrine derives directly from the classical military-theoretical writings of Clausewitz' (Lambeth, 1974, pp.207–8).

[11] Dexter, 1960, Fuller, 1962, describe Communist strategy of peace as a direct application of Clausewitzian ideas of identity of goals pursued by states in peace and war. The distinguishing characteristic in Soviet warfare is the inter-changeability of political and military means. 'A "peace offensive" in Moscow, a cultural conference in Warsaw, a strike in France, an armed insurrection in Czechoslovakia, the invasion of Greece and Korea by fully equipped troops – all are instruments of one war, turned on and turned off from a central tap' (Byron, p.41); Fuller writes that the theory of a total and global unified war – a war in all dimensions – directed by a supreme central intelligence is in accord with Clausewitzian ideas (pp.202 ff.); he contends, as many other Sovietologists do, that Lenin inverted Clausewitz's well known dictum that 'war is a continuation of state-policy by other means' and substituted for it a formula holding that 'state policy is a continuation of war by every means' ('Our War Problems', *Marine Corps Gazette*, Nov. 1960, p.10). Garthoff contends that Soviets regard all policy, international as well as internal, as one unified process of permanent struggle, where military and non-military means are used according to circum-stances. 'Except that they are phases of policy with a different component of

armed force, no distinction between peace and war is meaningful in Soviet doctrine' (1954, pp.11–12).

[12] Matthäus, 1967; Donald E. Davis, 'Marxism and People's Wars', *Orbis*, Winter 1972; Johnson, 1973; Wallach, 1972, ch. 13: 'Der subversive Volkskrieg: Mao Tse Tung, Giap, Guevara'; W. Hahlweg, *Typologie des modernen Kleinkrieges*, Wiesbaden 1972; his: *Guerilla. Krieg ohne Fronten*, Stuttgart 1968.

[13] Lenin called the Clausewitzian formula 'dictum', 'saying', 'maxim', etc. In his quotations of Clausewitz he underlined the brilliant form in which the old truth was expressed, he didn't, however, regard Clausewitzian ideas as a quite new theory.

16   Wars within states

The subject

Civil war is a term that has long been used to denote armed struggle between citizens of the same country. Civil wars seldom figured in treatises on war, however, because such works were written by military men for whom the only natural and acceptable use of military forces was in war against the military forces of another state. In the main stream of the study of war, civil war was a treasonable activity, more a topic of interest to politicians and political philosophers than loyal military theorists. And it was naturally more as political philosophers and revolutionaries than practical military men that Marx and Engels used the term civil war ('Bürgerkrieg') parallel to the orthodox term war ('Krieg') to convey quite another concept from the standard one, namely that of a class war between bourgeoisie and proletariat. When Lenin later developed a theory of war, he adopted the term civil war in the above sense. Since non-Marxists continued to use the term in its traditional sense, a good deal of confusion existed as to what the term really meant.

Gradually the term was replaced in some non-Marxist approaches in favour of internal war, or complemented by it. This covers a great variety of armed clashes whose only common feature is that they occur within the confines of one country. The political goals of the belligerents may thereby be very diversified, the point of contention ranging from the rights of a religious or ethnic minority, to the degree of local autonomy, the political standing of a particular person, or a particular policy; it may reflect the ambition of some groups within the society to reform the system or even, as in Marxist–Leninist civil war, to transform it radically. In addition, the main actors in internal war may represent one of a great many divisions of the society, for example, a military élite in a *coup d'état*, or a national movement in a struggle for independence. External participation in internal war is considered to be an almost normal phenomenon.

It has only been recently that Soviet scholars have also allowed internal war as a general term to cover civil wars, some wars of national liberation, and wars for ethnic, tribal, or religious autonomy. However they do not recognise that the concept has any theoretical significance. The essential difference between wars lies in the political goals for which they are fought, not in the outwardly visible features of the hostilities. Of the main types of war conceived by orthodox Marxism, and in most non-Marxist theories classified as internal two remain in primary focus of the Soviet theory. The most important of the two is civil war. The meaning of the concept, virtually unchanged since its initial formulation, is armed struggle between antagonistic classes within a country for state power;

for the working class the goal is therefore either social liberation or the defence of socialistic gains, whereas for the bourgeoisie (or the imperialists, to use the more modern term) it is to consolidate its political domination or, in a post-revolutionary situation, to regain power. More recently, the concept has been extended to include armed struggle of the popular masses for general democratic aims in opposition to monopolistic circles that attempt to establish a dictatorship by using armed violence (these are called democratic civil wars).

The second main category of war, sometimes regarded as internal in Soviet literature, but often classified with international wars are wars of national liberation. Other kinds of internal armed violence are more or less ignored since they are not considered to be representative of the main contradictions of the epoch. Excluded from the concept of internal war are all internal clashes between members of antagonistic classes whose object is not related to control over the state apparatus, for example, an armed clash between demonstrators or strikers and troops sent to disperse them.

Since each approach has retained its original conceptual scheme, the gap between the two remains very great. Yet within both approaches have occurred shifts of emphasis that indicate a tendency towards convergence. Ever since the October Revolution non-Marxist scholars have come to recognise the particular importance that wars of social revolution have in our time. What the Marxist–Leninists call civil war is thus one of the most studied topics of non-Marxist writers on internal wars. [1] On the other hand, the Soviet introduction of the term internal war, however trivial it is in their esteem, can be interpreted as an admission that not all internal armed struggles fit exactly into the class concept of war.

In line with the course of historical events, both approaches have moved their attention away from developed countries and concentrate now almost entirely on armed struggles in under-developed countries; but they do not agree on the reason for the historical development they both observe. Soviet scholars say that civil wars are simply unnecessary for the future victory of socialism in the developed countries since the great growth of the revolutionary movement has brought about a favourable correlation of forces there (or according to the newest terminology, because the bourgeoisie does not dare to apply counter-revolutionary armed violence). Non-Marxist observers maintain that revolution is now unlikely in Western developed countries because the introduction of social welfare politics has made it meaningless.

The emphasis on under-developed countries has led both approaches to introduce sub-categories into their major classifications. Each follows its orthodox line, however: Soviet scholars begin with the assumption that all wars have a class base and try to fit new kinds of civil war within that framework; non-Marxists take the variety of internal wars as further evidence of the folly of trying to reduce all wars to class struggles.

Roots, continuation, consequences, and assessment

It is as true that there exists a Marxist–Leninist theory of civil war as that in the West there is no general theory of internal war but rather a loosely united collection of observations and hypotheses of varying degrees of generality. A major consequence of this inequality is that Marxist–Leninists have a clear idea of the causes (the fundamental ones at least) and the consequences of civil wars, whereas non-Marxist scholars and political realists among them do not.

According to Marxism–Leninism, civil wars are primarily rooted in the economic contradictions that divide society into antagonistic classes permanently engaged in civil struggle against each other. Since this intense struggle is for Marxist–Leninists a given, a fact of life, civil wars are thus a normal, inevitable, or at least highly probable, means by which these contradictions are resolved. Since the class struggle is moreover the quintessence of politics, civil war is the natural continuation of class politics. If the progressive classes win a civil war (which is the usual case), it is deemed a functional war, for social progress requires that old class formations make way for new until socialism ultimately triumphs.

Soviet scholars accuse the non-Marxist study of two sins, of burying the true civil war (i.e. war for socialist revolution) in the mass of civil violence, and of masking the economic-class roots of the great social upheavals. The reason they do so is to weaken the class struggle. [2]

Political realists often accept that there are important insights in the Marxist–Leninist theory; intolerable socio-economic conditions may well give rise to some kinds of internal war. But they would argue that it lacks that refinement which would make it truly useful. Economic inequalities are not by themselves sufficient causes of revolution since they are virtually universal and almost universally tolerated. [3] The reason people begin to find economic inequalities intolerable is usually connected with the inability of the government to act in such a way that the people can feel confident of future improvement. A further condition necessary for revolution must therefore be that the government is for some reason prevented from acting in a way that inspires the confidence of its people. Political realists deny that there is any *a priori* reason why governments in countries with market economies must sooner or later become incapable of keeping sufficiently abreast of popular demands. Indeed the success of the Western democracies in achieving greater economic equality is ample evidence that the Soviet theorists are mistaken. It is another matter, however, that a government may for one reason or another find itself in a situation it cannot cope with. The origins of such a situation may be many and varied, from faults in the decision-making system (lack of accurate information, corruption, time-consuming procedures, etc.) to structural weaknesses of the society (historical cleavages between minorities etc.) and also to external disturbances beyond the control of the government (war, world inflation, etc.). The political controversy at the heart of an internal war may be one of a similarly broad range of issues. But for none of these issues is war a natural means of resolution (the political controversy with the Communists is perhaps an exception, but only because it is one of the goals of the Communists to instigate revolutionary wars). The general attitude of political realists is that internal wars are dysfunctional: they arise from failures in the political system that could in principle be avoided; as a means of resolving problems they are more expensive in terms of social costs than non-violent methods. Some good may come of internal wars, but in general they check progressive evolution more than they foster it.

On some points recent theoretical developments suggest a slight tendency for the two approaches to meet each other. Soviet scholars have always been prepared to accept that particular circumstances, including external factors, may affect the outbreak and course of civil war if only incidentally. Now Soviet scholars as well as Western scholars stress the importance of relating all struggles to the global conflict between the two ideological camps; they accuse each other of intentionally fomenting aggression to suit their own purposes. [4]

Two recent modifications of the Soviet view of civil war do not seem to have been noted in the West yet. In the first place, it is considered to be possible that revolutionary changes may in favourable circumstances be brought about without open war. Secondly, it is declared that the working classes resort to armed violence only in response to the counter-revolutionary violence of the bourgeoisie, that is if it tries to prevent social reform. The prevailing Western view keeps to the orthodox position that Marxist–Leninists believe armed uprising to be necessary for revolution and that they deliberately strive to initiate war. It seems, however, that the shifts in the Soviet position constitute a less hostile interpretation of what to expect from the bourgeoisie. As Western scholars come to realise the import of this change, the debate may become less charged with accusations of bad faith and thereby gain in constructiveness.

Strategy and the question of new kinds of war

Both approaches emphasise that the strategic problems of internal wars are quite distinct from those of interstate war.

Western students of the strategic problems of guerilla revolutionary war usually point out the following distinctive features: the initial asymmetry of forces, the disparity in available resources, the irregular character of the forces and the methods they use, the great emphasis placed on political activity, and the total character of the goals. Some scholars contend that these features are so greatly different from the traditional strategy of war that the whole concept of war needs to be redefined.

Soviet scholars also underline features specific to civil wars: the extreme violence of battles, the rapid change of methods and forms of the armed struggle to keep pace with changes in the concrete situation, and the initial inferiority in the number of trained troops comprising the revolutionary forces, a feature which compels the revolutionaries to act energetically to win popular support in order to make possible a rapid recruitment of troops. They emphasise moreover that in civil war conciliation is precluded by the fact that the incompatibility of the antagonists' interests is very deeply rooted.

There can be no doubt that wars for socialistic revolution have strategic features all their own, especially when guerilla warfare is adopted. That this should be reason enough for redefining the whole concept of war does not seem to be a very convincing argument, however. The forms in which wars have been fought have constantly been changing, and it may very well be that each new war in the Third World will diverge from textbook models in one respect or another. To treat each war as a unique type, with its own designation and definition, does not seem to be a very meaningful enterprise. It can only lead to an obliteration of the distinction between peace and war, which requires a general concept of war to be expressible. Moreover, when many Western scholars propose that the concept of war be redefined, it is the line of interpretation of Clausewitz that excludes civil war they have in mind. The Marxist–Leninists never accepted this interpretation, however. They substituted a concept of politics, built up logically upon the basic class struggle within states, for the Clausewitzian emphasis on foreign policy. Civil war has therefore always been an integral part, indeed a fundamental part, of the Soviet theory of war, and of all kinds of war it is that which best corresponds to the war/policy formula; in this respect the Western views differ from the Soviet.

The Soviet interpretation of the formula that war is a continuation of policy thus accommodates the problems now pointed out by Western scholars.

The relation between internal and interstate war

The debate on the adequacy of the Clausewitzian formula in accounting for modern forms of war thus touches upon the way each of the two theories treats the relation between internal and interstate war.

The political realist tradition begins from the premise that wars occur between states because their basic interests to some extent compete and because the anarchical nature of the international system compels each state to fend for itself as best it can. The international legal order is very rudimentary; at all events it is completely unable to aid a state in the defence of its vital interests if they should conflict with those of another. There the ultimate arbiter is the relative power of the competing states. The foreign policy of states thus defines the national interests and defends them in the context of the international balance of power. Within states, the interests of citizens compete, as the interests of states do internationally, but war between them is prevented because the sovereign power imposes order by means of a monopoly of legal force. Only when this order breaks down does civil war arise. Since disorder is ubiquitous between states but rare within them, the study of internal war is merely an appendix to the main theory of war. For the most part the two kinds of war occur independently of each other, although some relation between their occurrence may arise. A country engaged in international war may become unable to maintain domestic order with the result that internal war breaks out. Conversely, if a foreign power has a vital interest in the outcome of a clash between rivalling groups in a second country, it may intervene and thereby initiate an international war. Sometimes an international war stems from an attempt to avoid civil war, to redirect the dissatisfaction of the population to an alleged external threat. But whether there is a link between actual internal and international wars is an empirical question dependent upon circumstances; there is no regular logical connection.

Marxism–Leninism turns this whole view on its head by postulating that all wars are rooted in the division of society into economic classes whose interests are irreconcilable. The legal order maintained within states is not an impartial arbiter treating all citizens equally as the non-Marxist theories would have it, but an instrument by which the dominant class protects its interests and suppresses the interests of other classes. Thus it leads to internal clashes and wars. International war breaks out when a reactionary ruling class wants to pursue its interests on the international arena by the use of armed force or when it tries to stave off the possible collapse of its political power by easing the contradictions inherent in the economic system. The relation between instances of international and civil war is therefore not simply empirical: they have the same roots, pursue the same goals; they are conceptually the same basic conflict despite differences in their manifest features.

Western scholars have begun to take an interest in the domestic sources of foreign policy, and thereby of international war. They have also begun to search for some regularities and stable links between them, which is a step towards the Soviet approach.

On the whole, however, the research strategies adopted by each approach continue the respective conceptual tradition. Many Western scholars try to determine whether there is a significant correlation between various types of internal and international war but leave open the question of whether wars of both types can have the same basic cause.

Soviet scholars on the other hand are completely uninterested in such correlations. Their task is to interpret the way in which the main contradictions of the epoch and the correlation of forces between the two antagonistic camps affect the occurrence and course of both international and internal wars; it being taken for granted that they must. Moreover, the outcome of each potential war, internal or international, will have an impact on the international struggle. The arguments *pro* and *contra* in each case of possible war must be carefully weighed since particular circumstances make it difficult to formulate regularities.

Notwithstanding these theoretical dissimilarities, however, there appears to be agreement on both sides that in the present strategic situation there are close ties between all internal and international conflicts; no war can remain isolated from the global conflict. For the Soviets, this state of affairs is quite natural, given that the main contradiction of the epoch is the struggle between the two social systems. Imperialism can be expected to try to stop progressive movements by applying counter-revolutionary force. It may engage in military adventures, but the vigilance of peace-loving forces together with the military might of the socialistic countries should manage to prevent major wars. The ideological struggle is intense; one example was the refusal of Western science to recognise civil wars as fully-fledged wars until it was compelled to do so by the wave of revolutions that came after the Second World War.

Some political realists have expressed much the same ideas, but with a crucial shift of emphasis. The reason all wars are so intimately linked together in the present situation is that the Communists pursue a deliberate strategy of international civil war. [5] This strategy has two main features: militarily, it is based on the tactics of prolonged war – attrition, 'salami-tactics', exploitation of dissent, avoidance of major confrontations, etc.; politically it is designed to fulfil the Communist aspiration for control of the world. The Soviet theory of war provides the rationalisation for interventions that in international law are acts of aggression.

The mutual critique

As we have seen above (as well as in chapters 13 and 14), a recurrent theme in the debate between Soviet and Western scholars over the nature of war has concerned the proper interpretation of internal (or civil) war. It is perhaps not surprising that with radically unlike departures, the one approach should find the other unscientific and claim that insofar as it is exploited in the ideological struggle, it is dangerous.

That the two theories are unscientific seems to me to be an exaggerated claim, except perhaps with respect to the most extreme positions that, on the one hand, deny any significance to pressures arising from contradictions in the socio-economic structures, or, on the other, deny the necessity of taking into consideration national, religious, or other non-economic factors in explaining the genesis of armed violence.

The difference in views should be seen as a natural consequence of the more general difference of view concerning the nature of state and society. In the Marxist–Leninist model, social development requires that progressive forces control the state apparatus; this can only be accomplished by revolutionary violence, and war is considered to be a natural instrument of such revolution, even if it can nowadays be avoided. In many non-Marxian models it is the function of the state to reconcile opposing social forces; the outbreak of civil war is therefore a symptom of failure that normally impedes social development.

A good deal of the criticism is directed not so much at the substance of what the opposite side says about internal war, but more at the way it studies the phenomenon. Soviet scholars deplore the Western preoccupation with the apparent features of war, which prevents a more penetrating analysis of the fundamental class contradictions at the root of all wars. Western scholars take Soviets to task for their infatuation with the idea that there is a single primary source of all social conflicts; such a perspective is only a hindrance since it precludes what must always be one of the central questions in the study of war. [6]

On balance, both approaches have contributed to a better understanding of war. Marxism–Leninism, with its focus on deep internal contradictions was the first to treat internal war as a fully-fledged type of war, and it was the forerunner of all theories of war, which search for its roots in the structure of society. Although political realists have focused on the processes leading up to international war, their method of analysis has also cast light on many factors that may influence the outbreak of internal war.

Notes

[1] The new trend has resulted in the inclusion of internal war into many definitions and descriptions of war. Coats, 1966, points out that 78 of the 278 wars for the period 1480–1941 were civil wars (p.18); Trythall, 1973, includes internal war in the definition of war (p.51); Beaufre, 1972, devotes to revolutionary war a large part of his analysis of contemporary war. In the summary prepared by the members of the pre-Congress conference 'War: Its Causes and Correlates' (at the IXth. ICAES) it is announced: 'Conference participants could not agree on a definition of war which restricted it only to conflict between nation-states' and: 'Both civil and international war may be considered the result of absence of stable social structures. Between nations an adequate structure had never been achieved' (Nettleship, et al., eds, pp.775, 776).

[2] For the critique of Western theories of civil war see *Marxism–Leninism on War and Army*, Ch. Two. 3; Khmara, 1971.

[3] Rapoport argues that Marxist theory which puts allocation of resources at the basis of all major social conflicts, both endogenous and exogenous, is invalid, since conflicts are guided not by abstract issues but rather by the identification of the participants with the contemporary social groups, and the immediate criterion of such identification may be several stages removed from the original issue. Rapoport recognises two important sources of social conflicts besides allocation of resources: the struggle for power, and the need for autonomy. He presents two examples of conflicts where armed violence is occasionally used and in which the identification of the parties cannot be connected with class division of the society, the current strife in Northern Ireland, and the racial strife in the United States (1974, pp.187–8).

[4] Strausz-Hupé contended that Lenin converted the abstract doctrine of

324

conflict between classes into a highly effective instrument of conflict between nations. He projected the class struggle from capitalist society into world politics and exploited it in the ideological struggle ('Protracted Conflict: A New Look at Comunist Strategy', *Orbis,* Spring 1958, pp.13–38).

[5] The main theme of Strausz-Hupé, 1959. Cf. his 'Protracted Conflict: A New Look at Communist Strategy', *Orbis,* Spring 1958. 'Lenin's singular contribution to communist theory was the conversion of Marxism from an abstract doctrine of conflict between classes into a highly effective instrument of conflict between nations'.

[6] Michael Haas contends, that 'there is some ambiguity' in Marx's two ideas: (a) that ruling classes use war for their internal and external interests, and (b) that war is generated by definite types of economic systems (these ideas correspond to two approaches, to the functional and prerequisite approach) (1974, pp.168 ff.).

This critical comment doesn't seem to be sufficiently substantiated: Marx assumes that ruling classes decide to use war for their interests because of the nature of the exploiting system: such a decision is caused by the system. War is generated by the system as functional for its ruling classes.

17 War and conflict

Both Soviet and Western scholars have devoted much effort to the analysis of the relation between wars and the conflicts that have preceded and led up to them. The traditions have taken different paths, however.

Pro and contra a general theory of conflict

Many Western social scientists, whose research was briefly reviewed in chapter 5, have attempted to create a general theory of conflict in the hope that they could thereby explain some essential features of war, especially its outbreak. They contend that to study war within the framework of a general theory of conflict is both possible and useful: [1] possible because war is a kind of conflict, and useful because the process of escalation of conflict and its transformation into war might be better understood if analogies are drawn with other types of social conflicts that share the same dynamics. For many the decision to study the general problem of conflict resolution was based on the conviction that the traditional disciplines had said all they could about war yet without offering a solution as to how to avoid it. A fresh perspective was needed if war and total annihilation were to be prevented.

Although Soviet scholars also study such problems as the genesis of different types of war in political conflicts, the interrelation of political, economic, ideological, and cultural conflicts, and the cause and effects of interpersonal and intrapersonal conflicts, they deal with each subject separately and do not try to establish from their findings general rules of conflict considered as a category of human behaviour. More than this Soviets view attempts to create a general theory of conflict, including war, as both impossible and undesirable. Conflicts differ in so many essential respects: the field of human activity (political, economic, cultural, psychological, ideological, etc.); the social level (interstate, interclass, interpersonal, intrapersonal); the method of resolution (peaceful or by armed violence); and the social setting (capitalistic or socialistic). Thus, so various are the kinds of conflict that no regularities common to all can be found nor even a common methodology by which they can all be investigated. Moreover, it is undesirable to try to develop a general theory for the following reasons:[2]

(a)   Listing a multitude of sources, causes, factors, variables, correlates, and suchlike conditions affecting the occurrence of a great variety of conflicts would amount to equating the primary with the secondary, the deep with the superficial, the direct with the remote. It would be impossible to identify the primary roots of social conflict.

(b)   Concentration on the general features of all conflicts would divert attention from identifying the characteristics of the main types of conflict that may now lead to war. All universal schemes obscure the concrete contents of actual conflicts and wars.

(c)   A general theory would obscure qualitative distinctions between various kinds of conflict, as peaceful conflict and war.

(d)   Mixing all conflicts together without due differentiation obscures the fact that the economic class struggle is the mainspring of all violent social and political conflicts and that imperialism is the source of all wars in our epoch.

Soviet scholars contend that the purpose of incorporating war into a general theory of conflict is to direct the attention of the popular masses away from the real sources of conflicts and wars, namely imperialism. With the alleged regularities they discover, Western scholars hope to convince the masses that all governments, social systems, lands, nations share the responsibility for the occurrence of conflicts and wars.

Methods of research

Having set themselves different tasks, the two groups of scholars naturally go about their study of conflict in quite different ways. In the West many social scientists have adopted the inductive method. They devote most of their resources to the empirical study of particular conflicts and wars. From these they hope to generate statements about conflicts of increasing generality, and ultimately to formulate a general theory of conflict. Soviet scholars begin with the concepts of historical materialism, which might be called a general theory of social relations, and apply these general ideas to particular wars. Western research on conflict therefore takes the character of a search for concepts, while Soviet writing describes the true reality in the light of an accepted conceptual framework. Non-Marxist studies are explorative: Soviet studies are expositive. The distinction can be traced through all important aspects of the subject.

Much of the literature in the West on the theory of conflict resolution consists of a discussion regarding the most useful conceptual elements. No general agreement has yet been reached, however, whether conflict should be defined to include everything from the first expression of contradiction to the ultimate elimination or whether the term should be reserved for certain types of behaviour. It is also unclear what is meant by war, whether it is a sub-class of conflict or a certain stage of conflict. The terms actor and social group are widely used, but there is little unanimity concerning the meaning of the concepts they denote.

By applying dialectical materialism in their analysis of society, Soviet scholars derive the theory of class struggle of which the theory of war forms a part. All conceptual difficulties are solved logically. All social conflicts are based on the division of society into classes, which are the main actors, states being merely an arm of the ruling class. The basic conflict is economic but is reflected in political and ideological conflicts. When these become so intense that peaceful means cannot resolve them, war breaks out. War is thus the most acute form of conflict.

As to the source of social conflicts, Western scientists have generally adopted the view that this springs from the division of human society into many groups. Social groups can be defined in psychological terms on the basis of the feeling of identification experienced by each member of society towards the others, and/or in sociological terms on the basis of the relative amount of interaction each person has with the others. Such groups can spring up in any field of human activity: economic production or consumption, politics, religion, social status, etc. The extent to which these various divisions overlap and thereby reinforce each other, or on the other hand, cut through society in many different ways is important for social equilibrium. In some cases, conflicts rooted in these divisions may lead to physical clashes, and even to various kinds of armed violence. Western scholars criticise the Soviet reduction of all divisions to one primary division based on class economic characteristics, as well as of all conflicts to the economically rooted ones.[3][4]

By analogy, international conflicts and war are traced to the inclination of states to group themselves according to their identifications and interests, and in structural inequalities of power and resources. Various propositions have been put forward to relate the impact of the character of different political systems, the peculiarities of cultural development, and the differences in techno-logical, economic, and ecological resources of states on the international system.

These and other theories are countered by one simple explanation: the division of society into two antagonistic classes is the primary root of all internal and international conflicts and wars. All other features of society and the inter-national system may be treated as factors impeding or facilitating the outbreak of war. They cannot be their primary source, however, for these factors are themselves conditioned by the fundamental class conflict, although they may also reflect peculiarities of the epoch and the particular country in which they occur. The political system of a given country is, for example, only part of the superstructure built upon the socio-economic base: by itself it cannot explain the political activity carried on within the state. The properties of the inter-national system that influence the outbreak of conflicts and wars reflect the concrete conditions of the epoch, of which one of the most important is the internal structure of states. Models based on a bi- or multi-polar balance of power may describe the international system, but they cannot explain the occurrence of modern conflicts and wars; neither are the various quantitative analyses capable of providing a suitable explanation. [5] An understanding of this problem can only be gained by analysing the two competing socio-political systems and the war-generating role of imperialism, Soviet scholars maintain.

As to the value of social conflict, Western social scientists are divided between those who emphasise the destructive consequences of conflicts and those who point to its dynamic benefit to society. Even more marked is the lack of unanimity regarding the general assessment of the role of war. Those who believe that war is in general dysfunctional recognise that in various historical epochs particular wars have had some favourable consequences. Opinions on this matter reflect to some extent differing views of social development: in one, society is normally in a state of equilibrium and is continually subjected to evolutionary changes; in the other, to a certain extent, Marxian-inspired view, society is normally in a state of disequilibrium and is periodically transformed through revolutionary shocks. The prospect of nuclear war makes a clearcut opinion about all war in modern times very difficult to hold.

Something of the same ambiguity can be seen in the double-edged answer Soviet scholars give to the problem: revolutionary wars are functional while counter-revolutionary wars are dysfunctional; international wars that weaken the rule of exploiting classes or their domination over other nations are functional, others are not. Marxist–Leninist ideology provides fairly clear criteria for determining the political characteristics of war on the basis of which the assessment is made. Conditions in the nuclear era add to the social cost of war and devalue even functional wars. Since the cost of war seems to outweigh its benefits in each instance the general tendency of the Soviet position seems to be similar to that of Western social scientists who regard war as dysfunctional. In theory, the general assessment of war as a social phenomenon is still dependent upon its political characteristics.

Some comments

Soviet scholars are highly critical of the expressed aim of the Western scholars who are working towards a general theory of conflict resolution. According to their polemics, they see the enterprise as a manoeuvre in the ideological struggle on the part of imperialistic science to hide the truth about war. Despite their critique of a general theory of conflict, however, it can be argued that just such a theory is implicit in Marxist–Leninist dialectics. This posits that all things, phenomena, and processes consist of a unity of internal contradictions. From the tension between these contradictions comes the motive force that is the source of the object's development. It is the task of dialectics to discover the laws of such development. When the dialectical theory of conflict is mixed with materialistic epistemology and applied to human society, it yields a theory of social conflict based on contradictions in the relations of men to the material foundation of society, that is, the economy. Social conflict, political conflict, revolutionary war, and many other related concepts are all derived logically from this basic idea, and for this reason they constitute a consistent whole. Soviet scholars can therefore state with conviction that all conflicts and wars are primarily caused by the contradictions in the structure of the socio-economic system. In this sense the Marxist-Leninist theory seems to be the more general, at least at the moment.

On the other hand, however, Western scientists who search for a general theory of conflict try to be as open as they can to new ideas. By going beyond those conflicts which lead to war, they may have something to say about societies free of revolution and war. They may find conflicts intrinsic in society that are sources of continuous social development and yet are not confined to class society. If they succeed in their enterprise, they would have a theory more general than Marxism–Leninism.

Notes

[1] The basic theme of many articles in *Journal of Conflict Resolution.* For an analysis of the first twelve years see: 1968: 'Special Review Issue', esp. the review by Clinton F. Fink.

[2] For a sharp critique of the Western attempts to create a general theory of conflict see: *Mezhdunarodnye konflikty,* 1972, pp.27 ff. The authors hold that some extremist adherents of the general theory of conflict are ready to apply it to all natural phenomena and to present the general theory of conflict as a new general theory of evolution (pp.27–8).

However, when Soviet scholars participate in international conferences they don't *a priori* repudiate all attempts to create a general theory conflict; they even state, that it would be purposeful to investigate some general problems of conflict as a category. At the same time they underline that all such attempts in the West have up to now been unsuccessful (D.V. Yermolenko, 'Mezhdunarodnyi konflikt kak obiekt sotsiologicheskogo issledovaniya', in *Sotsiologicheskie problemy mezhdunarodnykh otnoshenii,* Izd. 'Nauka', Moscow 1970, pp.104 ff).

[3] For a critique of the Soviet 'economic reductionism', see Dahrendorf, 1959; Coser, 1956, 1967 (ch.7: 'Karl Marx and contemporary Sociology'); Rapoport, 1974. Coser writes: '. . . it seems necessary to reject Marx's reductionist tendencies, his propensity, for example, to see in the political order only a reflection of the economic order' (1967, p.150).

[4] 'What is basically wrong is the very attempt to reduce causes of such an intricate phenomenon as war to the correlation of power. As a matter of fact this initial position rests on the notion which struck root in the bourgeois science that wars are inevitable, that the aggressiveness of all states is immanent and is displayed every time a definite correlation of their power makes it possible. . . . An approach to the problem with the only criterion of the correlation of material power, for all its importance, makes it impossible to disclose the true origins of wars, their economic, social and ideological roots and leads to wrong conclusions' (A. Filyov, 'In The Labyrinth of Numbers', *International Affairs,* Moscow, 1974:7, p.147).

Fuller states that the Soviets treat civil war as an instrument of international war, of the war against the capitalist countries: '. . . the highest economy of force is to be sought in transforming an international or imperialist conflict into a civil war – that is, into war, in which the enemy destroys himself' (1962, p.204).

[5] For a comprehensive critique of the quantitative analysis of the essence and occurrence of conflicts and wars see Filyov, 1974 cf. note 4. Cf. *Metodologicheskie problemy voennoi teorii i praktiki.*

For a critique of the quantitative analysis in the sociological study as a discipline, see: *Marksistskaya i burzhuaznaya sotsiologiya segodniya,* Izd. 'Nauka', Moscow 1964; for a sharp critique of futurology as based on quantitative research, see: A. Sergeyev, 'Bourgeois Pseudo-Science About the Future' *International Affairs* (Moscow), 1972:2. For evaluation of the game theory see: Gennadi Gerasimov, 'War and Peace Problems as Viewed from the Angle of the Game Theory', *Coexistence,* 1971:9.

The attempts to present war as the result of a certain war-generating situation being a sum total of diverse causes are criticised; in such concepts war breaks out by accident, and not by a deliberate political decision. The contention that one can forecast the origination of such situations by quantitative analysis and quantitative methods and means of forecasting are also critisised: 'It is vital to separate the factors into the main, determining ones [that may be discovered only by a socio-political analysis-JL] and into derivative, secondary ones [that may be discovered by a quantitative analysis], as well as to determine a correct ratio between them' (Barsegov, Khairov 1973, p.76).

18   The historical character of war

The past and the future of war

The question of whether or not war is a social phenomenon that will always plague mankind is a vital one for both approaches, although they disagree over the correct answer. For Soviet scholars – unlike political realists or other non-Marxists – it is almost equally as important to ask if, in the past, war has always been a feature of man's existence. The reason for this is that the whole conceptual scheme of Marxism–Leninism hangs on the assumption that war is a class phenomenon: if wars can be shown to have occurred before the evolution of economic classes, then Marxist–Leninist ideology would be severely shaken. Gone would be a part of the grounds for the claims that capitalism is inherently aggressive and that socialism will bring an end to the war epoch in man's history.

One might suppose that Western scholars engaged in the ideological struggle would try to disprove the Marxist–Leninist contention that war first occurred when class societies emerged. The research of Western anthropologists has resulted in a variety of opinions about the earliest wars, however, and much of the issue has boiled down to quarrels over the proper choice of minimum criteria of war and speculation about when they may first have been met. Political realists have disdained such discussion; for them it is enough to be able to state without fear of contradiction that wars have been an element in all major civilisations for which we have historical evidence. [1]

Attitudes towards the future of war are correspondingly consistent. Political realists are on the whole pessimistic about eliminating war. The permanence of the conflicts in interests between men and states are political realities that militate against replacing war as the ultimate means of resolution. Cognisant of the risks entailed in modern wars, realists admit the value of striving for new methods of solving political conflicts peacefully, but they do not believe that radical changes in either the international or internal political structures of states can be accomplished in the foreseeable future. [2] Political realists are therefore sceptical about the feasability of many of the proposals put forward by other Western researchers to outlaw war or to assert collective control over military resources. These they scorn as utopian and naive. What may be achieved, they believe, is a reduction of the incidence of war. Their programme is one of pragmatic measures towards gradual *détente*, international cooperation through international organisations, arms limitation and development aid to meet the wave of rising expectations. At the same time, they state, in the struggle for the minds of men, a tireless effort must be made to defend inalienable human rights and to prevent the spread of Communist ideology by disclosing the distortions of Communist propaganda. The defence of the free-world from Communist totalitarianism requires military power sufficient to deter the Soviet Union from putting its expansionistic policy into action.

Soviet scholars, who state that wars first occurred with the emergence of class society, have a much more optimistic attitude towards the future of war. Wars will disappear with the demise of class society. Moreover, it is towards such a fate that men are inexorably moving according to the laws of social development. In a world without economic classes there will be no irreconcilable conflicts of interest; in a world of socialistic states, there will be no need to maintain armed forces, and there will be no war.

The programme of the socialistic states, called by the Soviets the policy of peaceful coexistence, has two main elements: one is to take preventive action against imperialism, the source of all wars, by promoting cooperation between states and by taking all possible organisational and psychological measures to reduce the risk of war; the other is to consolidate the international class struggle in order to diminish the influence of the imperialistic states on the course of world affairs and to further socialistic transformations in an ever greater number of countries.

The mutual critique

Each side is very critical of the attitudes and ambitions of the other.

The Soviets strongly censure non-Marxists for their scepticism about the historical origins of war. By throwing doubt on the contemporaneous appearance of war with class society, Western scholars make it more difficult to eliminate war, indeed they serve to perpetuate it. Even in concrete situations, their lack of clarity about the genesis of wars makes more difficult the task of unmasking the forces responsible for armed violence, and the war-mongers can more safely carry on with their preparations. [3]

As to the future of war, Soviet scholars regard the pessimism of political realists on this point as symptomatic of the quandary of capitalistic science in the transitional era. The realists attempt to save war as an instrument of politics, since they recognise that war is essential to the capitalistic way of life; at the same time they realise that war has become an inpracticable instrument for meeting the main political challenge to capitalism because of the strength of the socialistic forces making it. The schemes of other Western theorists to eliminate war are either utopian or treacherous. To the latter category belong: all theories of world government, which in fact aim at eliminating the independence of nations and at subjecting them to the hegemony of the United States; [4] many theories of arms control, which are designed to make other states defenceless or to put their armed forces under the control of the imperialistic powers; and all plans for the creation of international armed forces, which are intended to be used to deprive the socialistic states of independence.

The main features of the political realist critique of Soviet ideas might be summarised in the following way. The optimism radiated by Communists about being able to make an end to war is based on two premises: firstly, that Communism will ultimately win over capitalism throughout the world; and secondly, that no conflicts can occur in a purely communistic world. The validity of both can be questioned. Ample evidence refuting the second can already be put forward; communistic countries are troubled with internal conflicts that must be put down by armed force, and between communistic states military force is used, either latently or openly, to ensure that the Soviet line is followed in 'brotherly disagreements'. [5]

Whether Communism will one day win the world remains a moot point. It is clear, however, that the Communists are prepared to foment war until it does. Since the perfect society, whether Communist or other, has always proved an elusive dream, it might be argued that the revolutionary strategy of Communism amounts to a declaration of perpetual war. [6] The fine-sounding phrases of Communist propaganda should be seen as part of a deliberate deception by which the Soviet Union hopes to gain control of the world.

Similarities

The differences are thus considerable and the tone of the polemics is sharp. Given the fact that the attitudes towards the historical character of war stem from divergent ideologies that put contrasting interpretations on historical events and political values, no other conclusion would have been possible.

But the elimination of war is not only an abstract problem of interest to theoreticians and ideologists; it is also a matter of urgent and immediate concern to entire peoples whose lives are at risk if an actual war were to break out tomorrow. The policies followed by those charged with the everyday conduct of government are therefore very similar. Both sides act on behalf of peaceful co-existence based on a balance of power. They take cooperative measures to damp the arms race, to prevent unwanted wars, and to limit hostilities that may break out. Despite the Western belief that Soviet ideas about eliminating wars are utopian, Western governments have nevertheless cooperated to realise Soviet proposals designed to prevent interstate wars. Despite the Soviet critique of balance of power theories, the Soviet government continually acts to maintain such a balance.

To sum up, the long-range views of Soviet and Western scholars are radically opposed, but the policies they approve in the short-term are much alike. The prospect of a war-free world of socialistic states with harmonious interests is set against the prospect of a world of states with competing interests in which conflicts are normally resolved by peaceful means. Against a programme of revolutionary social transformation is set a plan of gradual, pragmatic reform of the existing system. In the short-term, however, both sides agree to the need to reduce the general risks of war, to avert imminent wars, and to create an atmosphere in which a better mutual understanding of peoples and governments is encouraged.

Notes

[1] It seems that in contrast to the Soviet views, even scholars who associated the origin of war with the origin of the states emphasised either that (a) war was one of the conditions of the origin of state; or that (b) war was instrumental for society (not for the governing class): the origin of state created material conditions for waging wars; or (c) both.

[2] One of the main themes of Raymond Aron's treatise on Clausewitz (1976) is that in an unforeseeable future the world society will remain to consist of nations which don't understand each other, of states which want to preserve their independence, and of incompatible ideologies.

[3] Common to all Soviet analyses and reviews of Western theories of causes of war is the accusation that all such theories treat war as an eternal phenomenon, and by treating it so, they contribute to its eternisation (cf. previously quoted writings); Filyov, 1974 (note 4 in ch.17) deals with political realists' 'eternization' of war.

[4] For a critique of views of a 'world government', see: E. Kuz'min, *Mirovoe gosudarstvo: illuzii ili realnost?*, Izd. 'Mezhdunarodnye Otnosheniya', Moscow 1969; *Sotsiologicheskie problemy mezhdunarodnykh otnoshenii*, Izd. 'Nauka', Moscow 1970.

Barsegov, Khairov, present the following arguments against a 'world government': (a) such projects ignore the objective laws of development of modern society, the powerful upsurge of national liberation struggle and the general strengthening of national feelings; (b) they ignore such obstacles as the existence of opposite social systems and the resistance in the capitalist world itself; (c) such projects are harmful, because they undermine the mainstays of the UNO and hinder its normal development (1974, p.74).

[5] See: Richard J. Barnet, *Roots of War. The Men and Institutions Behind U.S. Foreign Policy*, Penguin Books, Baltimore-Maryland 1972, p.207. In his view, the assertion that war will not exist in a socialist society has been formulated in a very general form, it appears as a kind of faith proclaimed even without any interpretation.

[6] Marxists view war as springing from the nature of the society, which is a theory of a 'social inevitability' of war, because it is impossible to change the whole social system so radically, as Marxists propose (Margaret Mead, 'Warfare Is Only an Invention – Not a Biological Necessity', in Bramson, Goethals, p.269).

19 The assessment of wars

The traditional positions

Within European civilisation there has over the centuries emerged a theory of just war by which instances of war can be assessed in relation to various principles, some ethical, some political. Both Soviet scholars and political realists have rejected this theory, however, although for different reasons.

Marxists–Leninists have abandoned it because in their view it is too saturated with the reactionary values of class society: for example, the traditional theory condemns as unjust popular uprisings for liberty and it condones wars by which the ruling classes suppress the rest of society. As part of the ideology of capitalism it has been replaced by a theory of just war based on the values of classless society. This socialistic theory is given a prominent place in the Soviet analysis of particular wars, and it is presented as forming the basis for the political attitude adopted by the Communist Party and the Soviet state. Two inter-related criteria are used to decide the justness of a given war: its political characteristics, i.e. the political forces represented by the belligerents and their goals, and its role in the socio-historical progression towards socialism.

Political realists have adopted the position that a moral assessment of wars is either impossible or meaningless although such an assessment is often officially presented. When going to war, states always justify their actions with the best of motives and intentions. Cynicism is warranted here, for in war the winner is invariably proven to have been on the side of justice. The difficulty with trying to assess wars in relation to their consequences for social evolution is that these cannot be known in advance. Political realists therefore prefer to adopt what they call an amoral posture. They maintain that the dynamics of international relations can be adequately understood by assuming that states act in the pursuit of their national interests. Whether these interests are just or not is not a question the analyst of international affairs is required to answer in his capacity as analyst. This does not preclude him from making value judgements of course, most realists support the values of democratic liberalism or conservatism, but any such statements are ancillary.

It could be argued that the distance between the Soviet scholars and political realists is in one sense much less than it might appear. Political realists stress the importance of being clear about the concrete political goals of state adversaries, goals which will be rooted in state interests whatever the idealistic phrases in which they are couched. Soviets, as we have seen, have always given first priority to the political characteristics of war, which are said to determine the socio-historical and ethical assessments. Wars between progressive and reactionary forces are simultaneously just and unjust, depending upon whether the war is being discussed with reference to the progressive or the reactionary side. The question 'which side is progressive?' which in turn wholly depends on the political characteristics of both warring parties is thus prior to the question 'is the war just?' In this way their ethical assessment is conditioned by their political analysis.

Modifications in the nuclear era

Having discardeed, for different reasons, the traditional theory of just war, both approaches are now exhibiting tendencies to reintroduce one of its ideas, namely the just means. The reason is the revolutionary implications of the destructive power of nuclear weapons for the nature of war.

The potential use of nuclear weapons has occasioned considerable controversy amongst political realists. On one side are those for whom it is inconceivable that a nuclear conflagration could possibly lie in the national interest; they argue that nuclear war should at least be condemned. Opposite them are political realists who argue that the only way of preventing a nuclear war is to prepare for one seriously. Moral qualms about using nuclear weapons are out of place since they may not be shared by a potential enemy. In the middle are realists who argue that the free world cannot rely solely on nuclear weapons in the struggle against Communist aggression, for these can by their nature only be used in an utter calamity. Since it has become the Soviet strategy to avoid major confrontations and to advance its position in the world gradually through limited wars, the West must also be well equipped to counter such threats by conventional means. Underlying all three positions can be discerned an awareness of the radical change in the instrumentality of war that has occurred with the development of nuclear missile technology, a change which also has an impact on the value of full-scale war.

In the Soviet view, the advent of nuclear weapons has not altered the grounds for appraising wars. Their socio-political content would be similar to that in the pre-nuclear era, which also determines the socio-historical and ethical assessments. An unjust war – one waged by a capitalistic state – would be even more unjust if the state used nuclear weapons. But it would not be unjust of a socialistic state to use nuclear weapons in a just war. [1] On the other hand, Soviet scholars do argue that nuclear wars, even just ones, are undesirable because of their high social costs and that they should therefore be prevented. In other words, the importance of the nuclear revolution in military technology is not admitted in the theory of just war, which has been the orthodox basis for policy decisions, but despite this fact, the change in the means of warfare has been allowed to make an impact on Soviet policy through the expedient of giving greater weight to criteria outside the theory of just war.

This shift implies that Soviet policy has become more pragmatic: the theory of just war amounts to the view that the end justifies the means; now it is recognised that the value of the means may to some extent be independent of the ends for which it is used. The traditional position of realists, that states will behave as if the ends justify the means, has undergone a similar change, for it is difficult for them to conceive of how nuclear war could be a means to anything.

The mutual critique

Political realists take the stand that the value of war as an instrument of policy cannot be deduced *a priori* from some set of abstract principles. If a government believes it is in the national interest to obtain some goal by waging war, then it will go to war. And it will find plenty of reasons why it was in the right to do so. The Soviet theory of just war is an excellent example of this kind of moralising. Certainly as an ethical theory it is in some respects deficient: it fails to identify conclusively the cause of proletarian revolution with the cause of humanity as a whole; it does not link Communist ethics with universal moral principles; and it rejects widely accepted international moral and legal norms. As the Soviet scholars themselves point out, the justness of war depends upon its political characteristics. By this they mean that the important thing is which side represents the progressive forces (less important is the question of which side struck first; if it was the progressive side, the argument can always be made that it was provoked and that it was the reactionary forces that really started the war). The progressive forces are identical with those which defend the interests of the working class and strive towards socialism, that is to say the groups supported by the Soviet Union. It turns out therefore that what is just is what is in the interest of the Soviet Union in its quest for power. The terms just and unjust are only slogans designed to give Soviet policy greater appeal in world public opinion.

The critique of the Marxist-Leninist theory of just war proffered by political realists thus confirms them in their belief that moral ideas are used by politicians to justify whatever they want to do.

The Soviet critique of political realism is somewhat more complicated. The moral indifference so proudly proclaimed by political realists to be one of their virtues is really a great fault, for it conceals the morality of capitalistic ideology. Implied in their approach is the belief that might is right, a value that very well coincides with the interests of the ruling class. It is an analytical stance completely in accord with the whole ideology of the exploitive society in which each class fights for its own interests. [2]

Furthermore, the realists do not succeed in keeping their studies free from openly expressed values. They make reference to standards fetched from the traditional theory of just war. This is not a truly moral theory, however, since the main ground for assessing wars is their political utility for the existing social system. By this criterion all progressive wars for national or social liberation are regarded as unjust and reactionary wars as just. The Soviet critique of political realism on this point is thus two-sided: its professed amorality is a deception that conceals its deep ideological roots in class society; therefore, it is really an immoral theory, for the truly human morality must be anchored in the ideology of classless society.

Some comments

From this survey of the debate on the assessment of war, a few observations might be made. The first concerns the perceptions the political realists and Soviet scholars have of each other's way of thinking. Political realists maintain that war is an activity that cannot be understood or explained in moral terms; they therefore contend that because Soviets give high priority to political characteristics when assessing wars they do not really base their decisions about war on moral grounds. Soviets believe that the basic contradiction of the material base is reflected throughout society, even in ideology and morality. They therefore ascribe to political realists the values they must have as apologists of the capitalistic system. Each side thus sees the other within the bounds of its own conceptual framework: realists see the Soviet scholars as power seekers and the latter see the realists as ideological combatants.

Secondly, in both approaches to the assessment of wars, the political characteristics imply the socio-historical and moral assessments. General moral principles must be interpreted with reference to a concrete period and situation in order to be applicable to it. Such interpretation is one of the functions of ideology which relates the moral principles to the political values. At the same time the place of a concrete war in history depends on its assessment according to some ideological criteria. Thus, if the two ideologies clash in their political evaluation of wars, they do also in their socio-historical and moral appreciation. A conflict of political interests always implies a conflict in moral values and assessments.

Notes

[1] Western contentions that the nuclear missile war ought to be excluded from the category of just wars are criticised in the Soviet literature (see Dmitriev, 1975, p.14), since such a war waged by socialist states would be just. On the other hand, Western theories holding that such a war waged by the West would be just are criticised even more sharply, especially writings of Herman Kahn. 'Khan's gloomy, inhuman writings have, he is proud to admit, become indispensable books on military planning for the Pentagon' (*Problems of War and Peace*, p.88).

[2] For a comprehensive critique of the political realism views of ethics in connection with politics and war, see Aleksiei Karenin, *Filosofiya politicheskogo nasiliya*, Izd. 'Mezhdunarodnye Otnosheniya', Moscow 1971, ch.VII: 'O morali'.

20   The prospect of war in the nuclear missile era

Both schools of thought have come to the conclusion that all problems concerning the nature of war and its role in social development need to be reconsidered in the situation that has arisen since the Second World War. Different motives were put forward, however; in the West it was usually the revolution in military technology; [1] in the Soviet Union it was the radical new relation of forces in the world, although the impact of nuclear weapons was also mentioned. The topics discussed in the re-examination of each side were therefore not quite the same. In the West the discussion centred on the practical difficulties of waging limited and general war, the strategy of deterrence, the nature of the Cold War and crisis management, and the validity of the Clausewitzian formula. Soviet scholars focused on the instrumentality, desirability, and inevitability of war, the changing nature of war, and the meaning of peaceful coexistence. Despite the difference in the form in which the problems of the new age were dealt with by each side, the main arguments can be compared under four main headings: (a) the instrumentality and desirability of war; (b) the inevitability of war; (c) changes in the nature of war (the validity of the Clausewitzian formula); and (d) the new character of peace and the strategy for achieving political goals by peaceful means.

The instrumentality and desirability of war

It seems that both schools of thought have not only modified their opinions about the instrumentality of war but also changed the terms in which they discuss the problem. Previously it was enough for a war to be considered instrumental if it offered good prospects through a military victory of achieving the political goals for which it was fought. In the nuclear missile era the problem of the relation between total social costs, direct and indirect, and resultant gains was added.

The need was also accentuated to consider whether these same goals could be achieved by peaceful means at a lower cost. The problem of the instrumentality of war was thereby united with the problem of its desirability.

The two approaches arrived at somewhat different conclusions about the instrumentality of nuclear missile war between social systems. Political realists, and with them the overwhelming majority of Western scholars, regard such war as unusable in an absolute sense: the social costs incurred would exceed the benefits associated with any conceivable aims a party to war might have. In the Soviet view, war between the two social systems would be uninstrumental only for imperialism, which would be defeated: it would be instrumental for the socialistic camp since it would win. When consideration is taken of the social costs of such war for humanity and of the impact it would have on the development of the socialistic countries and on the cause of the international revolutionary movement, Soviet scholars conclude, however, that nuclear missile war would be undesirable. The views of the opposing schools of thought thus converge on this point, although the supporting arguments are different.

As to local and limited interstate wars, both sides argue that they should both be practicable in the sense that they can further political goals. Soviet scholars look much more unfavourably on such wars, however. In the theories of many political realists, limited wars appear as normal and even beneficial since they provide a means of resolving interstate conflicts that cannot now be resolved through major war. A change of views can be noted, however, after the war in Vietnam and in the face of prospects for a military victory in other places. Soviet scholars emphasise that such wars, like all wars of the epoch, are rooted in the politics of imperialism. They are the manifestation of imperialistic aggression and the means by which the stronger capitalistic countries attempt to subjugate the weaker peoples and to intervene in the internal affairs of Third World countries. Such wars are dangerous because they may escalate into world wars. This judgement was somewhat tempered by the reflection that in some circumstances a limited interstate war might result in victory for the forces of progress. Since, on the other hand, political realists were less enthusiastic about limited wars that could end in Communist gains, naturally, there was in practice much less separating the two approaches with regard to the merits of limited war as such as it could be concluded from general statements.

Wars for socialistic revolution or national independence were usually included by Western writers amongst limited and local wars but constitute a separate category in Soviet analysis. Both of the confronted approaches considered that they could be used to achieve political goals, but each appraised the desirability of such wars quite differently. For the most part, political realists were not favourably disposed to such wars on the grounds that the aims for which they were fought could be realised by administrative measures and socio-economic reforms in the case of social revolution or, with respect to wars of national-liberation, by granting independence once the colony had become ripe for it. The Soviet scholars upheld their traditional tenet that revolution is the only way social transformation and national independence can be achieved, but they now added that the possibility of carrying out such revolutions by peaceful means had greatly increased owing to the successes of the progressive forces. In the post-war world the desirability of violent revolution thus depended upon the particular circumstances.

It should perhaps be noted that there was quite a wide range of views in the West on the subject of local war. British and French scholars seemed on the whole to be more inclined to underline the shortcomings of such wars than their American colleagues, although the difference of opinions was as much individual as national. A distinction can often be drawn between the supposed advantages of local war for those directly involved on the one hand and its assessment by those who can be drawn into the conflict on the other: for example, war may be considered by the Arab states and Israel to be both instrumental and desirable as a means of resolving the conflict between them while the superpowers or even the European states would be more fearful of what consequences such a war would have for their own security.

To sum up, despite basic conceptual differences about the nature of war and its role in social development, when the comparative costs of war and peaceful means are taken into account, both sides come to the same conclusion – that in the majority of situations war is undesirable.

The inevitability of war

Traditionally both of the confronted approaches regarded war as inevitable. In the Marxist–Leninist view, revolutionary wars were as unavoidable as the basic contradictions of class society which gave rise to them; as long as capitalistic states existed, international wars would occur because of the internal contradictions of these states and because of the uneven development between them; and wars of national-liberation were a natural and inevitable consequence of the national oppression of colonialism. Political realists, whose focus was almost exclusively on interstate wars, seemed to argue that war was necessarily bound up with a system of independent states whose interests must at times conflict. While they thus implied that war was inevitable, Marxist–Leninists declared the idea outright in their political doctrine and propaganda.

Since both schools of thought came to the conclusion that most wars in the nuclear era are undesirable, moreover that they are undesirable for all parties concerned, it would seem strange to insist that wars were nevertheless inevitable. Owing to the vagueness of their traditional position, political realists were not required to change the exposition of their theory, even if they no longer considered some types of interstate war as inevitable under contemporary conditions. Nor were they obliged to reconcile such a view with the increase in the frequency of certain kinds of interstate war and mixed wars. On the other hand, because the Marxists–Leninists had previously emphasised the doctrine of the inevitability of war, they were now to a certain extent compelled to renounce it officially.

The change they have made is more a shift of emphasis, albeit an important one, than a modification in theory. Some kinds of war are no longer inevitable because the source of all wars, imperialism, has become weaker in relation to the peace-loving forces, which now stand in an unprecedented position to prevent war. War is still possible since its source has not yet been eliminated, but social development has moved forward one stage in the progression towards the ultimate victory of socialism when wars will completely cease to occur.

The interpretation given by political realists is also consistent with their own general view. That general war is no longer probable is accounted for by the development of military technology. The present balance of power is such that no state can hope to gain any political goals by war. However, the source of war, the system of independent states, remains, and will remain as far as it is possible to foresee. If some dramatic change were to occur in the balance of power – a technological breakthrough, the proliferation of nuclear weapons, etc. – the probability of major war may well rise again. In the meantime, there will remain a definite possibility that a world war will break out, but a low probability of its doing so.

Again there is great similarity in the conclusions reached by the two approaches. Both have abandoned the idea that war, or at least certain kinds of war, are inevitable. Moreover, their argument is in form the same: both believe that war can now be avoided because the conditions of the post-war world act as a check on the source of war. Since the root of war remains intact, however, there is still a possibility that war can break out.

Here the two approaches diverge. For political realists it is the balance of nuclear power that prevents international anarchy from leading to war. For Soviet scholars it is the favourable correlation of forces that keeps the imperialists in check. The balance of power, we have seen, is not the same thing as the correlation between progressive and reactionary forces, however. At the present juncture, there happens to be a certain correspondence between the two, but how long this will last is a moot point. Soviet scholars will say that the correlation between forces can only become more favourable to the progressive, peace-loving forces. Political realists tend on the other hand to be pessimistic about the possibilities of maintaining a balance of power for any length of time.

Is the nature of war changing?

The debate on these aspects of war led naturally to the more general question of the nature of war itself. Is war in the nuclear era still a continuation and instrument of policy? Or if it is, has the degree to which it is a continuation and instrument of policy changed?

In the West, the prospect of nuclear war led many scholars to question the primacy of politics. For those who followed the more militaristic tradition of interpreting Clausewitz, the new situation could be easily accommodated in their idea of war. Pure wars followed their own laws, but in reality political constraints normally interfered in their operation to some extent. In the nuclear era, the decision to launch war may still be a political one, but the conduct of war would be a purely military matter without political controls. What has happened in other words is that war has come to more closely resemble the Clausewitzian ideal of absolute war. To that extent the nature of war has changed and an adjustment of the political formula is in order. Many of those who have taken the opposite tack, that Clausewitz meant that war should be kept under political control, express the same conclusion more categorically: since war can no longer be controlled by policy, it has ceased to be an instrument of policy. The formula has become obsolete.

But the main core of political realism contends that despite these objections the formula has not been invalidated by modern developments. On the contrary, the evidence supports Clausewitz: nuclear war has not broken out, and its occurrence is widely believed by both sides to be very unlikely because neither can conceive of achieving any political goals by using it. Rather than deny the validity of the Clausewitzian formula on the grounds of a hypothetical war that no state wants, it would be more appropriate to regard a theoretically possible outbreak of nuclear war as a breakdown of political activity, not its continuation. Moreover, nuclear war is only one type of war. As the number of sovereign states has increased, limited wars have become more frequent. It remains part of the political reality of our world that states resort to war in the pursuit of political goals when this appears a practicable means of achieving them.

Some Western scholars have suggested that the concept of war, which in the West was traditionally applied to interstate wars, needed to be redefined to account for modern types of internal war, especially guerilla wars for socialistic revolution. These had become so frequent that they no longer could be regarded as less important than interstate war. Again, however, the general consensus amongst Western scholars seems to be that the Clausewitzian formula accommodates these adequately if it is given a more flexible interpretation to include internal politics, irregular warfare, and loosely organised troops.

In the Marxist–Leninist theory, of course, internal war never was a grounds for questioning the thesis that war is an instrument of policy. And the suggestion put forward by one or two scholars in the mid-1950s that the advent of the possibility of nuclear war had invalidated war as a socio-political phenomenon was sharply criticised. It can therefore be said that that formula was never officially questioned in the Soviet Union. The discussion on the problems of war in the nuclear era focused instead on the question of whether some of the essential elements of war could change, and if so, which. The first question was answered in the affirmative and three elements were recognised to change: the content of policy, the forms of warfare, and consequently the form of the relation between politics and war.

Soviet scholars conclude that what this means for the modern era is that the possibility of starting a war is now more limited, that it is in some cases more difficult to exert political control over warfare, and that the possibility of achieving desirable political goals through war has become greatly restricted. In other words, the nature of war has changed such that it is less useful as a political instrument than before, but it has not ceased to be a political phenomenon.

In sum, both Soviet scholars and political realists, each regarding the problem from their own perspective, conclude that war remains a continuation and instrument of politics although its value as an instrument has diminished.

The changing nature of peace

Parallel to these analyses of the nature of war in the nuclear era, an inquiry was conducted into the nature of modern peace.

For the political realists, the striking feature of peace in the postwar period was the intensity of the political conflict between the two main actors. The collision between their political interests was total and uncompromising, as in war. Moreover, military power was given a much greater role in diplomatic manoeuvres. This led some Western observers to talk of the war-like peace. Although Soviet scholars have emphatically denied that the postwar peace resembled war, they too have stressed that the danger of war has characterised the modern peace to an extent unprecedented in history.

The assessment of each side made of the character of peace, its main political features, the main socio-historical trends, and its ethical value, were even more similar in form, and even more contrary in content. Each contended that the opposite camp took advantage of the peace to prepare for war while great injustices remained to be rectified. The political realists pointed in this regard to the socio-political repression suffered by millions in the totalitarian communist states, and to the economic and social problems of the underdeveloped countries. Soviet scholars stressed the social inequalities existing throughout the non-socialistic world, and the socio-economic injustice imposed by neo-colonialism on the Third World.

More central to both than these general assessments was the appropriate strategy to adopt in order to progress, each according to its own lights. They agreed that the peace was a form of struggle, but they disagreed on whose policy was aggressive, whose defensive; and on who was to blame for the Cold War. To meet Western counter-revolutionary attempts to 'roll back Communism', Soviet scholars adopted the policy of peaceful coexistence and the prevention of war. In the West, the main political goal was to stop Soviet expansion, disguished under the mantle of social revolution, through a policy of containment and deterrence.

In both these political doctrines, the threat of war played an important role. Their common politico-military goal was to prevent war, especially total war, but to be prepared to win a war, should the other side start one. Dulles's strategy of massive retaliation resembled the Soviet warning that any local war would inevitably escalate into a total war. The strategy of flexible response was similar to the Soviet military policy of being prepared for all kinds of war. On both sides, the military power thereby played a crucial double role: for peace, by deterring aggression (or preventing war in the Soviet terminology), and for war should deterrence fail.

345

The two sides have accused each other of hiding their true aims behind their officially proclaimed policy. The following are the main points in this Soviet attack on the Western policy of containment and deterrence adopted by most Western governments and expounded by political realists.

(a)   It justifies the creation and build-up of anti-Communist alliances under the leadership of the United States, the establishment of military bases in a ring around the socialistic countries, and the continuation of the arms race. In short, it justifies preparations for war against the socialistic countries.

(b)   It justifies political crises, which are unleashed as a substitute for war and a prelude to armed interventions.

(c)   It justifies the use of violence to suppress social revolutions and national liberation movements.

(d)   It justifies the spread of anti-Communist propaganda and war propaganda under the pretext that war is already being waged. [2]

The Western analysis of the Soviet doctrine of peaceful coexistence also claims to find ulterior motives. [3]

(a)   It is in fact a strategy of conquest by which the Soviet Union aims to weaken the Western countries by a great number of small actions and to prepare a future crushing blow; it is a form of total war in which very sophisticated covert methods are used instead of conventional military means.

(b)   It stimulates the arms race in all kinds of weapons.

(c)   It aims at exploiting social discontent and injured national pride, which may indeed be legitimate, to stir up trouble and seize control of the revolutionary forces.

(d)   It makes use of psychological techniques to try to undermine the cohesion of Western societies with a bombardment of propaganda.

In all these charges there is some grain of truth but much more exaggeration. The political goals of the two camps are in many respects mutually exclusive, and it seems only natural that this be reflected in a tough ideological struggle. The claim that the true goal of the official policies is war seems extremely exaggerated, however. Now that the climax of the Cold War is past, peace seems to be the sincere aim of both opposing camps. Since it is one of the keystones of both policies that the opponent could unleash war if it so desired, and that it must therefore be deterred from doing so by the maintenance of a high level of military readiness, to ascribe to the opponent peaceful intentions would only give counter-arguments to domestic opponents of the level of military expenditures necessary to pursue the strategy of deterrence.

Deterrence and the prevention of war

In the West the idea of preventing wars has been developed into the highly sophisticated theory of deterrence. This theory first appeared in the late 1940s and early 1950s with the so-called 'big deterrent', a threat of massive nuclear attack against the Soviet Union should it make any military move against the Free World. In the late 1950s, it was complemented by other forms of nuclear deterrence, and in the 1960s by the idea of conventional deterrence in order to meet different forms of attack at an appropriate level of response. It reached its most developed form in the theory of graduated deterrence, which became one of the premises of the strategy of flexible response. [4]

Both the theory of deterrence and military doctrine have taken various forms. The latest doctrine promulgated in the winter of 1974 by the US Secretary of Defense, J. R. Schlesinger, includes the capacity to use nuclear weapons flexibly against selected military targets and is said to introduce flexibility both into nuclear employment planning and into the whole deterrence policy: the enemy will be deterred from making any kind of military challenge, even nuclear black-mail, because the United States will be able to respond not only with a massive countervalue attack, but also with blows which bear some relation to the provo-cation. [5]

With increased flexibility, deterrence will become effective on all conceivable situations, according to prominent exponents of the new doctrine.

While in the West the strategy of deterrence was widely discussed and even presented in official expositions of the politico-military doctrine, in the Soviet Union the full implications of the idea of a deterrent threat were nowhere expressed. Nevertheless, similar ideas are implied in strategic concepts relating to the prevention of war. Unlike the American theory of deterrence, the threat on which it was initially based was a massive conventional attack on Western Europe. Later the threat of a massive nuclear rocket blow was added. In its recent most developed form the strategy calls for the ability of the Soviet Union to threaten a would-be aggressor with crushing blows of different size and intensity using a full range of weapons from conventional to nuclear.

The kinds of war which it was the purpose of the respective strategies to deter or prevent were also in some respects similar. Western politicals and theorists were most anxious to deter three kinds of Soviet aggression: a direct attack on American territory, an attack on the territory of one of the American allies, and intervention of some sort in one of the world's trouble-spots, particularly in the Third World. The expressed purpose of the Soviet doctrine was to prevent an American attack supported by NATO forces on the socialistic countries in Europe, including the Soviet Union itself, and also to prevent Western armed intervention taken to put down a popular struggle for national- or social-liberation.

Despite these similarities the attitude officially adopted to the strategy of the opponent was highly critical. The Soviet Union had the following main objections to the strategy of deterrence. [6]

(a) Its purpose was not to prevent war but to compel other countries to act in accordance with American demands, deterrence meaning intimidation.

(b) The whole doctrine of deterrence was built up round the nuclear arsenal of the United States. It thus served as an instrument by which the United States could gain control over the military policy of its allies and compel them to fall in line with its policy of global Cold War.

(c) The doctrine was used to stimulate the arms race. Each development in the doctrine of deterrence called for further improvements in the American armed forces. The big deterrent justified the enormous build up of the American strategic nuclear arsenal; the conventional deterrent required large investments in tactical nuclear weapons and conventional weapons: and the need for a flexible response stimulated the development of anti-guerilla weapons and forces.

(d) The theory of deterrence glorified military means by making them the main tool of international politics in peace-time.

(e) It greatly increased the danger of the occurrence of wars. These objections have been repeated in response to the new American doctrine, which under the pretext of deterring and avoiding a general nuclear war enhances the danger of the outbreak of a so-called limited nuclear war (or limited strategic war). Such a war would in fact lead directly and inevitably to a general war. Under the pretext of deterring all possible challenges it aims at expanding the usability of nuclear weapons, e.g. it attempts to give the United States armed forces the capability to launch a massive counterforce attack. The critics summarise their comments with the charge that such changes in the military doctrine and deterrence policy contradict the policy of limiting strategic weapons and add impetus to the arms race; they run contrary to the tendency to *détente*. [7]

The critical remarks Western observers directed at the Soviet doctrine of preventing war have been in most respects analogous. Far from preventing war, the Soviet doctrine was used to prepare for war, to justify the production of newer kinds of nuclear missile armament, and to keep Soviet allies under the Soviet command.[8] In recent years, under the pretext of preventing war through a favourable correlation of forces, the Soviets have been increasing their global military power and, in particular, their counterforce capability; the new American doctrine is only a response to the Soviet build-up.

This was said to be a natural consequence of the different place which the ideas of deterrence (prevention of war) have occupied in respective doctrines. Whereas in the US doctrine deterrence has been the primary idea, and the emphasis has been on forces capable of inflicting massive punishment on the society of the opponent, in the Soviet doctrine, the prevention of war has never been presented as a separate task, a function of armed forces *per se;* on the contrary, the war-fighting function has always been emphasised. It has been maintained that deterrence, or the prevention of war can be achieved by having the ability to fight and win a war, and not by creation of forces assigned with the pure task of deterrence, i.e. with the countercity (counter-value) strategy. As Western critics maintain, such a position led to a steady expansion of the Soviet counterforce capabilities, i.e. strategic offensive forces, and to the Soviet resistance to the American propositions to establish a security system based on the mutual ability of offensive forces to survive attack.

As was pointed out at the beginning of this section, the doctrine of deterrence has quite a different status from the Soviet goal of preventing war. Indeed, as a theoretical construct it is much more elaborated, and many alternative models of deterrence have been presented. There are several reasons for this theoretical activity in the West. In the first place, the United States found itself both militarily and politically in an entirely new, and in some respects unique, position after the Second World War. It was the only state that possessed nuclear weapons, and it had succeeded France and Great Britain as the bulwark of Free World interests throughout the world. The new ideas generated by these circumstances were subjected to a debate to which contributions were made from many directions. Analysts in other Western countries could see Western aims and problems in a somewhat different light from the Americans. Moreover the competition between the various branches of the American armed forces caused each to develop theories to justify its role and to motivate its request for greater financial resources and armaments. Finally, owing to the openness of the debate, many arm chair strategists and scholars from various disciplines were able to present their ideas as well.

Therefore, within the West itself, the debate on the value of deterrence was often quite heated. For the most part the arguments belong to one of three kinds. The most common was to criticise the official view of deterrence and to present an alternative. Sometimes it was argued that the needs of deterrence were given greater priority than the needs of defence, in other words that too much emphasis was being placed on strategic nuclear weapons which were not intended to be used, at the cost of conventional weapons, for which there was a real need. [9] The third kind of argument maintained that deterrence was a dangerous concept that jeopardised peace. [10] The new flexible option deterrence strategy was criticised from this perspective.

Although some of these arguments are similar to the Soviet critique, Western critics of the theory of deterrence shared with its proponents the view that the danger of war lay in the expansionist aims of the Soviet Union.

Despite this basic difference in socio-political referents which more than anything else divided Soviet ideas from political realist ones, the doctrine of deterrence and the doctrine of the prevention of war have had much in common. Firstly, the official policy of both superpowers has been to prevent war, at least the most dangerous kinds of it. Since neither superpower has considered it advisable to engage in such a war itself, the task both superpowers have set themselves has been to prevent the other from making the first move. Both have also tried to restrain the opponent from participating in smaller wars.

Secondly, both doctrines are to some extent strategies of hypothetical war. The military actions by which the superpowers threaten each other are in principle those which they would take if war were to break out. Since both doctrines can therefore be used to justify a military build-up, there is some truth in the charge that they stimulate the arms race.

Thirdly, and most important of all, both superpowers have used their respective ideas of deterrence and prevention of war as something more than a purely military policy. It has been an important means by which they have justified actions to assert their pre-eminence within their own camp, to ensure the support of their allies for their particular political line, and to extend their influence in other parts of the world. It might be said that the prevention of war has become a means of policy. In the Cold War phase, all other political issues were subordinated to this central problem; nothing might be allowed to happen that would indicate discord within each camp, for that would invite a military move from the other side which it was their whole purpose of the policy to prevent. This was a period marked by fear of the opponent. – It was uncertain whether the policy adopted by each side was really having the desired effect or whether the absence of war had other immediate causes. The experiences of the 1950s and early 1960s are used to support the theory of deterrence, however. The deployment of American forces in Western Europe helped prevent the rise of communism there, just as the stationing of Soviet troops in Eastern Europe effectively stopped any counter-socialistic threats; the threat of a Soviet missile attack at the time of the Middle East War of 1956 helped bring a quick end to it; and quite clearly the action of American forces during the Cuban Crisis of 1962 compelled the Soviet Union to withdraw its missiles there.

Both sides admit that in the period of *détente,* war can be prevented by means of strategic military balance. The superpowers have thus felt secure enough to find ways of cooperating for common benefit. It has been vital that the military balance be preserved and this consideration has guided their policy towards each other and towards many of the world's problems.

One of the paradoxes of the nuclear era thus consists in the fact that deterrence has become not only a part of the general policy of competition between the two systems but also a condition of cooperation between them: in fact, it is included in both political-strategic doctrines by which the great countries rival each other for power and in collective arrangements that aim at increasing cooperation between nations.

What this means is that while a global war between systems is in practice no longer an instrument of policy in the socialist-capitalist confrontation, a military competition remains such an instrument. The role of the military has become less that of winning wars and more that of preventing them. Deterrence has given military power quite a new political dimension.

In sum, war between systems cannot achieve political goals, but military moves short of such war may help to do so. If this is true, the nature of such political-military action has become one of the central problems of the study of military affairs.

Notes

[1] It seems, however, that many Western scholars tacitly admit the great

importance of the changes in the international system structure, and some present this view in a categorical way: their approach differs from the Soviet one inasmuch as it puts emphasis on the emergence of the Third World and not on the socialist system. E.g. Arnold L. Horelick points out two main causes of the changes in instrumentality of war: reordering of the international system and revolution in military affairs ('Perspectives on the Study of Comparative Military Doctrines' in Horton III, et al., 1973, pp.194 ff.).

[2] Innumerable Soviet books and articles on the Western policy after World War II, especially on the NATO. I have written critical comments on policy of containment in 1962, 1963, 1965, 1967, 1968, 1969, 1970.

[3] Innumerable books and articles on the Soviet concept of peaceful coexistence; cf. the section on Cold War (in ch.8). A typical comment: The Soviet Union is convinced, that 'the imperialists' are getting weaker so quickly that with careful handling they may miss their moment and be induced in effect to surrender peacefully (Strachey, 1962, p.229).

Soviet policy of coexistence has been described by another scholar as aiming at: (a) retaining the bipolarity of the cold war; (b) closing Soviet sphere of influence to Western penetration; (c) opening the rest of the world to Soviet penetration, and the conclusion is the following: 'Détente Soviet-style would become a form of graceful historical capitulation of Western nations to the Soviet (or Russian) imperial will without the risks of war with a united West' (Richard B. Foster, 'The Emerging US Global Strategy: Its Implications for the U.S.-European Partnership' in Foster, Beaufre, Joshua, 1974, p.26).

[4] Some recent publications: Kaplan, 1973; Phil Williams, 'Deterrence' in Baylis et al., 1975; J. Barnett, 1974.

[5] Secretary of Defense James R. Schlesinger, Press Conference, 24 January 1974, Excerpts: 'Flexible Strategic Options and Deterrence', in *Survival,* March/April 1974; Barry Carter, 'Flexible Strategic Options No Need for New Strategy', *Survival,* Jan./Febr. 1975; Lynn Etheridge Davis, 'Limited Nuclear Options. Deterrence and the New American Doctrine', *Adelphi Papers no. 121,* The Institute for Strategic Studies, London, Winter 1975–6.

[6] Innumerable books and articles on Western political and military doctrines. The term to deter, or *Abschrecken* has been translated in the Soviet literature by *ustrashat* or *zapugivat* which means 'to intimidate' (instead of *otstrashat* or *otpugivat*).

[7] Michael A. Milstein, Leo S. Semeiko, 'Problems of the Inadmissibility of Nuclear Conflict', *International Studies Quarterly,* March 1976. Milstein is a well-known author of many studies on the Western military doctrines.

A very interesting passage of the above mentioned article is worth quoting: 'Of course the concept of nuclear deterrence, which assumes the existence of gigantic nuclear forces capable of "assured destruction" is not the ideal solution to the problem of peace and of the avoidance of nuclear conflict. But the question is, how do we move away from this concept which according to widely held opinion, expresses the actual situation at the moment?' (p.98). Naturally, the critical assessment of the nuclear deterrence is not abandoned here, and the path of *détente* is afterward praised; but the expression 'is not ideal solution' is very moderate, the words 'expresses the actual situation' stress the stabilising effect of the present strategic situation.

Cf. David Holloway, 'The Soviet Strategists Attack Schlesinger', *The New Scientist,* 5 Dec. 1974.

[8] For a detailed analysis of the Soviet deterrent policy see: Goure, Kohler, Harvey, 1974. The authors point out four components of the Soviet deterrence: (a) the passive deterrence i.e. deterrence of a Western or Chinese attack on the Soviet Union; (b) the active deterrence i.e. deterrence of an attack on Soviet allies; (c) deterrence of Western efforts to oppose and crush the liberation movement and the progressive countries either through a direct military action or 'by proxy'; (d) deterrence of efforts at exporting 'counter-revolution' (esp. pp.32–3).

For a profound analysis of the earlier stages of the history of the Soviet deterrence see writings of Dinerstein, Griffith and Wolfe.

[9] Of many books and articles, for a more detailed analysis see: Liddell-Hart 1960; Miksche 1955, 1958, 1965 (I have described their views in 1969a).

[10] Of the rich literature, see Philip Green 1966, Senghaas 1969, 1972. Following accusations are corollary: deterrence is immoral, as based on the threat of annihilating the innocent population; it accelerates the arms race and jeopardises the economic development. Cf. writings of Kenneth Boulding, Amitai Etzioni, Ralph Lapp and others.

21 Instead of conclusions: three observations

Approaches: different yet similar

The first and most important observation that can be made on the basis of this survey of what war is according to Soviet Marxism–Leninism and political realism is that however much the two views may be modified in consequence of the changing situation and policies, and however they may come closer to each other, they are founded upon two separate and irreconcilable concepts of society, social behaviour, and history. Although both consider war to be a political act and an instrument of politics, since each has its own idea of politics, each interprets the sources, role, and consequences of war in a fundamentally different way.

Concepts are not eternal and unequivocal truths, however, even if they are often presented as such, especially in Soviet theory, and even if they give a semblance of continuity to different succeeding doctrines.

Concepts of the nature of war have a double nature. They are not only, as scientific constructions, a part of ideology but they are also, as a guide to action, included in the political and military doctrine. In the first role they may change only with changes in the knowledge and interpretation of the world and society; in the second role, however, they may be interpreted very flexibly, especially if they are formulated in a very general way amenable to ambiguous conclusions.

Thus, to begin with the latter, in the Western interpretation the process of decolonisation took place because it was in the interests of the former colonial powers, and according to these interests (and with the use of violence if against them), thus it could be presented as voluntary: anticolonial wars were often described as either an outcome of political failures or as the product of the instigation of communist countries; armed intervention in the countries of the Third World could be presented as a counteraction against the communist attempts to change the world balance of power. On the other hand, in the Soviet exposition, military actions of the Soviet Union have always been justified by its role as the main force of the world camp struggling for progress, independence and peace among nations, as just wars in defence against the imperialist aggression.

To turn, however, to the first role, it would be mistaken to treat concepts of the nature of war only as convenient verbal structures elastically used according to the needs of the current policy. They are an important part of the belief system, of the attitudes and preconceptions which the governing elite brings to the evaluation of the national interest, of long-range policy, and of the course of current action; and this is at least to some extent shared by the population which has to be convinced of the necessity and desirability of a given policy, especially if it involves armed struggle.

To be useful, concepts cannot long remain static, of course, even in their general theoretical wording, i.e. as a part of science and consequently, ideology. Their elastic character cannot openly contradict changes in the environment. For example the theory of the inevitability of war cannot remain unchanged if the instrumentality of war has greatly and rapidly decreased.

Thus, as reality changes so too must the ideas with which men try to understand it. Since the Second World War three radical changes in the structure of the international system have had a bearing on problems of war and peace: the confrontation between the socialist and non-socialistic developed countries, the emergence of the Third World, with the dissolution of colonial empires, and the advent of nuclear rocket technology. All three have had a profound impact on attitudes towards war. This can be seen in the changes that have occurred in political doctrines as politicians have grappled with a new political reality, and in modifications in the orthodox theoretical views on the nature of war. By and large the adjustments made have tended to narrow the gap separating the two concepts of war.

1 Both Marxist–Leninists and political realists have altered their view of the political conflict of which a hypothetical world war would be the continuation. Political realists have come to recognise that something more is at stake than a conflict of interests between states. They may talk of a struggle between totalitarianism and democracy rather than social-ism and capitalism, but they treat it now as a conflict to some extent independent of any conflict between states. They also realise that a world war could exert an extremely great impact on the social order in the fighting countries.

   The peaceful road is chosen even when victory in war is believed to be assured because of the great social costs such a war would entail even for the socialistic states themselves. What this implies is that the Soviet Union or the socialistic states together have particular interests that are in some measure separate from the fundamental class-centred conflict. The difference between this concept and the state-centred concept of politics adopted by political realists has thereby diminished.

2 Related to the modifications that have occurred in the concepts of politics are shifts in the notion of the political actor who is a party to war. By broadening the scope of their theory to include internal war, political realists have dropped their traditional state-as-actor premise. Marxist–Leninists have in turn softened their class criterion in order to account for wars in which the direct actors were not classes. Since both also included irregular warfare and irregular forces within their theory of war, the combined effect of these modifications was to make more similar the kinds of hostilities each regarded as war.

3 Both sides are agreed that the usefulness of war as an instrument of policy has diminished. In the West, the role of the technological revolution in

this connection has been exaggerated in much of the literature. As an indication of the Western position this emphasis is somewhat misleading, however. It is recognised that the balance of power is more than a matter of weapons; it also has a political dimension that makes more difficult the use of violent means against revolutionary movements which can be supported by the Soviet Union. Moreover, Western politicians also realise that much of the pressure for change in the world is not manufactured by agitators but is a genuine result of a growing political awareness of people in all parts of the world. On the other hand, despite the emphasis Soviet scholars and politicians put on the strength of the socialistic system as the condition that makes war unusable for the imperialists, they give more weight than they admit officially to the preventive impact of the new weapons that have made most wars unprofitable. Both camps fear the socio-political consequences of a world war or of any major war even though each finds reassurance in the belief that the adversary would suffer more in a potential war.

4 Both Soviet scholars and political realists are aware of the great risk that war may break out accidentally and that irrational factors may exert much influence on the course a war takes and the level of violence at which it is fought. In other words, both sides recognise that the difficulty of maintaining rational political control over warfare has increased.

5 In neither approach is war any longer treated as an inevitably recurrent method of resolving political disputes. Both now assert that even the most difficult of conflicts, internal or international, can in theory be resolved by peaceful means.

To recapitulate, both approaches have adapted the basic formula that war is a continuation and instrument of politics to modern conditions by modifying their interpretation of its key concepts. In the process, the gap separating the Marxist–Leninist and political realist positions has been somewhat narrowed.

Such theoretical adjustments must be marginal, however. As long as the two ideologies retain incompatible concepts of society, social activity, and social history, the most essential aspects of their idea of the nature of war, its fundamental source, its role in social intercourse, and its future can never converge.

The formulae that had a career

What is all this interest in what Clausewitz wrote on the subject of war really about? Why do so many military theorists, and political scientists begin by acknowledging their intellectual debt to Clausewitz for the ideas they are about to expound? Why is there so much controversy about what Clausewitz really meant? Has 'what he really meant' the status of scientific truth, whereas 'what he might be interpreted to mean' can be a distorted nonsense?

There may be some historical value *per se* in trying to understand the ideas behind the words Clausewitz wrote, even if we can never know whether or not we have reached such an understanding. [1] I believe, however, that the scientific value of Clausewitz's contribution lies not in the truth of his ideas (whatever these may have been) but in generality of his tools of analysis, the formulae on war.

Clausewitz endeavoured to comprise two main features of war in his study of its character: its outward visible aspect as armed violence, and its inward, purposive aspect as an act in pursuit of political goals. Therefore he not only defined war as an act of violence by which we intend to compel our opponent to obey our will but to penetrate deeper into its nature he also offered two formulae, one about war as an act of violence pushed to its utmost bounds, the other about war as a continuation of politics. The first is a logical extension of the military aspect of war to an absolute extreme where there are no checks on military violence, and the second reminds us that war is one of that class of actions called political.

Many commentators of Clausewitz have treated these as two quite different concepts. Some have contended that while he conceived of absolute war as an ideal type in a philosophical sense he related the formula on war as a political phenomenon to real wars. [2] Others have tried to express a similar duality between a pure and an impure form of war by identifying absolute war with total (unlimited) war and regarding war governed and controlled by politics as inherently limited. I believe, however, that whatever Clausewitz himself may have meant by war it is not necessary to suppose that the two formulae represent two quite different ideas, or two quite different levels of generalisation; they can plausibly be interpreted as two complementary expressions of the same notion.

Absolute war

To begin with, it seems to me that Clausewitz conceived of absolute war not only as an abstraction, an archetype[3] or a complex of features common to all wars[4] but also as a kind of real war. He considered the Napoleonic wars to be absolute wars, for example, and he indicated that he believed absolute wars could occur again.[5] Secondly, many of the features of actual wars, without which they could hardly be called wars at all, he attributes to the concept of absolute war. In the third place, he writes that absolute war can spring from 'extremely powerful motives', which one might also expect to give rise to extreme political goals.[6] In this view, an absolute war is not one completely divorced from political aims, but rather what we should nowadays call a total war fought by unlimited military means for unlimited political goals. Fourthly, in a little known article from 1827 Clausewitz writes about the two kinds of war that occur in reality: the one aims at defeating the enemy by destroying him politically or by disarming him and compelling him to unconditional surrender; the aims of the other are limited. Since the degree of limitation may vary, there are many intermediate forms of war, but in principle their aims may be limited or unlimited. Fifthly, Clausewitz describes absolute war in such a way that its features appear indispensable to actual wars. Moreover, in describing real war as something distinct from absolute war, he uses such expressions as 'in general', 'often', etc., which seems to suggest that war may sometimes take another form, namely the form of absolute war. Finally, from Clausewitz's detailed presentation of absolute war and the strategy of winning it, the reader can hardly avoid getting the impression that he believed it actually possible to fight an absolute war.[7]

War as a continuation of politics

On the other hand, the Clausewitzian formula on war as a political phenomenon can hardly be meant as a strictly factual description of real wars. Neither of two conditions of subordination to politics – that the outbreak of war be a result of a rational decision to pursue the same political goals that had been pursued in peace by violent means, and that political goals should guide the whole course of war, is ever completely fulfilled in practice.

To begin with, the first condition can only be fully met if the decision to go to war is completely isolated from extraneous influences – issues that arise shortly before the outbreak of war, the actions of the opponent, fortuitous circumstances, personal or collective emotions and feelings, misinterpretation of the situation etc. Clearly, many such factors always influence the decision the leaders of a country come to about war.

Warfare can never be completely directed by political considerations in actual fact. Military action is conducted according to certain rules which cannot be followed or abandoned at will; once the decision to take military action is taken, a commitment is made to follow certain kinds of behaviour until the state of war ceases. Moreover, as long as military action seems to succeed, there will be little reason to question it; and if the war goes badly for one side, it may very well be too late for it to tell the opponent that it is satisfied and desires a ceasefire. There is thus a certain inertia about war that compels the combatants to keep fighting once they have started until one of them wins a total victory.

Another limitation to political control is the difficulty involved in assessing the political implications of unexpected military developments. Political assessment takes time, which in war is a very scarce resource. Since the military implications are usually more obvious, the next move is in general based more on military than on political considerations.

A second indication that Clausewitz recognised that real wars are never completely controlled by policy is his description of war as a 'wonderful trinity' of passion, chance, and purpose. How should rational political aims govern strong feelings of hatred, let alone chance?

Thirdly, Clausewitz's study consists in a detailed description of how to attain an absolute military victory. He does not say how political control might be asserted over warfare to ensure the achievement of limited political goals. He

357

neither describes how political means should be used to stimulate military action nor how to hamper it if it overreaches political limits. His whole manner of writing gives the impression that he considers efficient political control of war to be less a matter of fact than an analytical postulate.

The war-politics formula is therefore as much an abstraction as absolute war, and it can be treated as an archetype in relation to which all real wars may be measured. If absolute war represents what would happen if nothing prevented military logic from completely determining the course of action, so too 'war as a continuation of politics' represents what would happen if nothing hindered the logic of political intercourse from taking a fully deliberate decision to unleash war, and from steering the course followed in war.

To conclude, the war-politics formula expresses in an ideal form the fact that all real wars are more or less governed by political goals. There is nothing, however, in this formula that precludes the possibility of unlimited war, for the political goals that ideally guide war may themselves be unlimited, or they may become so in the course of war.

Two complementary perspectives

On this interpretation, the set of Clausewitzian formulae constitute a single concept of war with two main aspects, one military and the other political. The two formulae provide two complementary perspectives from which to view any given war: according to the means employed or the extent by which it is controlled by political aims.

Seen in this perspective, the formulae on war are analytical statements about the concept of war, for they say what is entailed in the idea of war. The social phenomenon we call war is a combination of armed violence and political activity: it is fought by military means to achieve certain political ends, regardless of whether the outcome actually meets expectations: it is always considered a means to those ends, or else it is not resorted to. In all situations the formulae can be applied to the analysis of a concrete war.

Since Clausewitz was living in a world different in significant ways from our own, his understanding of all the main concepts must have been significantly different from ours. To study Clausewitz's writings may help us better to understand the era in which he lived and give us a new perspective on the modern world. However, we should only misunderstand modern war if we were to suppose that the substance of Clausewitz's ideas could be uprooted from their nineteenth-century context and directly replanted in the present-day situation. For this reason, modern interpretations tend to say more about the interpreters' general ideas of war, society, and war than they do about what Clausewitz meant himself.

Clausewitz's lasting contribution, and it is a very great one, amounts to his having provided a framework within which war can be analysed. His formulae express the idea of war at such a high level of generalisation that they are still useful today. The formula on absolute war became famous after the Prussian military staff adopted it about a century ago. It subsequently became notorious in relation to the use to which it was put by militarists. Nowadays the war-politics formula is used as the point of departure for most military analyses in both East and West. [8]

This brings me back to the main theme of this section. Since the theory of war in both East and West adopts the analytical framework proposed by Clausewitz, it is possible to use his formulae as an analytical tool with which to compare these theories and thereby to reveal basic similarities and differences in the two underlying interpretations of socio-political reality. One of the preconditions to a meaningful exchange of views, which would further the cause of peace, is the perception that reality may be legitimately understood in several different ways. Thus, although the solution to the problem of doing away with war is only remotely related to the debate over the interpretation of the Clausewitzian formulae *per se*, it does depend in part upon reducing the risk inherent in the dialogue of the deaf.

War as two interrelated processes

Most past wars, including both world wars, have lent support to the thesis that war is a means by which political conflicts are resolved. However, the relationship between warfare (military activity aiming at military victory) and the broader purpose of war to achieve political goals, is not a simple one, and is worthy of some comment.

Politics, social activity aimed at achievement of principal goals of a political unit, usually has at its disposal both military and several non-military means: diplomatic, economic, ideological, and others. Ideally warfare should therefore relate to the whole political process of war, to the process of pursuing political goals through war, as a part to the whole, and as means to ends.

One implication of such a relation should be that some proportionality exists between the political goals at stake and the military goals adopted, and consequently, the amount of violence used. In other words, the more important the political ends, the more violent the means chosen and used.

Historically, this contention has been shown to be valid, at least in some respects and in regard to most wars. With the development of destructive power of warfare and its growth in scale and intensity, however, the number of exceptions to the ideal picture of war has increased. The two world wars in particular have compelled scholars to re-examine it.

The proportionality has always existed, or almost always, in the initial stages of war. Belligerents have usually chosen the level of military operations that they expect would be adequate to their political needs. But this proportionality has often tended to disappear during the hostilities. As a rule, once the decision is taken to go to war and the military process set in motion, political considerations tend to become secondary to whatever has been deemed necessary to military victory, and the level of violence begins to be conditioned more by military objectives than by the original political goal. This can be best observed when the termination of war is considered. If the political goals stand within reach, the successful party aims at bringing the military process to end. However, since it is virtually impossible to terminate war by negotiation as long as one of the parties holds any hope of gaining a better bargaining position by further fighting, war continues but at a higher level of violence. The logic of military operations calls for escalation until a decisive military victory is possible. The growth of the amount of violence beyond the level initially foreseen, and the consequent expansion of military goals, cannot fail to affect the political goals, which also expand and are redefined. As the level of violence escalates, and the amount of destruction, pain, and suffering increases, the societies fighting each other become more embittered towards each other and more intransigent in their policies.

In some wars, the military aspect had such an impact on the political that the original dispute was lost sight of. The logic of military strategy in dictating a level of violence out of proportion to the political goals, in pursuit of which a war was originally started, had not only led to a full-scale war but also created pressure for unlimited political goals.

In other words, the military process, or warfare, has more and more frequently influenced the political process by stimulating the expansion of political goals of more offensive character: the part had often changed the whole, the means had modified the ends. What is more, the military process had sometimes led an existence independent of the political process.

In the nuclear missile era, as the interrelation between the two processes, military and political, has undergone further changes, the ideal picture has become more confused than ever before. On the one hand, the revolutionary growth of military technology has created the possibility and the danger of an unlimited application of an unlimited amount of violence, i.e. of warfare completely free from political control, waged completely independently of the political process. On the other hand, however, politics has acquired quite new dimensions. Two of them seem to have the greatest impact on the relation between the military and political processes: first, the growing interaction between national policies and world policies, and second, the greater political awareness and resistance to war. Paradoxically, at a time when warfare has gained unlimited technological possibilities, political limitations on their use have become greater than ever.

The consequences for at least some kinds of war have been immediate and evident. More and more frequently the part-whole and means-end relation has failed to bring about desired results, which testifies to the diminished instrumentality of warfare and to growing importance of non-military means in political disputes.

What is more, non-military activity has sometimes paralysed the effects of military successes. Even when the Americans inflicted heavy damage on the Vietnamese economic and military potential, they could not exploit these effects; on the contrary, these measures added to a political setback. Israel suffered a sort of political defeat in 1973 despite military success in the final stage of war. In other words, even if the means have worked, they have not brought about desired ends, and the other parts of the whole, especially the diplomatic and economic activity have prevailed in the determination of results of war. (This may remind us of the fact that even in the ideal concept, military victory leads to desired political results only when the will of the enemy to resist is broken. If that will refuses to allow itself to be defeated, if a war can be continued by an increasing use of non-military means, then it is not terminated, nor is it won, for the political ends are not yet achieved.)[9] Reality has thus diverged even further from the ideal of the formula. In some wars, the military process has not only ceased to be subordinate to the political process but also become separate from it. The part is no longer a part of the whole: the means is no longer instrumental.

Furthermore, the great diversification in the kinds of war, from frequent guerilla wars to unthinkable nuclear missile war, makes it difficult to interpret the formula on war as a political act identically for all different kinds of war; the interrelation between the two processes, military and political, may be different in different wars.

These reflections are perhaps adequate to indicate that in our times the way in which war is an instrument of politics is by no means obvious, and the interaction between the two processes, warfare and political dispute, need to be re-examined. [10]

In this context, the discussion on the nature of war, reviewed in this study, as well as the search for regularities in the occurrence and conduct of wars, which is a topic beyond the aim of this study, is encouraging.

It is my impression, however, that some of the new methods of analysis that have gained vogue do not offer much promise of providing us with generalisations about the political aspect of war. The search for quantitative indicators may yield methods of describing more precisely the physical features of wars, but they say little about the political significance of war. The place of military methods in the whole arsenal of means used by politics in war and the means-ends relation in wars in the nuclear missile era must be clarified. Models of rational action cannot provide adequate explanations of the political side of war.

Whatever the value of such models as tools of analysis, they can neither describe the wide range of political goals that can conceivably be pursued in war by a great number of methods, nor adequately capture the political drama surrounding war, when mass killing and suffering stir up such powerful irrational forces as mass fear and hatred. Indeed, I doubt that it is possible to develop a model of presenting the relation between the military and political aspects of war as a set of regularities. Not only do military ends, means, and consequences (to take only three military variables) relate differently to different kinds of war, defined in political terms; in addition, the political character of war may change during the hostilities, and the political ends may become redefined. Initial and final ends, and political results of war (to take only three political variables) relate differently to different kinds of warfare. The task of reducing all these dynamic complexities to the sort of model mentioned above seems to me to be next to impossible.

If anything, the difficulty of all theories of war to show unambiguously how the military process is or can be kept subservient to the political process points to one important conclusion: when the consequences of the military process are as devastating as they are in the nuclear era, it should be one of the objects of policy everywhere to eliminate war as a phenomenon completely.

Notes

[1] It must be conceded that Clausewitz's manner of writing does not make his meaning very clear. He combines contrary views, sometimes without giving his own opinion, sometimes suggesting that the truth lies between the views presented. One might wonder why Clausewitz was so unclear about his own judgement. Perhaps he wished to avoid interrupting the flow of presentation in which after each argument he considers possible counterarguments; perhaps he found the nature of war to be too complex to form any definite opinion about it. It is also possible that the peculiarity of his presentation is a consequence of the disparity between his original intentions and the final study, which was never completed. In an undated notice written some time between 1816 and 1818, Clausewitz observes that he had initially planned to write short aphorismic chapters, 'grains of thought', which should stimulate men to think but that he had afterwards enlarged them into systematic lectures. The elaboration was never finished, however, the analysis of the political side of war being least developed.

Some assessments of Clausewitz's way of writing:

'Not one reader of Clausewitz's works in a hundred was likely to follow the subtlety of his logic and to preserve a true balance amid his philosophical juggery' (Liddell-Hart, 1967, p.355); 'He often starts out saying very nearly the opposite of what he concludes with, following the Hegelian pattern of thesis, antithesis, and synthesis' (Brodie, 1973, 1974, p.11); in Clausewitz's writings diverse followers found support for divergent interpretations by his habit of first stating an idea in its extreme form, in order to trace its full logical implications, and only then bringing forward considerations which modify the application of the pure theory (Kissinger, 1957, p.341).

[2] Dieter Senghaas who analysed the two Clausewitzian concepts of war

named the absolute war formula the philosophical concept of war, the war-policy formula the concept of real war (1972, pp.40 ff.).

[3] Rapoport, 1968 (1941), p.14. Most scholars quote Clausewitz's appraisal of the absolute form of war as a 'general point of direction' and 'the natural measure' of all that is done in war (*Sketches for Book Eight*, ch.II, p.370, ed. 1968).

[4] '. . . only through this kind of view (i.e. by keeping constantly in view the absolute kind of war) does war recover unity, only through it can we see all wars as things of one kind' (*Sketches for Book Eight*, ch.IV, p.403).

[5] Clausewitz writes that we might doubt whether the notion of the absolute character of the nature of war was founded in reality 'if we had not seen real warfare make its appearance in this absolute completeness even in our times' (*Sketches for Book Eight*, ch.II, p.369) and adds that 'in the next ten years there may perhaps be a war of that same kind'.

[6] The greater and more powerful the motives for a war, the nearer it will approach its abstract form, the more it will be directed towards the destruction of the enemy. (*Book One*, ch.I, pp.119–20).

[7] Some scholars consider total wars as Clausewitzian absolute wars. 'Total war is unlimited in character; it is what Clausewitz called "absolute war". It differs from that type of war which prevailed in the two centuries prior to World War I. Then war was "limited."' (Hans Speier, 'Class Structure and Total War', in Speier 1952, p.253). Antonio Pelliccia, who believes that such a war would not be 'apolitical' in the traditional sense attributed to this concept. Clausewitz changed his concept when he developed his philosophy of war: while he initially considered such a war as a conflict between forces left to themselves and obedient only to their own laws, he ended up by considering absolute war to be war waged with extreme violence in order to destroy the adversary's forces and thereby to attain political aims. In this light wars waged by Alexander the Great, by the Romans, and by Napoleon were total or absolute wars but they had clearly determined political aims. A thermonuclear war if it breaks out, could be such a war, and it would not contradict the principal primacy of policy, although the political control over the conduct of war would be very difficult (1970–71, p.62).

[8] Innumerable examples can be quoted of taking Clausewitzian ideas as point of departure for the presentation of own views. Some instances have been presented in note 57, ch.8 (Aron, Rapoport, Senghaas, Brodie). These ideas have also been used for supporting the own position. For instance, the description of Clausewitzian theory serves to Turner as an argument in his claim for a limited war strategy (Turner, Challener, 1960, ch.I); many other examples have been quoted in ch.8.

[9] Beaufre terms such non-military actions which are very important for attaining victory in war 'external maneuver; this is a frequent theme in his writings (1965, 1972, 1974).

[10] Up to now, the study of the nature of war has focused on why war breaks

out and quite naturally so, since the main problem of today is how to prevent the outbreak of war. But the problems of the escalation of war, and its termination should also be studied since they are intrinsic to the nature of war, and when war does occur to prevent its escalation, and to bring it to an end, is nowadays of primary importance.

Select bibliography

Books

Abshire, David M., Allen, Richard V. (eds), *National Security. Political, Military and Economic Strategies in the Decade Ahead*, Praeger, New York–London 1963.

Acheson, Dean, *Power and Diplomacy*, Harvard Univ. Press, Cambridge, Mass. 1958.

Afheldt, H. et al., *Kriegsfolgen und Kriegsverhütung*, von Weizsäcker (ed.), Hanser, Munich 1971 (2nd ed.).

Ahlander, B. *Krig och fred i atomåldern*, Gebers, Stockholm 1965.

Alcock, Norman Z., *The War Disease*, CPRI Press, Oakville, Ontario 1972.

Andreski, Stanislav, *Elements of Comparative Sociology*, Univ. of Calif. Press, 1965.

Andreski, Stanislav, *Military Organization and Society,* Routledge and Kegan Paul, London, and the Univ. of Calif. Press, 1968 (2nd ed.).

Angell, Norman, *The Great Illusion: A Study of the Relation of Military Power to National Advantage*, Putnam, New York 1910.

Aptheker, Herbert, *American Foreign Policy and the Cold War*, New Century Publishers, New York 1962.

Ardrey, Robert, *African Genesis*, Atheneum, New York 1961.

Ardrey, Robert, *The Territorial Imperative*, Atheneum, New York 1966.

Ardrey, Robert, *Social Contract*, Atheneum, New York 1970.

Arendt, Hannah, *On Violence*, Harcourt, Brace & World Inc., New York 1969.

Arnold, Theodor, *Der revolutionäre Krieg*, Pfaffenhofen 1961.

Aron, Raymond, *On War*, Secker and Warburg, London 1958 (Doubleday Anchor Books, Garden City New York 1959).

Aron, Raymond, *War and Industrial Society*, August Comte Memorial Lecture No.3, Oxford Univ. Press, London 1958.

Aron, Raymond, *Le Grand Debat: Initiation à la stratégie atomique*, Calman-Levy, Paris 1963 (transl. *Einführung in die Atomstrategie*, Kiepenheuer und Witsch, Köln-Berlin 1964).

Aron, Raymond, *Penser La Guerre: Clausewitz*, t.I 'l'Age européen', t.II 'l'Age planetaire', Gallimard, Paris 1976.

Aron, Raymond, *Paix et Guerre entre les nations*, Calman-Levy, Paris 1962 (transl. *Peace and War. A Theory of International Relations*, Weidenfeld and Nicolson, London 1966; Doubleday Co., New York 1966).

Azovtsev, N. N. *V. I. Lenin i sovetskaya voennaya nauka (V. I. Lenin and the Soviet military science)*, Izd. 'Nauka', Moscow 1971.

Baldwin, Hanson W., *Strategy for Tomorrow*, Harper and Row, New York 1970.

Barbera, Henry, *Rich Nations and Poor in Peace and War*, Lexington Books, Lexington, Mass., 1973.

Barnett, Richard J., *Intervention and Revolution*, World, New York 1968.

Barnett, Richard J., *Roots of War. The Men and the Institutions Behind U.S. Foreign Policy*, Penguin Books, Baltimore-Maryland 1971, 1972.

Barringer, Richard E., *War: patterns of conflict*, MIT Press, Cambridge, Mass. 1972.

Bauman, Zygmunt, *Zarys marksistowskiej teorii społeczeństwa* (An outline of the Marxist theory of society), PWN, Warsaw 1964.

Baylis, John, Booth Ken, Garnett John and Williams Phil, *Contemporary Strategy. Theories and Policies*, Croom Helm, London 1975.

Beaton, Leonard, *The Struggle for Peace*, Allen and Unwin, London 1966.

Beaufre, André, *Introduction à la Stratégie*, Colin, Paris 1963 (transl. *Introduction to Strategy*, Faber and Faber, London 1965).

Beaufre, André, *Dissuasion et Stratégie*, Colin, Paris 1974.

Beaufre, André, *Stratégie Pour Demain*, Plon, Paris 1972 (trans. *Strategy for Tomorrow*, Crane, Russak, New York 1974).

Beaufre, André, *La Guerre Révolutionaire. Les formes nouvelles de la guerre*. Fayard, Paris 1972 (transl. *Die Revolutionisierung des Kriegsbildes. Neue Formen der Gewaltanwendung*, Seevald, Stuttgart 1975).

Beaumont, Roger A., Edmonds Martin, (eds), *War in the Next Decade*, Macmillan, London and Basingstoke 1975.

Beitz, Charles R., Herman, Theodore (eds), *Peace and War*, Freeman, San Francisco 1973.

Bell, Coral, *The Conventions of Crisis. A Study in Diplomatic Management*, Oxford Univ. Press, London-Oxford-New York 1971.

Bell, J. Bowyer, *The Myth of the Guerilla. Revolutionary Theory and Malpractice*, Knopf, New York 1971.

Bennett, John C., (ed.), *Nuclear Weapons and the Conflict of Science*, Lutterworth Press, London 1962.

Berkowitz, Leonard, *Aggression: A Social-Psychological Analysis*, McGraw-Hill, New York 1962.

Berkowitz, Leonard, *Roots of Aggression: re-examination of the frustration – aggression hypothesis*, Atherton Press, New York 1969.

Bernard, Jessie, Pear, T. H., Aron, R. *The Nature of Conflict: Studies of the Sociological Aspects of International Tensions*, Unesco, Paris 1957.

Bernard L. L., *War and Its Causes*, Holt, New York, 1944 (2nd ed. 1946).

Bidwell, Shelford, *Modern Warfare: A Study of Men, Weapons and Theories*, Allen Lane, The Penguin Press, London 1973.

Black, Cyril E., Thornton, Thomas P. (eds), *Communism and Revolution*, Princeton Univ. Press, Princeton-New York 1964.

Blackett, P. M. S., *Studies of War*, Oliver and Boyd, Edinburgh-London 1962.

Blainey, Geoffrey, *The Causes of War*, Macmillan, London 1973.

Blix, Hans, *Sovereignty, Aggression and Neutrality*, Almquist and Wiksell, Stockholm 1970.

Bloomfield, L., Leiss, A. *The Control of Local Conflict*. (US Arms Control and Disarmament Agency), Government Printing Office, Washington 1967.

Bloomfield, L., Lincoln, P., Lewis, Amelia C. *Controlling Small Wars. A Strategy for the 1970s*, A. Knopf, New York 1969 (Allen Lane, The Penguin Press, London 1970).

Bohannan, P. (ed.), *Law and Warfare; Studies in the Anthropology of Conflict*, Doubleday, Garden City, New York, 1967.

Borisow, O. B., Koloskov, B. T. *Sino-Soviet Relations*, Progress Publishers, Moscow 1975.

Boulding, K. E., *Conflict and Defense: a General Theory*, Harper, New York 1962.

Bouthoul, Gaston, *Les Guerres: Elements de Polemologie*, Payot, Paris 1953.

Bouthoul, Gaston, *La guerre*, Presses universitaires de France, Paris 1963 (transl. *War*, Walker, New York 1962).

Bramson, L., Goethals, G. W. (eds). *War: Studies from Psychology, Sociology, Anthropology*, Basic Books, New York 1964 (Rev. ed. 1968).

von Bredow, Wilfried (ed.), *Zum Charakter internatioaler Konflikte. Studien aus West- und Osteuropa*, Pahl-Rugenstein, Köln 1973.

Brodie, Bernard, *Strategy in the Missile Age*, Princeton Univ. Press, Princeton, N.J., 1959.

Brodie, Bernard, *Escalation and the Nuclear Option*, Princeton Univ. Press, Princeton, N.J. 1966.

Brodie, Bernard, *War and Politics*, Cassell, London 1974.

Brucan, Silviu *The Dissolution of Power. A Sociology of International Relations and Politics*, Knopf, New York 1971.

Buchan, Alastair (ed.), *Problems of Modern Strategy*, Chatto and Windus, ed.), London 1963.

Buchan, Alastair, *War in Modern Society*, Harper and Row, New York 1968.

Buchan, Alastair, *The End of the Post-war Era. A New Balance of World Power*, Weidenfeld and Nicolson, London 1974.

Buchan, Alastair (ed.), *Problems of Modern Strategy*, Chatto and Windus, London 1970.

Buchan, Alastair, Windsor, Philip *Arms and Stability in Europe*, London 1963.

Bull, Hedley, *The Control of the Arms Race*, The Institute for Strategic Studies, London, New York 1961.

Burnham, Colin, *War or Peace?*, Batsford, London 1972.

Burnham, James, *Containment or Liberation? An Inquiry into the Aims of United States Foreign Policy*, Day, New York 1953.

Burt, Richard, *New Weapons Technologies. Debate and Directions*, Adelphi Papers no. 126, Summer 1976.

Burton, J. W., *Systems, States, Diplomacy and Rules*, Cambridge Univ. Press, Cambridge, Mass., 1968.

Butenko, A. P., *Voina i revolutsiya* (War and revolution), Gospolitizdat, Moscow 1961.

Butterfield, Herbert, Wight, Martin (eds), *Diplomatic Investigations*, Harvard Univ. Press, Cambridge, Mass. 1968.

Calvert, Peter, *Revolution*, Pall Mall Press, London 1970.

Carr, Edward Hallett, *The Twenty Years' Crisis 1919–1939*, Macmillan, London 1951 (Harper and Row, New York 1964).

Carthy, J. D., Ebling, F. J. (eds), *The Natural History of Aggression*, Academic Press, New York 1964.

Cerf, Jay H., Pozen Walter (eds), *Strategy for the 60's*, Praeger, New York 1961. *Civil Violence and the International System*, I–II, IISS Adelphi Papers nos. 82–83, London 1971.

Clark, Robin, *The Science of War and Peace*, Cape, London 1971.

Clarkson, J. D., Cochran T. C. (eds), *War as a social institution: the historian's perspective,* American Historical Association, Columbia Univ. Press, New York 1941.

Claude, Inis L. Jr, *Power and International Relations*, Random House, New York 1962.

von Clausewitz, Carl, *Vom Kriege*, Dümmler, Achtzehnte Ausgabe, Bonn 1972.

von Clauswitz, *On War*. Penguin Books, Harmondsworth, Middlesex, 1968, 1971.

Clough, Ralph N., Barnett, A. Dook, Halperin Morton H., Kahan, Jerome H. *The United States, China, and Arms Control*, The Brookings Institution, Washington 1975.

Coats, Wendell J., *Armed Forces as Power: The Theory of War Reconsidered*, Exposition Press, New York 1966.

Coffey, J. L., *Strategic Power and National Security*, Univ. of Pittsburgh Press, London 1971.

Cohen, S. B., *Geography and politics in a world divided*, Random House, London 1963.

Collins, John M., *Grand Strategy, Principles and Practices*, US Naval Institute, Naval Institute Press, Annapolis, Maryland 1973.

Corning P. A., Corning Constance H., *An evolutionary-adaptive theory of aggression*, American Political Science Association Meeting, Chicago 1971.

Coser, L. A., *The Functions of Social Conflict*, Free Press, New York 1956.

Coser, L. A., *Continuities in the Study of Social Conflict*, Free Press, New York 1967.

Craig, Gordon A., *War, Politics and Diplomacy. Selected Essays*, Praeger, New York 1966.

Crane, Robert Dickson (ed.), *Soviet Nuclear Strategy. A Critical Appraisal*, The Center for Strategic Studies, Georgetown University, Washington 1963.

Crozier, Brian, *The Rebels: A Study of Post-War Insurrections*, Chatto and Windus, London 1960.

Cummins, David E. et al., *Accidental War: Some Dangers in the 1960's*, Ohio State Univ. Press, Columbus 1960.

Dahrendorf, Ralph, *Soziale Klassen und Klassenkonflikt in der industriellen Gesellschaft*, Stuttgart 1957 (trans. *Class and Class Conflict in Industrial Society*, Stanford Univ. Press, Stanford 1959).

Dahrendorf, Ralph, *Essays in the Theory of Society*, Stanford Univ. Press, Stanford 1968.

Davies, James, *Human Nature in Politics*, Wiley, New York 1963.

Davis, Jack, *Political Violence in Latin America*, IISS, Adelphi Papers no. 85, London 1972.

Deborin, G. A., *O kharaktere vtoroi mirovoi voiny* (The character of the Second World War), Voenizdat, Moscow 1960.

Deitschman, Seimour J., *Limited War and American Defense Policy*, The MIT Press, Cambridge, Mass., 1964.

Derevyanko, P., *Revolustiya v voennom dele* (The revolution in military affairs), Voenizdat, Moscow 1967.

Deutsch, Karl W., *The Analysis of International Relations*, Foundation of Modern Political Science Press, Prentice-Hall Englewood Cliffs, New York 1968.

Dinerstein, H. S., *War and the Soviet Union: Nuclear Weapons and the Revolution in Soviet Military and Political Thinking*, Rev. ed., Praeger, New York 1962.

Dollard, J., Doob, L. W., Miller, N. E., Mowrer, O. H., Pears, R. R. *Frustration and Aggression*, Yale Univ. Press, New Haven 1939.

Dougherty, James E., Pflatzgraff, Robert L. Jr, *Contending Theories of International Relations*, Lippincott, Philadelphia, New York, Toronto 1971.

Dulles, Eleanor Lansing, Crane Robert Dickson (eds), *Détente. Cold War Strategies in Transition*, The Center for Strategic Studies Georgetown Univ., Praeger, New York-Washington-London 1965.

Durbin, E. F. M. et al., *War and Democracy*, Kegan Paul, Tench, Trubner and Co., London 1938.

Earle, Edward Meade (ed.), *Makers of Modern Strategy: Military Thought from Macchiavelli to Hitler*, Princeton Univ. Press, Princeton 1943.

Easton, D., *A Framework for Political Analysis* Prentice-Hall, Englewood Cliffs, NJ, 1965.

Eccles, Henry E., *Military Concepts and Philosophy*, Rutgers Univ. Press, New Brunswick, New Jersey 1965.

Eckstein, Harry Horage (ed.), *Internal War, Problems and Approaches*, The Free Press, New York-London 1964.

Edwards, David V., *International Political Analysis*, Holt, Rinehart, and Winston Inc., New York 1969.

Elliot-Bateman, Michael, Ellis, John, Bowden, Tom (eds), *Revolt to revolution. Studies in the 19th and 20th century*, European Experience, Manchester Univ. Press, Bowman and Littlefield, Manchester 1974.

Ellul, Jacques, *Autopsy of Revolution*, Knopf, New York 1971 (transl. from *Autopsie de la Révolution*, Calmann-Lévy, Paris 1969).

Emme, Eugene, (ed.), *The Impact of Air Power*, Princeton, New York 1959.

Erickson, John, *Soviet Military Power*, Royal United Service Institute, London 1971.

Erickson, John, *The Military-Technical Revolution*, Praeger, New York-London 1966.

Etzioni, Amitai, *Winning Without War*, Doubleday and Co., Garden City, New York 1964.

Fairbairn, Geoffrey, *Revolutionary Guerilla Warfare*, Penguin Books, Harmondsworth, Middlesex, 1974.

Falk, Richard A., Mendlowitz, Saul (eds), *Toward a Theory of War Prevention*, World Law Fund, New York 1966.

Falls, Cyril, *A Hundred Years of War*, Duckworth, London 1953.

Fanon, Frantz, *The Wretched of the Earth*, Penguin Books, Harmondsworth, Middlesex, 1967.

Finletter, Thomas K., *Foreign Policy: The Next Phase. The 1960s*, Praeger, New York 1960.

Fleming, D. F., *The Cold War and Its Origins, 1917–1960*, (2 vols.). Allen and Unwin, London 1961.

Foot, M. R. D., (ed.), *War and Society: Historical Essays in honour and memory of J. R. Western, 1928–1971*, Paul Elek, London 1973.

Fornari, Franco, *The Psychoanalysis of War*, Indiana Univ. Press, Blomington and London 1974 (transl. from the Italian).

Foster, Richard B., Beaufre, André, Joshu, Wynfred (eds), *Strategy for the West,* Stanford Research Institute, Macdonald and Jane's, London 1974.

Fox, William T. R., (ed.), *Theoretical Aspects of International Relations*, Univ. of Notre Dame Press, Notre Dame, Ind., 1959.

Frank, Jerome D., *Sanity and Survival. Psychological Aspects of War and Peace*, Random House, New York 1967.

Frankel, Joseph, *International Politics: Conflict and Harmony*, Penguin Books, Harmondsworth, Middlesex, 1969, 1973.

Frankel, Joseph, *International Relations* Oxford Univ. Press, London-Oxford-New York, 2nd ed. 1972.

Fried, Morton, Harris, Marvin, Murphy, Robert, (eds), *War. The Anthropology of Armed Conflict*, Garden City, New York 1968.

Fuller, J. F. C., *A Military History of the Western World*, (3 vols.) Funk and Wagnalls Co., New York 1954, 1955, 1956.

Fuller, J. F. C., *The Conduct of War, 1789–1961*, Rutgers Univ. Press, Brunswick, NJ 1961, (transl. *Die entartete Kunst, Krieg zu führen, 1789–1961*, Köln 1964).

Furniss, Edgar S., *American Military Policy: Strategic Aspects of World Political Geography*, Rinehart and Co., New York 1957.

Gallois, Pierre M., *Stratégie de l'age nucleaire*, Calmann-Levy, Paris 1960.

Gallois, Pierre M., *Paradoxes de la paix*, Paris 1967.

Galula, David, *Counterinsurgency Warfare. Theory and Practice*, Pall Mall Press, London–Dunmow 1964.

Gantzel, Jürgen, *System and Akteur*, Bertelsmann Universitätsverlag, Düsseldorf 1972.

Garnett, John (ed.), *Theories of Peace and Security. A Reader in Contemporary Strategic Thought*, Macmillan, London 1970.

Garnett, John (ed.), *The Defence of Western Europe*, Macmillan, London 1974.

Garthoff, Raymond L., *Soviet Military Doctrine*, Free Press, Glencoe, Ill., 1953 (*How Russia Makes War*, Allen and Unwin, London 1954).

Garthoff, Raymond L., *Soviet Strategy in the Nuclear Age*, Praeger, New York 1958, rev. ed. 1962.

Garthoff, Raymond L. (ed.), *Sino-Soviet Military Relations*, Praeger, New York 1966.

Gavin, James M., *War and Peace in the Space Age*, Harper and Row, New York 1958.

George, A. L., Hall, David K., Simons, William L., *The Limits of Coercive Diplomacy*, Little, Brown and Co., Boston 1971.

Giap, N. V., *People's War, People's Army*, Foreign Languages Publishing House, Peking 1961.

Ginsberg, Robert (ed.), *The Critique of War*, Regnery Co., Chicago 1969.

Ginsburgh, Robert N., *U.S. Military Strategy in the Sixties*, Norton, New York 1965.

Goldmann, Kjell, *International Norms and War Between States: Three Studies in International Politics*, Esselte Studium, Stockholm 1971.

Goldmann, Kjell, *Tension and Detente in Bipolar Europe*, Esselte Studium, Stockholm 1974.

Gorshkov, Sergei G., *Red Star Rising at Sea*, ed. by Herbert Preston, United States Naval Institute, 1974.

Gouré, Leon, Kohler, Foy D., Harvey, Mose L., *The Role of Nuclear Forces in Current Soviet Strategy*, Center for Advanced International Studies, Univ. of Miami, Washington 1974.

Grechko, A. A., *Na Strazhe Mira i Stroitelstva Kommunizma* (On Guard for Peace and the Building of Communism), Voenizdat, Moscow 1971.

Grechko, A. A., *Vooruzhennye Sily SSSR* (The Armed Forces of the U.S.S.R.), Voenizdat, Moscow 1 ed. 1974, 2 ed. 1975.

Grechko, A. A. (ed.) *Liberation Mission of the Soviet Armed Forces in the Second World War*, Progress Publishers, Moscow 1975.

Green, Philip, *Deadly Logic. The Theory of Nuclear Deterrence*, Ohio Univ. Press, Columbus 1966.

Greene, T. N. (ed.), *The Guerilla and How to Fight Him*, Praeger, New York 1962.

Groom, A. J. R., *British Thinking About Nuclear Weapons*, Pinter, London 1974.

Grudinin, I. A., *Dialektika i Voennaya Nauka* (Dialectics and Military Science), Voenizdat, Moscow 1971.

Guevara, Che, *Guerilla Warfare*, Monthly Review Press, New York 1961.

Gurr, Ted Robert, *Why Men Rebel*, Princeton Univ. Press, Princeton 1969.

Haas, Michael, *International Conflict*, The Bobbs-Merville Co., Indianopolis-New York 1974.

Hahlweg, Werner, *Typologie des modernen Kleinkrieges*, Institut für Europäische Geschichte, Wiesbaden 1967.

Hahlweg, Werner, *Lehrmeister des Kleinen Krieges. Von Clausewitz bis Mao Tse-tung und Che Guevara*, Darmstadt 1968a.

Hahlweg, Werner, *Guerilla, Krieg ohne Fronten*, W. Kohlhammer Vg., Stuttgart-Berlin-Köln-Mainz, 1968b.

Hahn, Walter F., Neff, John C. (eds), *American Strategy for the Nuclear Age*, Anchor Books, Doubleday and Co., New York 1960.

Halperin, Morton N., *Contemporary Military Strategy*, Little, Brown, Boston-Toronto 1967.

Halperin, Morton N., *Limited War in the Nuclear Age*, Wiley, New York-London 1963.

Halperin, Morton N., *Defence Strategies for the Seventies*, Little, Brown, Boston 1971.

Hamon, Léo, *La Stratégie contre la Guerre,* Gresset, Paris 1967.

Handelsman, John R., Vasquez, John A., O'Leary, Michael K., Coplin, William D., *Color it Morgenthau: A Data-Based Assessment of Quantitative International Relations Research*, Prince Research Studies, Syracuse Univ., Syracuse-New York 1973.

Haushoffer, Karl, *Weltpolitik von heute*, 'Zeitgeschichte', Verlag und Vertiebs-Gesellschaft, Berlin 1934.

Hermann, Charles F. (ed.), *International Crises: Insights from Behavioral Research*, Free Press, New York 1972.

Hessler, William H., *Operation Survival. America's New Role in World Affairs*, Prentice-Hall, Inc., New York 1949.

Higham, Robin (ed.), *Civil Wars in the Twentieth Century*, The Univ. Press of Kentucky, Lexington 1972.

Hobson, J. A., *Imperialism: A Study*, Allen and Unwin, London 1902.

Hoffmann, Stanley (ed.), *Contemporary Theory in International Relations*, Prentice-Hall, Inc., Englewood Cliffs, New Jersey 1960.

Hoffmann, Stanley (ed.), *The Acceptability of Military Power*, IISS, Adelphi Papers no. 102, London 1973.

Holsti, K. J., *International Politics. A Framework for Analysis*, Prentice-Hall, Inc., New Jersey, 1967, 1972.

Holsti, K. J., *Crisis, Escalation. War*, McGill-Queens Univ. Press, Montreal 1972.

Holsti, K. J., *Decisions on International War and Peace: The Role of Domestic Interests and National Decision System,* Introductory Remarks to the work of the Commission, IPSA IX World Congress, Montreal 1973.

Hopkins, Ramond F., Mansback, Richard W., *Structure and Process in International Politics,* Harper and Row, New York–Evanston–San Francisco–London 1973.

Horowitz, David, *Imperialism and Revolution,* Pelican Books, Harmondsworth, Middlesex 1971.

Horton III, Frank B., Rogerson, Anthony C., Warner III, Edward, L. (eds), *Comparative Defense Policy,* Johns Hopkins Univ. Press, Baltimore-London 1974.

How Wars End ?, Annals of the American Academy of Political and Social Sciences, 1970.

Howard, Michael (ed.), *The Theory and Practice of War*; Essays presented to Captain B. H. Liddell-Hart, Cassell, London 1965.

Howard, Michael (ed.), *Studies in War and Peace,* Marce Temple Smith, London 1970.

Howard, Michael (ed.), *Order and Conflict at Sea in the 1980s,* IISS, Adelphi Papers no. 124, London 1976.

Howard, Michael (ed.), *War in European History,* Oxford Univ. Press, London-Oxford-New York 1976.

Huntington, Samuel P. (ed.) *Changing Patterns of Military Politics,* Free Press, New York 1952.

Huntington, Samuel P., *The Soldier and the State. The Theory and Politics of Civil-Military Relations,* Belknap, Harvard 1957.

Huntington, Samuel P., *The Common Defense: Strategic Programs in National Politics,* Columbia Univ. Press, New York 1961.

Huntington, Samuel P., *Political Order in Changing Societies,* Yale Univ. Press, New Haven 1968.

Huszar, George B. (ed.), *National Strategy in an Age of Revolutions,* Praeger, New York 1959.

Hyde, Douglas, *On Roots of Guerilla Warfare,* The Bodley Head, London-Sydney-Toronto 1968.

Istoricheskii materializm i sotsialnaya filozofiya sovremennoi burzhuazii (Historical materialism and the social philosophy of the modern bourgeoisie), Akademiya Nauk SSSR, Institut Filosofii, Izd. Sotsialno-Ekonomicheskoi Literatury, Moscow 1960.

Ivanova, I. M., *Mirnoe sosushchestvovanie i krizis vneshnepoliticheskoi ideologii imperializma* (Peaceful coexistence and the crisis of the imperialist ideology of foreign policy), Izd. 'Mezhdunarodnye Otnosheniya', Moscow 1965.

Jacobsen, C. G., *Soviet Strategy–Soviet Foreign Policy,* R. Maclehose and Co., The Univ. Press, Glasgow 1974 (2nd ed.).

Janowitz, Morris, *The Professional Soldier,* Free Press, Glencoe, Ill., 1960.

373

Janowitz, Morris, *Sociology and the military establishment (in collab. with Roger W. Little), SAGE, Beverly Hills, London 1972 (3rd ed.).*

Janowitz, Morris, *Military Conflict.* Essays in International Analysis of War and Peace, SAGE, Beverly Hills, London 1975.

Johnson, Chalmers, *Revolution and the Social System*, Stanford Univ. Press, Stanford 1964.

Johnson, Chalmers, *Revolutionary Change*, Little, Brown, and Co., London 1966.

Johnson, Chalmers, *Autopsy on People's War*, Univ. of California Press, Berkeley, Los Angeles, London 1973.

Jouvenel, Bertrand de, *On Power: Its Nature and the History of its Growth*, Viking, New York 1949.

Kahan, Jerome H., *Security in the Nuclear Age. Developing U.S. Strategic Arms Policy*, The Brooking Institution, Washington 1975.

Kahn, Herman, *On Thermonuclear War*, Princeton Univ. Press, Princeton 1960.

Kahn, Herman, *On Escalation: Metaphors and Scenarios*, Praeger, New York 1965.

Kahn, Herman, *Thinking About the Unthinkable*, Horizon Press, New York 1962.

Kaplan, Morton A., *System and Process in International Politics*, Wiley, New York 1957.

Kaplan, Morton A. (ed.) *Strategic Thinking and Its Moral Implications*, The Univ. of Chicago, Chicago 1973.

Kaplan, Morton A. (ed.) *NATO and Dissuasion*, The Univ. of Chicago, Chicago 1974.

Kaplan, Morton A., *Great issues of international politics: the international and national systems*, Alsine, Chicago 1974.

Kardelj, Edward, *Socialism and War. A Survey of Chinese Criticism of the Policy of Coexistence*, Methuen and Co., London 1961.

Karenin A., *Filosofiya politicheskogo nasiliya* (The philosophy of the political violence), Izd. 'Mezhdunarodnye Otnosheniya', Moscow 1971.

Kaufmann, W. W. (ed.), *Military Power and National Security*, Princeton Univ. Press, Princeton 1956.

Kecskemeti, Paul, *Strategic Surrender: The Politics of Victory and Defeat*, Stanford Univ. Press, Stanford 1958.

Kelly, George A., Miller, Linda B., *Internal War and International System. Perspectives on Method*, The Center of International Affairs, Harvard Univ., 1969.

Kende, Istvan, *Local Wars in Asia, Africa, and Latin America 1945–1969*, Center for Afro-Asian Research of the Hungarian Academy of Science, Budapest 1972.

King-Hall, Stephen, *Defense in the Nuclear Age,* Victor Gollancz, London 1958.

Kingston-McCloughry, E. J., *Global Strategy*, Cape, London 1957, Praeger, New York 1957.

Kingston-McCloughry, E. J., *Defence: Policy and Strategy*, Praeger, New York 1960.

Kingston-McCloughry, E. J., *The Spectrum of Strategy: A Study of Policy and Strategy in Modern War*, Cape, London 1964.

Kintner, William R., *Peace and the Strategy Conflict*, Praeger, New York-Washington-London 1967.

Kintner, William R., and Scott, Harriet Fast (eds), *The Nuclear Revolution in Soviet Military Affairs*, Univ. of Oklahoma Press, Oklahoma 1968.

Kintner, William R., and Pfaltzgraff, Robert L. Jr. (eds), *Strategy and Values: Selected Writings of Robert Strausz-Hupé*, Lexington Books, Lexington, Mass. 1974.

Kissinger, Henry A., *Nuclear Weapons and Foreign Policy*, Harper, New York 1957.

Kissinger, Henry A., *The Necessity for Choice*, Chatto and Windus, London 1960, Harper and Bros., New York 1967.

Kissinger, Henry A., *The Troubled Partnership*, McGraw-Hill, New York–London–Toronto 1965.

Kissinger, Henry A., *American Foreign Policy*. Three Essays, Norton, New York 1969.

Kjellén, Rudolf, *Der Staat als Lebensform*, Hirzel Vg., Leipzig 1917.

Knorr, Klaus, *On the Uses of Military Power in the Nuclear Age*, Princeton Univ. Press, Princeton 1966.

Knorr, Klaus, *Military Power and Potential*, Heath, Lexington, Mass. 1970.

Knorr, Klaus, *Power and Wealth. The Political Economy of the International Power*, Basic Books, New York 1973.

Knorr, Klaus, and Rosenau, James N. (eds), *Contending Approaches to International Politics*, Princeton Univ. Press, Princeton 1969.

Kohler, F. F., *Soviet Strategy for the Seventies: from cold war to peaceful co-existence*, Center for Advanced International Studies, Miami Univ., Cort Gables, Florida 1973.

Konstantinov, F. V., Sladkovskii, M. I. (chief ed.), et al., *Kritika teoreticheskikh kontseptsii Mao Tsze-duna* (The critique of the philosophical conceptions of Mao Tse-tung), Izd. 'Mysl', Moscow 1970.

Kotzsch, Lothar, *The Concept of War in Contemporary History and International Law*, Droz, Geneve 1956.

Kovalev, A. M. (ed.), *Sovremennaya epokha i mirovoi revolutsionnyi protsess*, (The contemporary epoch and the world revolutionary movement), Izd. Moskovskogo Universiteta, Moscow 1970.

Krippendorff, Ekkehart, (ed.), *Friedensforschung*, Kiepenheuer – Witsch, Köln-Berlin 1968.

Lang, Kurt, *Military Institutions and the Sociology of War*, SAGE Publications, Beverly Hills, London 1972.

Lapp, Ralph E., *Kill and Overkill. The Strategy of Annihilation*, Basic Books, New York 1962.

Larson, Arthur (ed.), *A Warless World*, McGraw-Hill Book Co., New York-Toronto-London, 1962, 1963.

Larson, Arthur (ed.), and Kaplan, Abraham, *Power and Society*, Yale Univ. Press, New Haven 1950.

Lea, Homer, *Day of the Saxon*, Harper, New York 1912.

Legault, Albert, Lindsey, George, *Dynamik der nuklearen Gleichgewichts*, Metzner, Frankfurt M., 1973.

Leiden, C., Schnitt, K. (eds.), *The Politics of Violence: Revolution in the Modern World*, Englewood Cliffs, Prentice-Hall, New Jersey 1968.

Lenin, V. I., *State and Revolution*, International Publishers, New York 1932.

Lenin, V. I., *Imperialism, The Highest State of Capitalism*, International Publishers, New York 1939.

Leonard, Roger Ashley, *A Short Guide to Clausewitz On War*, Weidenfeld and Nicolson, London 1967.

Lepavsky, A., et al. (eds.), *The search for world order: studies by students and colleagues of Quincy Wright*, Appleton-Century-Crofts, New York 1971.

Levi, Werner, *International Politics, Foundations of the System*, Univ. of Minnesota Press, Minneapolis 1974.

Levine, Robert A., *The Arms Debate*, Harvard Univ. Press 1963.

Liddell-Hart, B. H., *Strategy. The Indirect Approach*, Rev. ed. Faber and Faber, London 1967.

Liddell-Hart, B. H., *Deterrent and Defence*, Stevens, London 1960.

Liddell-Hart, B. H., *Why don't we learn from history?* Allen and Unwin, London 1972.

Lider, Julian, *Pogadanki o dialektyce i materializmie* (Studies on dialectics and materialism), PiW, Warsaw (6th ed.) 1952a.

Lider, Julian, *Czynniki zwyciestwa* (Factors of victory), Wyd MON, Warsaw 1952b.

Lider, Julian, *NATO. Szkice o historii i doktrynie*, (NATO. Studies on history and doctrine), Wyd MON, Warsaw 1962.

Lider, Julian, *Doktryna wojenna Stanów Zjednoczonych 1945–1962* (Military doctrine of the US 1945–1962), KiW, Warsaw 1963.

Lider, Julian, *Doktryna wojenna i polityka wojskowa Wielkiej Brytanii* (The military doctrine and military politics of Great Britain), Wyd MON, Warsaw 1964.

Lider, Julian, *Zachód a rozbrojenie* (The West and the disarmament), ZAP, Warsaw 1966a.

Lider, Julian, *Myśl polityczno-wojskowa NRF 1949–1965* (The political military thought of GFR), KiW, Warsaw 1966b.

Lider, Julian, *Wojny i doktryny wojenne XX wieku* (Wars and doctrines of the XXth century), W. P. Warsaw 1966c.

Lider, Julian, *Co dalej NATO?* (The future of NATO), Czytelnik, Warsaw 1967.

Lider, Julian, *Problemy integra cji wojskowej na Zachodzie* (The problems of the military integration in the West), Instytut Zachodni Warsaw-Poznań 1968.

Lider, Julian, *Ludzie i doktryny* (Men and doctrines), Iskry, Warsaw 1969a.

Lider, Julian, *Pax Americana*, KiW, Warsaw 1969b.

Lider, Julian, *Nowe tendencje w myśli polityczno-wojskowej NRF 1966–1969* (New trends in the political-military thought of GFR), Instytut Slaski-Ossolineum, Wroclaw-Warsaw 1971.

Little, Richard, *External Involvements in Civil Wars*, Robertson, London 1975.

London, Kurt, *The Soviet Impact on World Politics*, Hawthorn Books, New York 1974.

Lorenz, Konrad, *On Aggression*, Harcourt, Brace and World, New York 1966.

Lowe, Georg E., *The Age of Deterrence*, Little, Brown, Boston-Toronto 1964.

Luard, Evan, *The Cold War. A Reappraisal*, Thomas and Hudson, London 1964.

Luard, Evan, *Conflict and Peace in the International System*, Little, Brown, Boston 1968.

Luard, Evan (ed.), *The International Regulation of Civil Wars*, Thomas and Hudson, London 1972.

Ludendorff, Erich von, *Der totale Krieg*. Ludendorff Vg., München 1937.

Mackenzie, W. J., *Politics and Social Science*, Penguin Books, Baltimore 1967.

Mackinder, Haldorf, *The Democratic Ideals and Reality*, Norton, New York 1962.

Mahan, Alfred Thayer, *The Influence of Sea Power Upon History, 1660–1783* Little, Brown, and Co., Boston 1867; Hill and Wang, New York 1960.

Makhalow, V. S., Beshentsev, A. V. (eds.), *Voina, Istoriya, Ideologiya*, Izd. Politicheskoi Literary, Moscow 1974.

Mao Tse-Tung, *On the Protracted War*, Foreign Language Press, 1954.

Marksistskaya i burzhuaznaya sotsiologiya segodnya (The Marxist and bourgeois modern sociology), Akademiya Nauk SSR, Institut Filosofii, Izd. 'Nauka'. Moscow 1964.

Marksizm–Leninizm o voine i armii, Voenizdat (1—5 eds.), Moscow 1957–68.

Marxism–Leninism on War and Army, Progress Publishers, Moscow 1972.

Martin, L. W., *Arms and Strategy*, Weidenfeld and Nicolson, London 1973.

Marwick, Arthur, *War and Social Change in the Twentieth Century*, Macmillan, London and Basingstoke 1974.

McArdle Kelleher, Catherine, *Political-Military Systems, Comparative Perspectives Research Progress Series on War, Revolution, and Peace-keeping*, SAGE, Beverly Hills, London 1974.

McClelland, C. A. (ed.), *Nuclear Weapons, Missiles, and Future War: Problem for the Sixties*, Chandler, San Francisco, 1960.

McClelland, C. A., *Theory and International System*, Macmillan, New York 1966.

McDougall, William, *An Introduction to Social Psychology* (14 ed.), Barns and Noble, New York 1960.

MacGwire, Michael, *Maritime Strategy and the Super-Powers*, IISS, Adelphi Papers, No. 123, London 1976.

McNamara, Robert S., *The Essence of Security*, Harper and Row, New York 1968.

McNeil, Elton B. (ed.), *The Nature of Human Conflict*, Englewood Cliffs, Prentice-Hall, New York 1965.

Melman, Seymour, *The Peace Race*, Ballantine Books, New York 1961.

Melman, Seymour, *A Strategy for American Security*, Lee Offset, Inc., New York 1963.

Melzer, Jehuda, *Concepts of Just War*, Sijthoff, Leyden 1975.

Metodologicheskie problemy voennoi teorii i praktiki, Voenizdat, Moscow 1969 (2nd ed.).

Mezhdunarodnye Konflikty (International conflicts), Akademiya Nauk SSSR, Institut Mirovoi Ekonomiki i Mezhdunarodnykh Otnoshenii, Izd. Mezhdunarodnye Otnosheniya Moscow 1972.

Middleton, Drew, *Can America Win the Next War?*, Charles Scribner's Sons, New York 1975.

Midlarsky, Manus I., *On War. Political Violence in the International System*, The Free Press, New York 1975.

Miksche, Ferdinand O., *Atomic Weapons and Armies*, London 1955.

Miksche, Ferdinand O., *La fallité de la stratégie atomique*, Paris 1958.

Miksche, Ferdinand O., *Kapitulation ohne Krieg, 1970–1980*, Stuttgart 1965.

Miksche, Ferdinand O., *Vom Kriegsbild*, Seewald Vg., Stuttgart 1976.

Miller, Linda B., *World Order and Local Disorder. The United Nations and Internal Conflicts*, Princeton Univ. Press, Princeton, N.J. 1967.

Millis, Walter, *A World Without War*, Center for the Study of Democratic Institutions, Santa Barbara, Calif., 1961.

Mills, C. Wright, *The Power Elite*, Oxford Univ. Press, New York 1956.

Mills, C. Wright, *The Causes of World War III*, Ballantine Books, New York 1963.

Milovidov, A. S., Kozlov, V. G. (eds), *Filosofskoe nasledie V. I. Lenina i problemy sovremennoi voiny* (The philosophical heritage of V. I. Lenin and problems of contemporary war), Voenizdat, Moscow 1972.

Mitchell, William, *Winged defense*, Putman, New York 1925.

Modelski, George, *Principles of World Politics*, The Free Press, New York 1972.

Momboisse, R. M., *Riots, Revolts, and Insurrections*, Thomas, Springfield, Ill., 1967.

Montagu, M. F. (ed.), *Man and Aggression*, Oxford Univ. Press, New York 1969.

Morgenthau, Hans J., *In Defense of the National Interest*, Knopf, New York 1951.

Morgenthau, Hans J., *Politics in the Twentieth Century*, Univ. of Chicago Press Chicago 1962.

378

Morgenthau, Hans J., *Scientific Man vs. Power Politics*, Univ. of Chicago Press, Chicago 1946, Phoenix 1965.

Morgenthau, Hans J., *Politics Among Nations,* Knopf, New York 1967, 4th ed.

Morris, Desmond, *The Human Zoo*, McGraw-Hill, New York 1969.

Morris, Desmond, *Imperial Animal,* Holt, Rinehart and Winston, New York 1971.

Moulton, Harland B., *From Superiority to Parity. The United States and the Strategic Arms Race 1961-1971*, Greenwood Press Inc., Westport-Connecticut-London.

Moulton, J. L., *Defence in a Changing World*, Eyre and Spottiswoode, London 1954.

Muir, Richard, *Modern Political Geography*, Macmillan, London and Basingstoke, 1975.

Mulley, F. W., *Politics of Western Defence*, Thames and Hudson, London 1963.

Murray, Thomas E., *Nuclear Policy for War and Peace*, World, Cleveland, Ohio 1960.

Naroll, R., Bullough, V. L., Naroll, F., *Military Deterrence in History: A Pilot Cross-historical Survey*, State Univ. of New York Press, Albany, New York 1974.

Nash, Henry T., *Nuclear Weapons and International Behavior*, Sijthoff, Leyden 1975.

Nauchno-Tekhnicheskii Progress i Revolutsiya v Voennom Dele (Scientific-Technical Progress and the Revolution in Military Affairs), Voenizdat, Moscow 1973.

Nef, John U., *War and Human Progress: an essay on the rise of industrial civilization*, Harvard Univ. Press, Cambridge 1952.

Nettleship, Martin A., Dalegivens, R., Nettleship, Andersson, (eds), *War, Its Causes and Correlates*, Mounton Publishers, The Hague-Paris 1975.

Neumann, J. V., Morgenthau, O., *Theory of Games and Economic Behavior*, Princeton Univ. Press 1964 (2nd ed.).

New Dynamics in National Strategy. The Paradox of Power. Foreword by General Maxwell D. Taylor, Crowell, New York 1975.

Nicholson, Michael, *Conflict Analysis*, The English Universities Press, London 1970.

Niezing, Johan, *Sociology, War and Disarmament*, Studies in Peace Research, Rotterdam Univ. Press, 1970.

Nikolskii, N. M., *Osnovnoi vopros sovremennosti* (The basic problem of the present time), Voenizdat, Moscow 1964.

Noel-Baker, Philip, *The Arms Race*, Calder, London 1958.

Northedge, F. S., Donelan, M. S., *International Disputes: The Political Aspects*, Europa Publications, London 1971.

Obermann, Emil (ed.), *Verteidigung der Freiheit*, Stuttgarter Verlagskontor, Stuttgart 1966.

O'Neill, Robert (ed.), *The Strategic Nuclear Balance*, The Australian National University, Australia 1975.

Osanka, Franklin M., (ed.), *Modern Guerilla Warfare: Fighting Communist Guerilla Movements 1945–1961*, Free Press of Glencoe, New York 1962.

Osgood, Charles, *An Alternative to War or Surrender*, Univ. of Illinois Press, 1962.

Osgood, Robert Endicott, *Limited War. A Challenge to American Strategy*, The Univ. of Chicago Press, Chicago 1957.

Osgood, Robert Endicott, *NATO. The Entangling Alliance*, The Univ. of Chicago Press, Chicago 1961.

Osgood, Robert Endicott, et. al., *America and the World*, Hopkins, Baltimore-London 1970.

Osgood, Robert Endicott, and Tucker, Robert W. *Force, Order, and Justice*, Hopkins, Baltimore 1967.

Palmer, Norman D. (ed.), *Design for International Relations Research: Scope, Theory, Methods, and Relevance*, The American Academy of Political and Social Science, Philadelphia 1970.

Paret, Peter, *Clausewitz and the State,* Oxford Univ. Press, Clarendon Press, Oxford 1976.

Paret, Peter, Shy, John W., *Guerillas in the 1960s*, Praeger, New York 1962.

Parkinson, Roger, *Clausewitz: A Biography,* Stein and Day, New York 1971.

Pear, T. H. (ed.), *Psychological Factors of Peace and War*, Hutchinson and Co., London 1950.

Peeters, Paul, *Massive Retaliation. The Policy and Its Critics,* Regnery, Chicago 1959.

Picht, Werner, *Vom Wesen des Krieges und vom Kriegswesen der Deutschen,* Stuttgart 1952.

Pomeroy, William J. (ed.), *Guerilla Warfare and Marxism,* Lawrence and Wishart, London 1968.

Possony, Stefan T., *A Century of Conflict, Communist Techniques of World Revolution,* Regnery, Chicago 1953 (transl. *Jahrhundert des Aufruhrs,* Isar Vg., Munich 1956).

Power, Thomas S., *Design for Survival,* New York 1964–5.

Power at Sea, I. The New Environment; II. Super-powers and Navies; III. Competition and Conflict, IISS, Adelphi Papers nos. 122–4, London 1976.

Prescott, J. R. V., *The Political Geography of the Oceans,* David and Charles, London-Vancouver 1975.

Problems of Modern Strategy, With a foreword by Alastair Buchan, Chatto and Windus, London 1970.

Problems of Modern Strategy, I–II, IISS, Adelphi Papers nos. 54–5, 1969.

Problemy voiny i mira. Kritika sovremennykh burzhuaznykh sotsialnofilosof-skikh kontseptsii, Izd. 'Mysl' Moscow 1967; Problems of war and peace. A Critical Analysis of Bourgeois Theories, Progress Publishers, Moscow 1972.

Prokop'ev, N. P., *O voine i armii,* Voenizdat, Moscow 1965.

Pruitt, Dean G., Snyder, Richard C. (eds), *Theory and Research on the Causes of War,* Prentice-Hall, New Jersey, 1969.

Puchala, Donald James, *International Politics Today,* Dodd, Mead, New York-Toronto 1971.

Pukhovskii, N. V., *O mire i voine,* Voenizdat, Moscow 1965.

Quester, George H. (ed.), *Power, Action, and Interaction. Readings in International Politics,* Little, Brown, Boston 1971.

Randle, Robert, *The Origins of Peace: A Study of Peacemaking and the Structure of Peace Settlements* Free Press, New York 1973.

Rapoport, Anatol, *Fights, Games, and Debates,* Univ. of Michigan Press 1960.

Rapoport, Anatol, *Strategy and Conscience,* Harper and Row, New York 1964.

Rapoport, Anatol, *Conflict in man-made environment,* Penguin Books-Pelican, Harmondsworth, Middlesex, 1974.

Ratzel, Friedrich, *Antrophogeographie,* Englehorn, Stuttgart 1899 (2nd ed.).

Ratzenhover, Gustav, *Die Staatswehr. Wissenschaftliche Untersuchung der öffentlichen Wehrangelegenheiten,* Stuttgart 1881.

Raven, Wolfram von (ed.), *Armee gegen den Krieg,* Stuttgart 1966.

Reinhardt, George C., *American Strategy in the Atomic Age,* Univ. of Oklahoma Press, Norman 1955.

Reves, Emery, *The Anatomy of Peace,* Harper and Row, New York 1945.

Richardson, Lewis F., *Statistics of Deadly Quarrels,* Stevens, London 1960.

Richardson, Lewis F., *Arms and Insecurity: A Mathematical Study of the Causes and Origin of War,* Ed. by Nicholas Rashevsky and Ernesto Trucco, Boxwood Press, Pittsburgh 1960.

Rosecrance, R. N. (ed.), *The Dispersion of Nuclear Weapons: Strategy and Politics,* Columbia Univ. Press, New York 1964.

Rosecrance, Richard, *International Relations: Peace or War?,* McGraw-Hill Co., New York 1973.

Rosenau, James N. (ed.), *International Aspects of Civil Strife,* Princeton Univ. Press, Princeton 1964.

Rosenau, James N. (ed.), *International Politics and Foreign Policy,* rev. edition, The Free Press, New York; Collier-Macmillan, London 1969.

Rosinski, Herbert, *The Evolution of the Conduct of War and Strategic Thinking,* Naval War College, Newport 1955.

Röling, Bert V. A., *Einführung in die Wissenschaft von Krieg und Frieden,* Neukirchener Vg., Neukirchen-Vluyn 1960.

Ruehl, Lothar, *Machtpolitik und Friedensstrategie,* Hoffmann und Campe, Hamburg 1974.

Ruge, Friedrich, *Politik, Militär, Bündnis,* Stuttgart 1963.

Ruge, Friedrich, *Seemacht und Sicherheit,* Bernard und Graefe, Frankfurt M. 1968.

Ruge, Friedrich, *Politik und Strategie. Strategisches Denken und politisches Handeln,* Bernard und Graefe, Frankfurt M., 1967.

Russett, Bruce M. (ed.), *Peace, War, and Numbers,* SAGE, London 1972.

Russett, Bruce M. (ed.), *Power and Community in World Politics,* Freeman, San Francisco 1974.

Rybkin, Y., *Voina i politika* (War and politics), Voenizdat, Moscow 1959.

Rybkin, Y., *Voina i politika v sovremennuyu epokhu* (War and politics in the contemporary epoch), Voenizdat, Moscow 1973.

Schelling, Thomas C., *The Strategy of Conflict,* Harvard Univ. Press, Cambridge, Mass., 1960.

Schelling, Thomas C., *Arms and Influence,* Yale Univ. Press, New Haven 1966.

Schelling, Thomas C., Halperin, Morton H., *Strategy and Arms Control,* Twentieth Century Fond, New York 1961.

Schilling, Warner, Hammond, Paul, Snyder, Glenn, *Strategy, Politics and Defence Budgets,* Columbia Univ. Press, New York 1962.

Schmidt, Helmut, *Verteidigung oder Vergeltung,* Seewald, Stuttgart-Dagerloch 1961.

Schmidt, Helmut, *Strategie des Gleichgewichts,* Seewald, Stuttgart-Dagerloch 1969.

Schramm, Wilhelm, *Clausewitz, Leben und Werk,* Bechtle, Esslingen am Neckar 1976.

Schwarz, Urs, *Strategie: Gestern, Heute, Morgen,* Econ, Düsseldorf 1965.

Schwarzenberger, G., *Power Politics,* Stevens, London 1964 (3rd ed.).

Seabury, Paul (ed.), *Balance of Power,* Chandler, San Francisco 1965.

Senghaas, Dieter, *Abschreckung und Frieden. Studien zur Kritik organisierter Friedlosigkeit,* Europa Verlagsanstalt, Fischer Taschenbuch V., Frankfurt M. 1969/1972.

Senghaas, Dieter, *Rüstung und Militarismus,* Suhrkamp, Frankfurt M., 1972.

Seversky, Alexander P. de, *America: Too Young to Die!* McGraw-Hill, New York-Toronto-London 1961.

Shinn, Robert I., *Wars and Rumours of Wars,* Abingdon Press, Nashville, Tenn. 1972.

Simmel, G., *Conflict,* Free Press, New York 1955.

Singer, David J. *Deterrence, Arms Control, and Disarmament: Toward a Synthesis in National Security Policy,* Ohio Stage Univ. Press, Columbus 1962.

Singer, David J. (ed.), *Quantitative International Politics: Insights and Evidence,* The Free Press, London 1968.

Singer, David J., Small, Melvin *The Wages of War 1816–1965. A Statistical Handbook,* Wiley, New York-London-Sydney-Toronto 1972.

Sino-Soviet Dispute, documented and analysed by Hudson, George F., Lowenthal, Richard and MacFarquhar, Roderick, Praeger, New York 1961.

Slessor, John, *The Great Deterrent,* Praeger, New York 1957.

Slessor, John, *Strategy for the West,* Cassell, London 1954.

Slessor, John, *What Price Coexistence? A Policy for the Western Alliance,* Praeger, New York 1961, Cassell, London 1962.

Smith, Dale O., *U.S. Military Doctrine,* Duell, Sloan and Pearce, New York 1955. Little, Brown, London 1955.

Snyder, Glenn H., *Deterrence and Defence. Toward a Theory of National Security,* Princeton Univ. Press, Princeton N.J. 1961.

Sokol, Anthony, E., *Seapower in the Nuclear Age,* Public Affairs Press, Washington 1961.

Sokolovskii, V. D. (ed.), *Voennaya strategia* (Military strategy), Voenizdat, Moscow 1968 (3rd ed.).

Sorel, Georges, *Reflections on Violence,* Free Press, New York 1950; Collier Books, New York 1961.

Sorokin, Pitirim A., *Contemporary sociological theories,* Harper, New York 1928.

Sorokin, Pitirim A., *Social and Cultural Dynamics,* American Book, New York 1937 (3 vols.).

Sotsiologicheskie problemy mezhdunarodnykh otnoshenii (Sociological problems of the international relations), Izd. 'Nauka', Moscow 1970.

Sovetskaya vneshnaya politika i evropeiskaya bezopasnost (The Soviet foreign policy and the European security), Izd. 'Mezhdunarodnye Otnosheniya', Moscow 1972.

Spanier, John, *Games Nations Play: Analysis of International Politics,* Nelson, London 1972.

Speier, Hans, *Social Order and the Risks of War, Papers in Political Sociology,* The MIT Press, Cambridge, Mass., 1952.

Speier, Hans, *Force and Folly. Essays on Foreign Affairs and the History of Ideas,* MIT Press, Cambridge, Mass., 1969.

Steinhoff, Johannes, *Wohin treibt die NATO? Probleme der Verteidigung Westeuropas,* Hoffmann und Campe, Hamburg 1976.

Steinmetz, S. R., *Soziologie des Krieges,* Barth, Leipzig 1929.

Sterling, Richard W., *Macropolitics, International Relations in a Global Society,* Knopf, New York 1974.

Strachey, John, *On the Prevention of War,* Macmillan, London 1962.

Strausz-Hupé, Robert, *Geopolitics: The Sruggle for Space and Power,* Putnam, New York 1942.

Strausz-Hupé, Robert, Possony, Stefan T., *International Relations,* McGraw-Hill, New York 1954.

Strausz-Hupé, Robert, et al., *Protracted Conflict,* Harper, New York 1959.

Strausz-Hupé, Robert, Kintner, William R., Possony, Stefan T., *A Forward Strategy for America,* Harper, New York 1961.

Strausz-Hupé, Robert, Dougherty, James E., Kintner, William R., *Building the Atlantic World,* Harper and Row, New York-Evanston-London 1963.

Tanter, Raymond, Ullman, Richard H. (eds), *Theory and Policy in International Relations,* Princeton Univ. Press, Princeton, N.J., 1972.

Tarr, David W., *American Strategy in the Nuclear Age,* Macmillan, New York 1967.

Taylor, Maxwell, *The Uncertain Trumpet,* Harper, New York 1960.

Taylor, Maxwell, *Responsibility and Response,* Harper and Row, New York-London 1967.

Taylor, Maxwell, *Precarious Security,* Norton, New York 1976.

Thompson, Dennis L. (ed.), *Politics, Policy, and Natural Resources,* The Free Press, New York 1972.

Thompson, Robert, *Defeating Communist Insurgency,* Chatto and Windus, London 1967.

Thompson, Robert, *Revolutionary War in World Strategy, 1945–1969,* Taplinger, New York 1970.

Timasheff, Nicholas S., *War and revolutions,* Sheed and Ward, New York 1965.

Toynbee, Arnold, *A Study of History,* An abridgment in 2 vols. by D. C. Somervell, Oxford Univ. Press, New York 1946, 1957.

Triska, Jan F., Finley, David D., *Soviet Foreign Policy,* Macmillan, New York 1968, Collier-Macmillan, London 1968.

Tucker, Robert W., *The Just War,* Hopkins, Baltimore 1960.

Turner, Gordon B., Challener, Richard D. (eds), *National Security in the Nuclear Age. Basic Facts and Theories,* Praeger, New York 1960.

Turney-High, H. H., *Primitive War: Its Practice and Concepts,* Univ. of South Carolina Press, Columbia 1949.

Twining, Nathan, *Neither Liberty Nor Safety,* Holt, Rinehart and Winston, New York 1966.

Ulam, Adam B., *Expansion and Coexistence. The History of Soviet Foreign Policy 1917–1967,* Praeger, New York 1968.

Vagts, A., *A History of Militarism,* Meridian Books, New York 1959 (2nd ed.).

Vigor, P. H., *The Soviet View on War, Peace, and Neutrality,* Routledge and Kegan Paul, London-Boston 1975.

Vilmar, Fritz, *Rüstung und Abrüstung in Spätkapitalismus,* Europäische Verlagsanstalt, Frankfurt M., 1970 (5th ed.).

Voennaya sila i mezhdunarodnye otnosheniya (Military force and international relations), Akademiya Nauk SSSR, Institut Mirovoi Ekonomiki i Mezhdunarodnykh Otnoshenii, Izd. 'Mezhdunarodnye Otnosheniya', Moscow 1972.

Wallach, Jehuda L., *Kriegstheorien. Ihre Entwicklung im 19. und 20. Jahrhundert,* Bernard und Graefe Vg. für Wehrwesen, Frankfurt M., 1972.

Wallensteen, Peter, *Structure and War. On International Relations 1920–1968,* Publications of the Political Science Association in Uppsala, Stockholm 1973.

Waltz, Kenneth N., *Man, the State and War; a Theoretical Analysis,* Columbia Univ. Press, New York-London 1959.

Waskow, Arthur, J., *The Limits of Defence,* Doubleday Co., Garden City, New York 1962.

Wegley, Russell F., *The American Way of War. A History of United States Military Strategy and Policy,* Macmillan, New York 1973.

Wells, Donald A., *The War Myth,* Pegasus, New York 1967.
Whetton, Lawrence L. (ed.), *The Political Implications of Soviet Military Power,* Macdonald and Jane's, London 1977.
White, Wilbur W., *Political Dictionary,* World, Cleveland, New York 1947.
Wolfe, Thomas W., *Soviet Strategy at the Crossroads,* Harvard Univ. Press, Cambridge, Mass., 1964.
Wolfers, Arnold, *Discord and Collaboration,* Hopkins, Baltimore 1962.
Wood, David, *Conflict in the Twentieth Century,* IISS, Adelphi Papers no. 48, London 1968.
Wright, Quincy, *A Study of War,* The Univ. of Chicago Press, Chicago 1942; 2nd ed., with a commentary on war since 1942, Chicago 1965.
Wright, Quincy, *The Study of International Relations,* Appleton, New York 1955.
Wylie, J. C., *Military Strategy: A General Theory of Power Control,* Rutgers Univ. Press, New Brunswick 1967.
Yepishev, A. A., *Ideologicheskaya bor'ba po voennym voprosam* (The ideological struggle in the field of military problems), Voenizdat, Moscow 1974.
Yin, John, *Sino-Soviet dispute on the Problem of War,* Nijhoff, The Hague 1971.
Young, Oran R., *The Intermediaries: Third Parties in International Crisis,* Princeton Univ. Press, Princeton, N.J., 1967.
Young, Oran R., *The Politics of Force: Bargaining During International Crisis* Princeton Univ. Press, Princeton, N.J., 1968.

Articles

Amann, Peter, 'Revolution: A redefiniton', *Political Science Quarterly,* March 1962.
Amme, Cpt., 'Changing Nature of Power' *U.S. Naval Institute Proceedings,* 1963:3.
Andreski, Stanislav, 'Evolution and War', *Science Journal,* Jan. 1971.
Arendt, Hannah, 'Reflections on Violence', *Journal of International Affairs,* 1969:1.
Aron, Raymond, 'Clausewitz's Conceptual System', *Armed Forces and Society,* Autumn 1974.
Aron, Raymond, 'La notion de rapport de forces a-t-elle encore un sens à l'ère nucleaire?', *Défense nationale,* January 1976.
Atkeson, Edward B., 'Hemispheric Denial: Geopolitical Imperatives and Soviet Strategy', *Strategic Review,* Spring 1976.
Atkinson, Alexander, 'Chinese Communist Strategic Thought, The Strategic Premise of Protracted War', *Royal United Service Institution Journal,* March 1973.

Atkinson, Alexander, 'Social War – the Death of Classicalism in Contemporary Strategic Thought?', *Royal United Service Institution Journal*, March 1974.

Auton, Graeme P., 'Nuclear Deterrence and the Medium Power: A Proposal for Doctrinal Change in the British and French Cases', *Orbis,* Summer 1976.

Bailly-Cowell, G. M., ' "Detente" in Soviet Strategy' *NATO's Fifteen Nations,* Dec. 1975-Jan. 1976.

Barber, James Alden, 'Military Force and Nonmilitary Threats', *Military Review,* 1975:2.

Baritz, Joseph J., 'Soviet Military Theory, Politics and War', *Military Review,* Sept. 1966.

Baritz, Joseph J., 'The Soviet Strategy of Flexible Response' *Bulletin,* Institute for the Study of the USSR, 1969:4.

Barnett, Roger W., 'Trans-SALT: Soviet Strategic Doctrine', *Orbis,* Summer 1975.

Barrett, Raymond J., 'Geography and Soviet Strategic Thinking', *Military Review,* Jan. 1970.

Barsegov, Yuri, Khairov, Rustem, 'A Study of Problems of Peace', *Journal of Peace Research,* 1973:1–2.

Baskakov, V., 'O sootnoshenii voiny kak obshchestvennogo yavleniya i vooruzhennoi bor'by', *Kommunist Vooruzhennykh Sil,* 1971:1.

Battreall, Jr., Raymond R., 'Thesis: Massive Retaliation, Antithesis: Flexible Response, Synthesis: The Nixon Doctrine?', *Military Review,* 1975:1.

Baumann, Gerhard, 'Sicherheitspolitik zwischen "Friedlicher Koexistenz" und socialistischer Weltrewolution', *Wehrkunde,* 1973:6.

Bernard, Jessie, 'Parties and Issues in Conflict', *Journal of Conflict Resolution,* March 1957.

Blasius, Dirk, 'Carl von Clausewitz und die Haptdenker des Marxismus', I–II, *Wehrwissenschaftliche Rundschau,* 1966:5,6.

Boulding, Elise, 'The Study of Conflict and Community in the International System. Summary and Challenge to Research', *The Journal of Social Issues,* Jan. 1967.

Bouthoul, G., 'Fonctions sociologiques de guerres', *Revue Francais de Sociologie,* 1961:2.

Burin, Frederic S., 'The Communist Doctrine of the Inevitability of War', *American Political Science Review,* 1963:2.

Burns, Arthur Lee, 'From Balance to Deterrence', *World Politics,* July 1957.

Buzzard, Anthony, 'Unity in Defence and Disarmament', *Royal United Service Institution Journal,* 1959:8.

Carroll, Berenice A., 'Peace Research: The Cult of Power', *Journal of Conflict Resolution,* Dec. 1972.

Carroll, Berenice A., 'War Termination and Conflict Theory: Value Premises, Theories and Policies', *The Annals of the American Academy of Political and Social Science,* 'How Wars End', Nov. 1970.

Carter, Barry; 'Flexible Strategic Options. No Need For New Strategy', *Scientific American,* May 1974.

Choucri, N., North, R. L., 'The Determinants of International Violence', *Peace Research Society* (International) Papers, 1969:12.

Coffey, Joseph I., 'Strategies and Realities', *U.S. Naval Institute Proceedings,* Feb. 1966.

Cohen, Saul B., 'Geography and Strategy: Their Interrelationship', *Naval War College Review,* Dec. 1957.

Converse, Elizabeth, 'The War of All Against All: A Review of the Journal of Conflict Resolution, 1957–1968, *Journal of Conflict Resolution,* Dec. 1968.

Corning, Peter A., 'The Biological Bases of Behavior and Some Implications for Political Science', *World Politics,* April 1971.

Dahrendorf, Ralph, 'Toward a theory of social conflict', *Journal of Conflict Resolution,* June 1958.

Davies, James, 'Violence and Aggression: Innate or Not?' *The Western Political Quarterly,* Sept. 1970.

Deane, Michael J., 'The Soviet Assessment of the "Correlation of World Forces": Implications for American Foreign Policy', *Orbis,* Autumn 1976.

Deutsch, Karl W., 'Changing Images of International Conflict', *The Journal of Social Issues,* Jan. 1967.

Deutsch, Karl W., 'Imperialism and Neocolonialism', *Papers* of Peace Science Society (International), 1974.

Dmitriev, A., 'Marksistsko-leninskoe uchenie o voine i armii – vazhnyi element nauchnogo mirovozzreniya voennykh kadrov', *Kommunist Vooruzhennykh Sil,* 1975:13.

Dulles, John Foster, 'Policy for Security and Peace', *Foreign Affairs,* Apr. 1954.

Dulles, John Foster, 'Challenge and Response in United States Policy', *Foreign Affairs,* Oct. 1957.

Eckhardt, William, 'Primitive Militarism', *Journal of Peace Research,* 1975:1.

Eckstein, Harry, 'On the Ethiology of Internal Wars', *History and Theory,* 1965.

Ellsworth, Robert, 'Military Force and Political Influence in an Age of Peace', *Strategic Review,* Spring 1976.

Emerson, William R., 'American Concepts of Peace and War', *Naval War College Review,* May 1958.

Erickson, John, 'Detente: Soviet Policy and Purpose', *Strategic Review,* Spring 1976.

Erickson, John, 'European Security: Soviet Preferences and Priorities', *Strategic Review,* Winter 1976.

Fink, Clinton R., 'Some conceptual difficulties in the theory of social conflict', *Journal of Conflict Resolution,* Dec. 1968.

Finsterbush, K., Greisman, H. C., 'The unprofitability of warfare in the twentieth century' *Social Problems,* Feb. 1975.

Fisher, Ernest, 'A Strategy of Flexible Response', *Military Review,* Mar. 1967.

Foster, J. L., Brewer, G. D., 'And the clocks were striking thirteen: The termination of war', *Political Science,* June 1976.

Fuller, J. F. C., 'The Pattern of Future War', *Brassey's Annual* 1951.

Galtung, John, 'A Structural Theory of Aggression', *Journal of Peace Research,* 1964:2.

Garthoff, Raymond L., 'War and Peace in Soviet Policy', *Russian Review,* Apr. 1961.

Garthoff, Raymond L., 'Unconventional Warfare in Communist Strategy', *Foreign Affairs,* July 1962.

Garthoff, Raymond L., 'SALT and the Soviet Military', *Problems of Communism,* 1975:1.

Gjessing, Gutorm, 'Ecology and Peace Research', *Journal of Peace Research,* 1967:4.

Gray, Colin S., 'New weapons and the resort to force', *International Journal,* Spring 1975.

Guy, Mery, 'Comments', *Survival,* Sept/Oct. 1976.

Haas, Ernst, 'The Balance of Power: Prescription, Concept, or Propaganda?', *World Politics,* July 1953.

Haas, Michael, 'Societal Approaches to the Study of War', *Journal of Peace Research,* Dec. 1965.

Haas, Michael, 'Social Change and National Aggressiveness, 1900–1960'.

Halle, Louis J., 'Does War Have a Future?', *Foreign Affairs.*

Hammer, T., 'The Geopolitical Basis of Modern War', *Norsk Luftmilitaert Tidskrift,* Apr. 1955.

Haniotis, Constantin, 'Politicalization of War', *NATO's Fifteen Nations,* Oct.-Nov. 1970.

Heimann, Leo, 'Guerilla Warfare: An Analysis', *Military Review,* July 1963.

Hoffmann, Stanley, 'Notes on the elusiveness of modern power', *International Journal,* Spring 1975.

Holsti, K. J., 'Resolving international conflicts: a taxonomy of behavior and some figures on procedures', *Journal of Conflict Resolution,* Sept. 1966.

Holsti, O. R., Brody, R. A., North, R. C., 'International relations as a social science', *International Social Science Journal,* 1965:3.

Howard, Michael, 'The Relevance of Traditional Strategy', *Foreign Affairs,* Jan. 1973.

Howard, Michael, 'Military Science in an Age of Peace', *Royal United Service Institution Journal,* March 1974.

Howard, Michael, 'The strategic approach to international relations', *British Journal of International Studies,* Apr. 1976.

Hunt, K., 'Future Trends in Global Strategies', *NATO's Fifteen Nations,* Dec. 1970-Jan. 1971.

Huntington, Samuel P., 'Arms Races: Prerequisites and Results', *Public Policy,* 1958.

Hutchinson, Martha Crenshaw, 'The Concept of Revolutionary Terrorism', *Journal of Conflict Resolution,* Sept. 1972.

Jacobsen, C. G., 'The emergence of a Soviet doctrine of flexible response?', *Atlantic Community, Quarterly,* Summer 1974.

Jahn, Egbart, 'The Role of the Armament Complex in Soviet Society (Is There a Soviet Military-Industrial Complex?)', *Journal of Peace Research,* 1975:3.

Janssen, Alfred, 'Theorie and Praxis der Krise', *Sicherheitspolitik heute,* 1974:1.

Johnson, James Turner, 'Just War, The Nixon Doctrine and the Future Shape of American Military Policy', *The Year Book of World Affairs,* Stevens, London 1975.

Jones, Christopher D., 'Just Wars and Limited Wars. Restraints on the Use of the Soviet Armed Forces', *World Politics,* Oct. 1975.

Kára, Karel, 'On the Marxist Theory of War and Peace', *Journal of Peace Research,* 1968:1.

Khrushchev, N. S., 'Za novye pobedy mirovogo kommunisticheskogo dvizheniya', *Kommunist,* 1960:1.

Kintner, William R., 'The U.S. and the U.S.S.R.: Conflict and Cooperation', *Orbis,* Autumn 1973.

Kirschin, J., 'Die sowjetische und die westliche Militärwissenschaft', *Österreichische Militärische Zeitschrift,* 1973:2.

Kissinger, Henry A., 'The 1976 Alastair Buchan Memorial Lecture', *Survival,* Sept./Oct. 1976.

Koch, Hans, 'Ideologische Grundlagen der sowjetischen Politik und Strategie', *Wehrwissenschaftliche Rundschau,* 1959:1.

Kolkowicz, Roman, 'Strategic Elites and Politics of Superpower', *Journal of International Affairs,* 1972:1.

Kondratkov, T., 'Sotsialnyi kharakter sovremennoi voiny', *Kommunist Vooruzhennykh Sil,* 1972:21.

Kristof, Ladis, K. D., 'The Origin and Evolution of Geopolitics', *Journal of Conflict Resolution,* 1960:1.

Kupperman, R. H., Behr, R. M., Jones, jr., T. P., 'The deterrence continuum', *Orbis,* Autumn 1974.

Lambelet, John, 'Do Arms Race Lead to War?', *Journal of Peace Research,* 1975:2.

Laqueur, Walter, 'Revolution', *International Encyclopedia of Social Sciences,* Macmillan, New York 1968.

Laqueur, Walter, 'The Origin of Guerilla Doctrine', *Journal of Contemporary History,* 1975:3.

Laqueur, Walter, 'Interpretations of Terrorism – Fact, Fiction and Political Science', *Journal of Contemporary History,* 1977:1.

Levi, Werner, 'On the causes of war and the conditions of peace', *Journal of Conflict Resolution,* Dec. 1960.

Levi, Werner, 'On the causes of peace', *Journal of Conflict Resolution,* March 1964.

Levine, Robert A., 'Anthropology and the Study of Conflict: Introduction', *Journal of Conflict Resolution,* March 1961.

Luard, Evan, 'Conciliation and Deterrence: A Comparison of Political Strategies in the Interwar and Postwar Periods', *World Politics,* Jan. 1967.

Mack, R. W., Snyder, R. C., 'The analysis of social conflict – toward an overview and synthesis', *Journal of Conflict Resolution,* June 1957.

Mackinder, Halford, 'The Geographical Pivot of History', *Geographical Journal,* Apr. 1904.

Mackintosh, Malcolm, 'Soviet strategic policy', *The World Today,* July 1970.

Mainland, Edward A., 'Political Instability in Developing Areas', *Naval War College Review,* Feb. 1969.

Malinowski, Bronislaw, 'The Deadly Issue', *Atlantic Monthly,* Dec. 1936.

Malinowski, Bronislaw, 'An Anthropological Analysis of War', *American Journal of Sociology,* 1941.

Marantz, Paul, 'Prelude to Detente: Doctrinal Change Under Khrushchev', *International Studies Quarterly,* Dec. 1975.

Marshall, C. B., 'Unconventional Warfare as a Concern for American Foreign Policy', *Annals of the American Academy of Political and Social Science, 1962.*

Milovidov, A., 'Marksizm-Leninizm – teoreticheskaya osnova Sovetskoi voennoi nauki'. (Marxism-Leninism – the theoretical basis of the Soviet Military science), *Kommunist Vooruzhennykh Sil,* 1974:19.

Milshtein M. A., Semeiko L. S., 'Problems of the inadmissability of nuclear conflict', *International Studies Quarterly,* 1976:1.

Milshtein M. A., Slobodenko, 'Problema sootnosheniya politiki i voiny v sovremennoi imperialisticheskoi ideologii'. (The problem of the relationship between politics and war in the contemporary imperialistic ideology), *Mirovaya ekonomika i mezhdunarodnye otnosheniya,* 1958:9.

Monks, Alfred L., 'Evolution of Soviet Military Thinking', *Military Review,* 1971:3.

Montagu, M. F. Ashley, 'The Nature of War and the Myth of Nature', *Scientific Monthly,* Apr. 1942.

Morgenthau, Hans J., 'Has Atomic War Really Become Impossible?', *Bulletin of the Atomic Scientists,* 1956:12.

Morgenthau, Hans J., 'The Four Paradoxes of Nuclear Strategy', *American Political Science Review,* 1964.

Morozov, V., 'Politics and War', *Soviet Military Review,* 1975:11.

Möschel, Eberhard, 'Geographie, Geschichte und Ideologie in der sowjetischen Militärstrategie', I–II, *Wehrkunde,* 1975:8,9.

Münter, Otto, 'Das Primat der Politik. Geschichte einer Militärtheorie von Clausewitz bis zum Gegenwart', *Wehrkunde,* 1975:6.

Neumann, S., 'The International Civil War', *World Politics,* 1948:1.

Nevskii, N. A., 'Modern Armaments and Problems of Strategy', *World Marxist Review,* March 1963.

Ney, Virgil, 'Bibliography on Guerilla Warfare', *Military Affairs*, Autumn 1960.

Nieburg, H. L., 'The Threat of Violence and Social Change', *The American Political Science Review*, 1962.

Nieburg, H. L., 'Uses of Violence', *Journal of Conflict Resolution*, Dec. 1963.

Nitze, Paul H., 'Assuring Strategic Stability in an Era of Détente', *Foreign Affairs*, Jan. 1976.

North, Robert C., Koch, Jr., Howard E., Zinnes, Dina A., 'The integrative functions of conflict', *Journal of Conflict Resolution*, Sept. 1960.

Oppenheimer, J. R., 'Atomic Weapons and American Diplomacy', *Foreign Affairs*, July 1953.

Osanka, Franklin M., 'Guerilla Warfare', *International Encyclopedia of Social Services*, vol. 7, Macmillan, Free Press, New York 1968.

Osgood, Charles E., 'A Case for Graduated Unilateral Disengagement', *The Bulletin of Atomic Scientists*, Apr. 1960.

Osgood, Robert E., 'Stabilizing the Military Environment', *The American Political Science Review*, March 1961.

Osgood, Robert E., 'Limited War', *International Encyclopedia of Social Sciences*, vol. 9, Macmillan, Free Press, New York 1968.

Paret, Peter, 'Clausewitz: A Bibliographical Survey', *World Politics*, 1965:2.

Paret, Peter, 'The History of War', *Daedalus*, Spring 1971.

Paret, Peter, 'Clausewitz', *International Encyclopedia of Social Sciences*, 1968.

Pelliccia, A., 'The Evolution of the Philosophy of War', *NATO's Fifteen Nations*, Dec. 1970-Jan. 1971.

Pelliccia, A., 'Clausewitz and Soviet Politico-Military Strategy', *NATO's Fifteen Nations*, Dec. 1975-Jan 1976.

Pietzcker, Franz, 'Krieg und Gesellschaft', *Beiträge zur Konfliktforschung*, 1972:3.

Putney, S., Middleton, R., 'Some Factors Associated with Student Acceptance or Rejection of War', *American Sociological Research*, 1963.

Quester, G. H., 'Can Deterrence be left to the deterrent?', *Polity*, Summer 1975.

Rapoport, Anatol, 'Various Conceptions of Peace Research', *Peace Research Society (International) Papers*, 1972.

Rapoport, Anatol, 'War and Peace', *The Annals of the American Academy of Political and Social Science*, March 1974.

Rathjens, G. W., 'Flexible Response Options', *Orbis*, Autumn 1974.

Raymond, Jack, 'The Influence of Nuclear Weapons on National Strategy and Policy', *Naval War College Review*, Apr. 1967.

Reynolds, P. A., 'The Balance of Power: New Wine in an Old Bottle', *Political Studies*, 1975:2,3.

Richardson III, Robert C., 'The Fallacy of the Concept of Minimum Deterrence', *Air University Quarterly Review*, Spring 1960.

Russett, B. M., 'Cause, Surprise, and no Escape', *Journal of Politics*, 1962.

391

Rybkin, Y., 'O suschchestve mirovoi yaderno-raketnoi voiny', *Kommunist Vooruzhennykh Sil,* Sept. 1965.

Rybkin, Y., 'Voiny sovremennoi epokhi i ikh vliyanie na sotsialnye protsessy', *Kommunist Vooruzhennykh Sil,* 1970:1.

Rybkin, Y., 'V poiskakh vykhoda iz tupika (Kritika popytok "modernizatsii" imperialisticheskikh kontseptsii voiny)', *Kommunist Vooruzhennykh Sil,* 1973:1.

Rybkin, Y., 'Leninskaya kontseptsiya voiny i sovremennost', *Kommunist Vooruzhennykh Sil,* 1973:20.

Rybkin, Y., 'Sotsiologicheskii analiz istorii voin: osnovnye kategorii', *Voenno-Istoricheskii Zhurnal,* 1973:2.

Schelling, Thomas C., 'The Role of Deterrence in Total Disarmament', *Foreign Affairs,* Apr. 1962.

Schieder, Theodor, 'Friedenssicherung und Staatenpluralismus', *Europa-Archiv,* 1968:24.

Schmidt, Helmut, 'Deutschland und das europäische Sicherheitssystem der Zukunft', *Wehrkunde,* 1967:3.

Schmückle, Gerd, 'Die "Kriesenbeherrschung"', *Wehrkunde,* 1966:5.

Schmückle, Gerd, 'Krisenbeherrschung durch eine Allianz', *Wehrkunde,* 1967:8.

Schmückle, Gerd, 'Theorie über die Krise', *Wehrkunde,* 1969:7.

Schramm, Wilhelm Ritter von., 'Von der klassischen Kriegsphilosophie zur zeitgerechten Wehrauffassung', *Wehrwissenschaftliche Rundschau,* 1965:9.

Schramm, Wilhelm Ritter von., 'Clausewitz als politischer Klassiker. Von der Phänomenologie des Krieges zur aktuellen Wehrauffassung', *Wehrkunde,* 1973:4.

Schulte, Ludwig, 'Interdependenzen zwischen Sicherheits – und Entspannungspolitik', *Beiträge zur Konfliktforschung,* 1975:1.

Schütze, Walter, 'Praxis und Grenzen der Krisenbeherrschung', *Wehrkunde,* 1966:6.

Schwartz, Klaus Dieter, 'Die Entwicklung der Sowjetischen Militärstrategie 1945-1974', *Sicherheitspolitik heute,* 1974:3.

Senghaas, Dieter, 'Konflikt und Konfliktforschung', *Kölner Zeitschrift für Sociologie und Socialpsychologie,* 1968.

Seuberlich, H. E., 'Das strategische Grundkoncept der Sovjetunion', *Wehrkunde,* 1967:10.

Seuberlich, H. E., 'Atomstrategie der Sovjetunion', *Wehrkunde,* 1968:1.

Snyder, Glenn H., 'Balance of Power in the Nuclear Age', *Journal of International Affairs,* 1960:1.

Sokolovskii, V., Cherednichenko, M., 'Voennaya strategiya i ee problemy', *Voennaya Mysl,* 1968:10.

Sokolovskii, V., Cherednichenko, M., 'Nekotorye voprosy sovetskogo voennogo stroitelstva v poslevoennyi period', *Voenno-Istoricheskii Zhurnal,* 1965:3.

Somit, Albert, 'Review Article: Biopolitics', *British Journal of Political Science,* Apr. 1972.

Sommer, Theo, 'Detente and Security: The Options', *NATO's Fifteen Nations,* Dec. 1970-Jan. 1971.

Steinbrunner, John, 'Beyond Rational Deterrence', *World Politics,* Jan. 1976.

Strausz-Hupé, Robert, 'Nuclear Blackmail and Limited War', *The Yale Review,* Winter 1958.

Trager, Frank N., 'Wars of National Liberation: Implications for U.S. Policy and Planning', *Orbis,* Spring 1974.

Trythall, A. J., 'The Origins of Strategic Thought', *Royal United Service Institution Journal,* Sept. 1973.

Wallace, Michael, 'Power, status, and international war', *Journal of Peace Research,* 1971:1.

Walter, E. V., 'Power and Violence', *American Political Science Review,* June 1964.

Waltz, Kenneth N., 'The Stability of a Bipolar World', *Daedalus,* Summer 1974.

Weltman, John J., 'On the Obsolescence of War', *International Studies Quarterly,* Dec. 1974.

Wette, Wolfram, 'Revolution und Krieg: Wendepunkte in der Idee des Krieges', *Wehrwissenschaftliche Rundschau,* Feb. 1967.

Wheeler, Earle G., 'The Design of Military Power', *Military Review,* Feb. 1963.

Wilkenfeld, Jonathan, 'Domestic and Foreign Conflict Behavior of Nations', *Journal of Peace Research,* 1968:1.

Williams, P., 'The Decline of Academic Strategy – A Reappraisal', *Royal United Service Institution Journal,* Dec. 1972.

Wohlstetter, Albert, 'The Delicate Balance of Power', *Foreign Affairs,* Jan. 1959.

Wohlstetter, Albert, 'Nuclear Sharing: NATO and the N+1 Country', *Foreign Affairs,* Apr. 1961.

Wright, Quincy, 'The value for conflict resolution of a general discipline in international relations', *Journal of Conflict Resolution,* 1957:1.

Wright, Quincy, 'The escalation of international conflicts', *Journal of Conflict Resolution,* Dec. 1965.

Wright, Quincy, 'The Study of War', *International Encyclopedia of the Social Sciences,* vol. 16, Macmillan, New York 1968.

Yermolenko, D., 'Sotsiologiya i nekotorye problemy mezhdunarodnogo konflikta', *Mezhdunarodnaya Zhizn',* 1968:8.

Name Index

Vagts, Alfred 72n18 79n67, n69, 173n63
Vandenberg, Hoyt S. 84n114
Vasilenko, V. 286n14
Vasquez, John A. 79n66
Vayda, Andrew P. 34n48, n51, 35n52, 123n45, 131n3
Vega, Luis Mercier 104n47
Vilmar, Fritz 51, 72n18, 73n23
Visher, Stephen S. 36n64
Vishnev, S. M. 306n9
Vishnyakov 202n8
Vittoria, Francisco 145n5
Volkogonov, D. 264n78, 288n23
Vorwerck, Erich 96n8
Voslensky, Michael 304n4

Wagner, Klaus 42n128, 75n42
Wallace, Josiah A. 105n48
Wallace, Michael 45n154, n155, 46n159, 121n30
Wallensteen, Peter 122n32
Walters, Robert E. 35n59, 37n74, 38n80, 76n51, 176n80
Waltz, Kenneth W. 27, 29n3, 46n164, 74n33, 78n60, 121n29, 270, 285n7
Waskov, Arthur I. 134n21
Weber, Max 32n29
Weede, Erich 133n10
Weigert, Hans W. 290n36
Weigley, Russell F. 84n112, 174n68
Weizsäcker, Carl Friedrich von 145n8, 147n12, 166n12
Wesson, Robert G. 101n31, 289n34
Wetting, Gerhard 165n7, 166n14
Wheeler, Erich 133n10
White, Leslie 34n51
White, Thomas D. 73n31
White, W. W. 44n149, 70n5, 71n9, 95n5
Wiener, Anthony J. 170n40
Winowski, L. 145n6
Wiesner, Jerome 74n34
Wilkenfeld, Jonathan 106n54, 107n63, 121n31
Wilkinson, Paul 103n37
Williams, Phil 169n31, 170n40, n41, n42, n43, 171n44, n45, 351n4

Williams, Robin M. 46n164
Williams, W. A. 167n21
Withey, Stephen 70n4, 71n10, 165n4
Wohlstetter, A. 81n81
Wolfe, Thomas W. 39n89, 306n10, 352n8
Wolfers, Arnold 75n40
Wood, David 119n8, 120n15
Wright, Quincy 17–18, 32n30, 39n97–40n104, n106, n108, 43n138, 45n155, 70n4, 71n9, 83n100, n106, 84n118, 96n7, n8, 118n5, 119n9, 120n15, n16, 121n18, 131n2, 132n5, 134n20, 284n2, 285n8, 288n21, 289n29
Wright-Mills C. 72n18, n19, 79n67, n71
Wylie, J. C. 43n143

Ximenes 99n18

Yakovlev, Alexander 287n19
Yemashev, I. 259n31
Yermolenko, Dimitrii V. 204n30, n34, 206n48, 207n55, 228n20, 330n2
Yevgenev, V. 204n33
Yin, John 260n40, 286n11
York, Herbert 74n34
Young, Clarence W. 32n35, 84n117
Young, Leilyn M. 170n43
Young, Oran 124n54, 169n37
Yu Chao-Li 260n43, 261n45

Zadorozhnyi, Georgi 259n33, 260n39, 264n78
Zagoria, Perez 289n32
Zamkovoi, V. I. 256n18, 258n28, 287n19
Zawodny, J. K. 95n4, n5, 101n28
Zemskov, V. 203n17, n25
Zhilin, P. 306n9
Zhukov, Y. 253n23
Zhurkin, V. 206n48, 207n49, n53, 304n4
Zimmermann, Karl 42n129, 76n48, 171n46
Zinnes, Dina A. 71n8, 123n39
Zitzewitz, Horst von 169n38, 171n48
Zubarev, V. 258n29

Subject Index

404

Legal approach to war 17–19, 31n22, 32n33, n24, 284n3

Levels of analysis of war 5, 24, 27, 29n1, 32n33, 48, 70n5, n6, 269, 271, 272

Limited war; *see* War

Macrocosmic theories 270, 273

'Malthusianism' and war causation 13, 274

Massive retaliation, doctrine of 136, 152, 166n11, 167n26, 309, 313n2, 345

Microcosmic theories 270

Militarism 29n7, 160, 167n21, 239, 255n13

Militaristic school 157–58, 161, 163, 308–9, 312

Military theory 22, 158

Military (politico-military) doctrine 67, 157, 158, 181, 191, 195, 248, 249, 252, 279, 309, 313n1, 346, 347, 351n2, 353

Military establishment 17, 50, 166n11

Military force (strength) 10, 55, 58, 111, 136, 140, 149, 151, 154, 161, 163, 166n11, n19, 186, 192, 255n11, 317

Military-industrial complex 50, 70n6, 112, 303

Military posture 17

Military power, (might) 72n15, n16, 73n23, n24, 119n9, n13, 166n14, 243, 277–9, 283, 297, 332, 344, 348, 350

Military-technical approach to war 17, 22–23, 50, 284, n3

Moral (ethical) approach to war 19–22, 275

Multipolar system of balance of power, and war causation 56, 72n18, 100n23

Nation, and war 6, 10–13, 20–21, 29n4, 31n15, 32n30, n33, n36, 48–57, 70n4, n5, 71n11, 72n13, n14, 73n29, 95n2, 111, 118n5, 137, 140–2, 150, 157, 160, 168n28, 181–184, 189, 197, 208, 216–17, 225n4, 245, 285n7, 294, 298, 327

Nation-state as actor in war 9, 17–24, 32n33, 49, 70n2, n5, 72n19, 119n13, 131n2, 324n1.

National-democratic revolution 236

National-liberation wars, movement, struggle 87, 89, 90, 93, 95n4, 96n7, 97n9, 99n14, n16, 100n21, 101n24, 136, 137, 187, 189, 193, 194, 197–8, 200, 214n8, 215n13, 210, 216, 225n2, n3, 230, 234–236, 241, 245, 252, 254n9, 255n10,

n11, 256n16, 279, 298–9, 303, 304n3, 317, 318, 338, 342, 347

National Security Council 154

Natural selection 6, 11, 20

Nuclear age (era, epoch) 55, 61, 64–8, 72n21, 73n27, 117, 140, 149, 152, 160, 164, 187–92, 209, 221, 235, 240–1, 301, 328, 340–50, 360–1, 363n3

Nuclear blackmail 346

'Nuclear Paralysis' 94; stalemate 154

Nuclear war; *see* War

Nuclear weapons (thermonuclear weapons) 23, 29n2, 55, 61, 115, 128, 140–4, 145n3, n4, 150–3, 161–3, 165n7, 166n18, 168n26, 169n32, 221, 226n8, 230, 243, 255n11, 256n19, 353, 360–2; forces 67, 72n21, 73n29, 141; missiles 15, 31n21, 141, 153–5, 226n7, 281, 337, 348, 354

Oceans, control of, and causation of war 14–15

October Revolution (the Russian Revolution) 89–93, 199, 232, 240, 282, 318

'Optimist' view of the future of war 108, 110–12, 209, 212, 332

Pacifism, pacifists 53, 58, 143–4, 157–60, 164, 237–40, 309, 310, 312

Peace 18, 29n2, n6, 32n24, n26, n34, 55, 56, 57, 65, 68, 70n2, 71n6, 72n14, n16, n17, n18, n19, n20, 73n31, 91, 108, 137, 149, 154–6, 160–4, 181, 186, 193–7, 210, 222–3, 225n1, 226n8, n9, 234, 237, 245, 251–3, 254n7, 255n13, n14, 256n16, 258n26, 297, 301–2, 340–4, 345–6, 353

Peaceful co-existence 72n17, 149, 165n2, 188, 232–44, 251–3, 254n8, n9, 255n13, 256n16, 257n25, 311, 333–4, 340, 345–6, 350n1

People's war 89–90

'Permanent violence' 235

Polemology 28, 32n36

Policy, foreign 31n16, n21, 32n36, 48, 50, 57, 59, 69, 112, 136–7, 150, 152, 159, 162, 166n13, 183, 188, 201n4, 249–51, 257n23, 273, 282, 295, 304n2, 307–9, 321; domestic (internal) 48, 59, 69, 183, 188, 213n4, 251, 295, 298, 307

Political approach to war 5, 25, 48, 49, 51–3, 58–9, 70n2, n3, n4, 115, 150, 283

Political geography 17, 30n14, 31n22

Superpowers, and war 54, 71n8, 109–11, 136, 149–53, 195–6, 200, 244, 252, 256n16, 296–7, 349–50
System (structure): economic 72n16, 90; political 51, 90, 100n21, 111–12, 128–9, 181–4, 216, 241, 254n6, 279–80, 328, 332; social 90, 108, 110, 114, 126, 270–5, 298, 338, 340; socio-economic 51, 87, 183, 210, 216, 241, 248, 298, 328; socio-political 59, 164, 184, 202n10, 223, 291, 297, 328
Systems analysis 59, 113

Teleological outlook, and war 19–20, 270, 273, 293
Territory, and war causation 5–7, 12–13, 29n3, 30n12, 31n15, n17, n19, 117n5, 127, 138, 183, 191, 208, 211, 218, 226n8, 252, 282, 286n12, n13, n14, 347
Terrorism 90, 94
'Third partner' theory 153
Third World 13, 88, 90, 136, 195, 200, 209, 218, 244, 282, 341, 345–7, 350n1, 353–4
'Transitional epoch' 209–12, 217
Tribes and war 6, 9, 10, 11, 17, 28, 32n25, 108, 126, 131n2, n3, n4, 208, 213n2, n3, n4

'Uncalculated risk' 153
'Unconditional surrender' 68–9, 302, 356
Unit, political 12, 19, 48, 110; social 108, 110
'Utopian' tradition, and war 159, 293

Victory in war 66–9, 73n27, n29, n30, 74n32, 99n18, 100n20, 144, 150, 155, 158, 166n11, 186, 192, 200, 218, 230–1, 253n2, n3, 302–3, 307, 319, 340–1, 348, 354, 357–9, 363n4
Violence 18, 20, 22, 23, 32n27, n37, 57, 60, 69, 71n8, 72n12, n17, n20, 73n27, 89, 90–1, 94n1, 95n2, n4, 96n7, n8, 97n9, n10, 99n13, n15, n17, 100n20, n21, n23, 101n24, n26, 109–17, 118n5, n6, 119n8, n11, n12, 130–1, 131n3, n4, 140–4, 145n4, n5, n6, 149–50, 154–61, 164n1, 165n2, n3, 166n12, n14, n18, 167n22, 168n29, n30, 181, 185, 188, 192, 198, 213n1, n6, 221, 232–40, 245–51, 254n6, 255n12, n15, 257n21, 278, 280, 294, 301, 308, 310–11, 313n2, 319–20, 323n2, 326–7, 333, 353, 355, 358, 363n3
Vietnamese War 68, 88, 236, 341, 360

War concept, definition 5, 10, 17–18, 23, 29n6, 32n23, n31, n33, n34, n35, 48, 63, 70n3, n4, n5, 73n28, 86–7, 95n4, 109, 157, 181, 250; absolute 156–61, 167n22, 168n26, n27, 248–50, 308–11, 343, 356; accidental 24, 164, 168n30; catalytic 24; classifications, typologies 24, 86, 88, 90, 94, 95n4, 96n7, n8, 97n9, 98n12, 99n15, 136, 218–24, 202n11, 299–300; colonial 59; conventional 24, 66, 142, 151, 155; defensive 26, 193, in defence of socialist countries 195, and development (progress), culture 10–11, 13, 21, 109, economic 11, moral 20–1, social, socio-political 10–12, 21, 135–7, 144; general 64, 340, 342, 348; global, world 53, 137, 149–50, 187, 193, 196–7, 202n14, 210, 217, 221–3, 233, 244, 252, 253n1, n2, 254n5, n7, 341–2, 350, 354–5; guerilla 63–4, 86–92, 96n7, 99n15, n17, n18, 100n19, n20, n21, 153, 165n7, 169n31, 301–2, 320, 343, 361; 'impossible', 'unthinkable' 155, 160, 163, 361; internal 17, 30n13, 51, 59, 86–94, 94n1–101n26, 117, 128, 131, 136, 144, 145n5, 163, 184–5, 194–5, 197–8, 248, 252, 275, 283, 297, 299, 304n3, 317–23, 323n1, 343–4, 354; international, interstate 5, 17, 24, 28, 29n4, 30n13, 49, 51, 61, 86–91, 95n2, n4, n5, 96n8, 98n11, 100n23, 111–12, 128–9, 185, 189, 193–7, 210, 231–4, 244, 248, 253, 274, 280–1, 284n4, 297–9, 311–12, 328, 334n1, 340–3; internal and international, mutual relationship 59, 93–4, 100n22, n23; limited 24, 32n31, 64–70, 70n5, 72n17, n18, 129, 136, 149, 153–6, 162–3, 166n15, n19, 168n27, n28, n29, 193, 217, 230–2, 340–1, 348; local 26, 32n31, 94n1, 136, 143, 155, 197, 230–2, 254n5, 340–1; mixed 299, 342; modern 69, 73n22, n27, 164, 192, 216, 301, 328, 332, 358; nuclear, thermonuclear 23, 26, 51–3, 62–9, 73n30, 111, 140–3, 145n2, 149, 155–6, 161–4, 165n3, 166n15, n16, n19, 168n29, 193, 196, 210, 221–2, 269–70, 276, 302–3, 304n3, n4, 310, 337, 339n1, 340–3, 348, 361, 363n3; offensive 26, aggressive 60, 193; and politics 21–5, 32n30, 55, 58–69, 70n3, n4, 73n22, 74n32, 149, 155–65, 185–93, 209, 246–7, 250–1, 270–1, 275–8, 293–300, 338, 343–4, 354, 356–8; pre-emptive 52, 63;

preventive 52, 63, 161, 333; primitive 126, 131n1, n3, n4; protracted, prolonged 64, 69; real 157, 163, 249, 308, 356–7, 362n2; between socialist states 281; between systems 184, 200, 201n7, 210, 217, 225n3, 230–1, 237–45, 281, 299, 340, 345, 350; total 21–3, 32n26, n31, 64–6, 136, 141–2, 149, 167n23, 168n29, 231, 309, 313n2, 314n5, 309, 345–6; unlimited 345, 363n2

Western allies 302

World War I 7, 57, 68, 73n22, 89, 158, 167n22, 187, 232, 254n5, 281, 307, 359

World War II 7, 15, 26, 31n17, n21, 60, 69, 73n22, n31, 89, 111, 152, 158–63, 165n1, 166n15, 182, 197, 217–22, 232, 237, 281–2, 285n6, 302, 309, 314n5, 340, 349, 354, 359

'Zones of friction' 15